William Alexander Baillie Hamilton

Lectures on Metaphysics and Logic

Lectures on metaphysics. Vol. 1

William Alexander Baillie Hamilton

Lectures on Metaphysics and Logic
Lectures on metaphysics. Vol. 1

ISBN/EAN: 9783742854483

Manufactured in Europe, USA, Canada, Australia, Japa

Cover: Foto ©Thomas Meinert / pixelio.de

Manufactured and distributed by brebook publishing software
(www.brebook.com)

William Alexander Baillie Hamilton

Lectures on Metaphysics and Logic

LECTURES

ON

METAPHYSICS AND LOGIC

BY

SIR WILLIAM HAMILTON, BART.,

PROFESSOR OF LOGIC AND METAPHYSICS IN THE
UNIVERSITY OF EDINBURGH;

Advocate, A.M. (Oxon.) &c.; Corresponding Member of the Institute of France; Honorary Member of the
American Academy of Arts and Sciences; and of the Latin Society of Jena, &c.

EDITED BY THE

REV. H. L. MANSEL, B.D., LL.D.,

DEAN OF ST PAUL'S;

AND

JOHN VEITCH, M.A.,

PROFESSOR OF LOGIC AND RHETORIC IN THE UNIVERSITY OF GLASGOW.

IN FOUR VOLUMES.

VOL. I.

WILLIAM BLACKWOOD AND SONS,
EDINBURGH AND LONDON.
MDCCCLXXVII.

LECTURES

ON

METAPHYSICS

BY

SIR WILLIAM HAMILTON, BART.

EDITED BY THE

REV. H. L. MANSEL, B.D., LL.D.,
DEAN OF ST PAUL'S;

AND

JOHN VEITCH, M.A.,
PROFESSOR OF LOGIC AND RHETORIC IN THE UNIVERSITY OF GLASGOW.

VOL. I.

SIXTH EDITION.

WILLIAM BLACKWOOD AND SONS,
EDINBURGH AND LONDON.
MDCCCLXXVII.

PREFACE.

THE following Lectures on Metaphysics constitute the first portion of the Biennial Course which the lamented Author was in the habit of delivering during the period of his occupation of the Chair of Logic and Metaphysics in the University of Edinburgh.

In giving these Lectures to the world, it is due, both to the Author and to his readers, to acknowledge that they do not appear in that state of completeness which might have been expected, had they been prepared for publication by the Author himself. As Lectures on Metaphysics,—whether that term be taken in its wider or its stricter sense,—they are confessedly imperfect. The Author himself, adopting the Kantian division of the mental faculties into those of Knowledge, Feeling, and Conation, considers the Philosophy of Mind as comprehending, in relation to each of these, the three great subdivisions of Psychology, or the Science of the Phænomena of Mind ; Nomology, or the Science of its Laws ; and Ontology, or the Science of Results and Inferences.[a]

a See below, Lecture vii., p. 121 *et seq.*

The term *Metaphysics*, in its strictest sense, is synonymous with the last of these subdivisions; while, in its widest sense, it may be regarded as including the first also,—the second being, in practice at least, if not in scientific accuracy, usually distributed among other departments of Philosophy. The following Lectures cannot be considered as embracing the whole province of Metaphysics in either of the above senses. Among the Phænomena of Mind, the Cognitive Faculties are discussed fully and satisfactorily; those of feeling are treated with less detail; those of Conation receive scarcely any special consideration; while the questions of Ontology, or Metaphysics proper, are touched upon only incidentally. The omission of any special discussion of this last branch may perhaps be justified by its abstruse character, and unsuitableness for a course of elementary instruction; but it is especially to be regretted, both on account of the general neglect of this branch of study by the entire school of Scottish philosophers, and also on account of the eminent qualifications which the Author possessed for supplying this acknowledged deficiency. A treatise on Ontology from the pen of Sir William Hamilton, embodying the final results of the Philosophy of the Conditioned, would have been a boon to the philosophical world such as probably no writer now living is capable of conferring.

The circumstances under which these Lectures were written must also be taken into account in estimating their character, both as a specimen of the Author's powers, and as a contribution to philosophical literature.

Sir William Hamilton was elected to the Chair of
Logic and Metaphysics in July 1836. In the interval
between his appointment and the commencement of the
College Session (November of the same year), the Author
was assiduously occupied in making preparation for dis-
charging the duties of his office. The principal part of
those duties consisted, according to the practice of the
University, in the delivery of a Course of Lectures on
the subjects assigned to the chair. On his appointment
to the Professorship, Sir William Hamilton experienced
considerable difficulty in deciding on the character of the
course of Lectures on Philosophy, which, while doing
justice to the subject, would at the same time meet
the wants of his auditors, who were ordinarily composed of comparatively young students in the second
year of their university curriculum. The Author of the
articles on *Cousin's Philosophy*,[a] on *Perception*,[b] and on
Logic,[c] had already given ample proof of those speculative accomplishments, and that profound philosophical
learning, which, in Britain at least, were conjoined in an
equal degree by no other man of his time. But those
very qualities which placed him in the front rank of
speculative thinkers, joined to his love of precision and
system, and his lofty ideal of philosophical composition,
served but to make him the more keenly alive to the re-
quirements of his subject, and to the difficulties that lay
in the way of combining elementary instruction in Philosophy with the adequate discussion of its topics. Hence,

a *Edinburgh Review*, 1829.   β *Ibid.*, 1830.   γ *Ibid.*, 1833.

although even at this period his methodised stores of
learning were ample and pertinent, the opening of the
College Session found him still reading and reflecting,
and unsatisfied with even the small portion of matter
which he had been able to commit to writing. His first
Course of Lectures (Metaphysical) thus fell to be written during the currency of the Session (1836-7).   The
Author was in the habit of delivering three Lectures
each week ; and each Lecture was usually written on the
day, or, more properly, on the evening and night, preceding its delivery. The Course of Metaphysics, as it is
now given to the world, is the result of this nightly toil,
unremittingly sustained for a period of five months.
These Lectures were thus designed solely for a temporary purpose,—the use of the Author's own classes ; they
were, moreover, always regarded by the Author himself
as defective as a complete Course of Metaphysics ; and
they were never revised by him with any view to
publication, and this chiefly for the reason that he intended to make use of various portions of them which
had not been incorporated in his other writings, in the
promised Supplementary Dissertations to Reid's Works,
—a design which his failing health did not permit him
to complete.

The Lectures on Logic were not composed until the
following Session (1837-8). This Course was also, in
great part, written during the currency of the Session.

These circumstances will account for the repetition,
in some places, of portions of the Author's previously

published writings, and for the numerous and extensive
quotations from other writers which are interspersed
throughout the present Course. Most of these have
been ascertained by references furnished by the Author
himself, either in the manuscript of the present Lectures, or in his Commonplace-Book. These quotations,
while they detract in some degree from the originality
of the work, can, however, hardly be considered as
lessening its value. Many of the authors quoted are
but little known in this country ; and the extracts from
their writings will, to the majority of readers, have all
the novelty of original remarks. They also exhibit, in
a remarkable degree, the Author's singular power of
appreciating and making use of every available hint
scattered through those obscurer regions of thought
through which his extensive reading conducted him. No
part of Sir William Hamilton's writings more completely
verifies the remark of his American critic, Mr Tyler :
" There seems to be not even a random thought of any
value, which has been dropped along any, even obscure,
path of mental activity, in any age or country, that his
diligence has not recovered, his sagacity appreciated,
and his judgment husbanded in the stores of his knowledge."[a] Very frequently, indeed, the thought which
the Author selects and makes his own, acquires its
value and significance in the very process of selection ;
and the contribution is more enriched than the adopter ;

---

[a] *Princeton Review*, October 1855.
This article has since been republished
with the Author's name, in his Essay
on the *Progress of Philosophy in the
Past and in the Future*. Philadelphia,
1858.

for what, in another, is but a passing reflection, seen
in a faint light, isolated and fruitless, often rises, in the
hands of Sir William Hamilton, to the rank of a great,
permanent, and luminous principle, receives its appropriate place in the order of truths to which it belongs,
and proves, in many instances, a centre of radiation over
a wide expanse of the field of human knowledge.

The present volumes may also appear to some disadvantage on account of the length of time which has
elapsed between their composition and their publication.   Other writings, particularly the *Dissertations*
appended to Reid's Works,[a] and part of the new matter
in the *Discussions*, though earlier in point of publication,
contain later and more mature phases of the Author's
thought, on some of the questions discussed in the
following pages. Much that would have been new to
English readers twenty years ago, has, subsequently, in
a great measure by the instrumentality of the Author
himself, become well known; and the familiar exposi-
tions designed for the oral instruction of beginners in
philosophy, have been eclipsed by those profounder re-
flections which have been published for the deliberate
study of the philosophical world at large.

But, when all these deductions have been made, the
work before us will still remain a noble monument of the
Author's philosophical genius and learning. In many
respects, indeed, it is qualified to become more popular

a The *footnotes* to Reid were for   temporaneously with the present Lec-
the most part written nearly con-   tures.

than any of his other publications. The very necessity
which the Author was under, of adapting his observations, in some degree, to the needs and attainments of
his hearers, has also fitted them for the instruction and
gratification of a wide circle of general readers, who
would have less relish for the severer style in which
some of his later thoughts are conveyed. The present Lectures, if in depth and exactness of thought
they are, for the most part, not equal to the *Dissertations* on Reid, or to some portions of the *Discussions*,
possess attractions of their own, which will probably
recommend them to a more numerous class of admirers;
while they retain, in no small degree, the ample learning
and philosophical acumen which are identified with the
Author's previous reputation.

Apart, however, from considerations of their intrinsic
value, these Lectures possess a high academical and
historical interest. For twenty years,—from 1836 to
1856,—the Courses of Logic and Metaphysics were the
means through which Sir William Hamilton sought to
discipline and imbue with his philosophical opinions, the
numerous youth who gathered from Scotland and other
countries to his class-room; and while, by these prelec-
tions, the Author supplemented, developed, and moulded
the National Philosophy,—leaving thereon the ineffaceable impress of his genius and learning,—he, at the same
time and by the same means, exercised over the intellects
and feelings of his pupils an influence which, for depth,
intensity, and elevation, was certainly never surpassed by

that of any philosophical instructor. Among his pupils
there are not a few who, having lived for a season under
the constraining power of his intellect, and been led to
reflect on those great questions regarding the character,
origin, and bounds of human knowledge, which his teach-
ings stirred and quickened, bear the memory of their
beloved and revered Instructor inseparably blended with
what is highest in their present intellectual life, as well
as in their practical aims and aspirations.

The Editors, in offering these Lectures to the public,
are, therefore, encouraged to express their belief, that they
will not be found unworthy of the illustrious name which
they bear. In the discharge of their own duties as
annotators, the Editors have thought it due to the fame
of the Author, to leave his opinions to be judged entirely
by their own merits, without the accompaniment of
criticisms, concurrent or dissentient. For the same
reason, they have abstained from noticing such criticisms
as have appeared on those portions of the work which
have already been published in other forms. Their own
annotations are, for the most part, confined to occasional
explanations and verifications of the numerous references
and allusions scattered through the text. The notes
fall, as will be observed, into three classes :—

I. Original ; notes printed from the manuscript of the
present Lectures. These appear without any distinctive
mark. Mere Jottings or Memoranda by the Author,
made on the manuscript, are generally marked as such.

To these are also added a few Oral Interpolations of the
Author, made in the course of reading the Lectures,
which have been recovered from the note-books of
students.

II. Supplied; notes extracted or compiled by the
Editors from the Author's Commonplace - Book and
fragmentary papers. These are enclosed in square
brackets, and are without signature.

III. Editorial; notes added by the Editors. These
always bear the signature " ED." When added as supplementary to the original or supplied notes, they are
generally enclosed in square brackets, besides having the
usual signature.

The Editors have been at pains to trace and examine
the notes of the first and second classes with much care;
and have succeeded in discovering the authorities referred to, with very few and insignificant exceptions.
The Editors trust that the Original and Supplied Notes
may prove of service to students of Philosophy, as
indications of sources of philosophical opinions, which,
in many cases, are but little, if at all, known in this
country.

The Appendix embraces a few papers, chiefly fragmentary, which appeared to the Editors to be deserving
of publication. Several of these are fragments of discussions which the Author had written with a view to
the Memoir of Mr Dugald Stewart, on the editorship
of whose works he was engaged at the period of his

death.   They thus possess the melancholy interest which
attaches to the latest of his compositions.   To these
philosophical fragments have been added a few papers
on physiological subjects.   These consist of an extract
from the Author's Lectures on Phrenology, and communications made by him to various medical publications.   Apart from the value of their results, these
physiological investigations serve to exhibit, in a department of inquiry foreign to the class of subjects
with which the mind of the Author was ordinarily
occupied, that habit of careful, accurate, and unsparing
research, by which Sir William Hamilton was so eminently characterised.

CONTENTS OF VOL. I.

xviii CONTENTS.

LECTURES ON METAPHYSICS.

LECTURE I.

PHILOSOPHY—ITS ABSOLUTE UTILITY.

(A) SUBJECTIVE.

GENTLEMEN—In the commencement of a course of instruction in any department of knowledge, it is usual, before entering on the regular consideration of the subject, to premise a general survey of the more important advantages which it affords; and this with the view of animating the student to a higher assiduity, by holding up to him, in prospect, some at least of those benefits and pleasures which he may promise to himself in reward of his exertions.

And, if such a preparation be found expedient for other branches of study, it is, I think, peculiarly requisite in Philosophy,—Philosophy Proper,—the Science of Mind. For, in the first place, the most important advantages to be derived from the cultivation of philosophy, are not, in themselves, direct, palpable, obtrusive : they are, therefore, of their own nature, peculiarly liable to be overlooked or disparaged by the world at large ; because to estimate them at their proper value requires in the judge more than a vulgar complement of information and intelligence. But, in

[margin notes: LECT. I. Philosophy; its benefits and pleasures. The exhibition of these, why peculiarly requisite.]

the second place, the many are not simply by negative incompetence disqualified for an opinion ; they
are, moreover, by positive error, at once rendered incapable of judging right ; and yet, by positive error,
encouraged to a decision. For there are at present
afloat, and in very general acceptation, certain superficial misconceptions in regard to the end and objects
of education, which render the popular opinion of the
comparative importance of its different branches, not
merely false, but precisely the reverse of truth ; the
studies which, in reality, are of the highest value as a
mean of intellectual development, being those which,
on the vulgar standard of utility, are at the very
bottom of the scale ; while those which, in the nomenclature of the multitude, are emphatically,—distinctively denominated the Useful, are precisely those
which, in relation to the great ends of liberal education, possess the least, and least general, utility.

Utility of
a branch of
knowledge,
of two
grand kinds
—Absolute
and Rela-
tive.
In considering the utility of a branch of knowledge,
it behoves us, in the first place, to estimate its value as
viewed simply in itself ; and, in the second, its value
as viewed in relation to other branches. Considered
in itself, a science is valuable in proportion as its
cultivation is immediately conducive to the mental
improvement of the cultivator. This may be called
its Absolute utility. In relation to others, a science
is valuable in proportion as its study is necessary for
the prosecution of other branches of knowledge. This
may be called its Relative utility. In this latter
point of view, that is, as relatively useful, I cannot at
present enter upon the value of Philosophy,—I cannot
attempt to show how it supplies either the materials
or the rules to all the sciences ; and how, in particular,
its study is of importance to the Lawyer, the Physi-

cian, and, above all, to the Theologian. All this I must for the present pass by.

LECT. I.

In the former point of view, that is, considered absolutely, or in itself, the philosophy of mind comprises two several utilities, according as it, 1°, Cultivates the mind or knowing subject, by calling its faculties into exercise ; and, 2°, Furnishes the mind with a certain complement of truths or objects of knowledge. The former of these constitutes its Subjective, the latter its Objective utility. These utilities are not the same, nor do they even stand to each other in any necessary proportion. As the special consideration of both is more than I can compass in the present Lecture, I am constrained to limit myself to one alone ; and as the subjective utility is that which has usually been overlooked, though not assuredly of the two the less important, while at the same time its exposition affords in part the rationale of the method of instruction which I have adopted, I shall at present only attempt an illustration of the advantages afforded by the Philosophy of Mind, regarded as the study which, of all others, best cultivates the mind or subject of knowledge, by supplying to its higher faculties the occasions of their most vigorous, and therefore their most improving, exercise.

Absolute utility of two kinds —Subjective and Objective.

There are few, I believe, disposed to question the speculative dignity of mental science ; but its practical utility is not unfrequently denied. To what, it is asked, is the science of mind conducive ? What are its uses ?

Practical utility of Philosophy.

I am not one of those who think that the importance of a study is sufficiently established when its dignity is admitted ; for, holding that knowledge is for the sake of man, and not man for the sake of knowledge, it is necessary, in order to vindicate its value, that every science should be able to show what are the advantages which it promises to confer upon its student. I, therefore, profess myself a utilitarian; and it is only on the special ground of its utility that I would claim for the philosophy of mind, what I regard as its peculiar and pre-eminent importance. But what is a utilitarian? Simply one who prefers the Useful to the Useless—and who does not? But what is the useful? That which is prized, not on its own account, but as conducive to the acquisition of something else,—the useful is, in short, only another word for a mean towards an end; for every mean is useful, and whatever is useful is a mean. Now the value of a mean is always in proportion to the value of its end; and the useful being a mean, it follows that, of two utilities, the one which conduces to the more valuable end will be itself the more valuable utility.

The Useful.

So far there is no difference of opinion. All agree that the useful is a mean towards an end; and that, *cæteris paribus,* a mean towards a higher end constitutes a higher utility than a mean towards a lower. The only dispute that has arisen, or can possibly arise, in regard to the utility of means (supposing always their relative efficiency), is founded on the various views that may be entertained in regard to the existence and comparative importance of ends.

Two errors in the popular estimate of the comparative utility of human sciences.

Now the various opinions which prevail concerning the comparative utility of human sciences and studies have all arisen from two errors.[a]

The first of these consists in viewing man, not as

[a] With the following observations may be compared the author's remarks on the distinction between a *liberal* and a *professional* education in his article on the study of mathematics, *Edinburgh Review,* vol. lxii. p. 409, reprinted in his *Discussions,* p. 263.—ED.

an end unto himself, but merely as a mean organised for the sake of something out of himself; and, under this partial view of human destination, those branches of knowledge obtain exclusively the name of *useful*, which tend to qualify a human being to act the lowly part of a dexterous instrument.

The second, and the more dangerous of these errors, consists in regarding the cultivation of our faculties as subordinate to the acquisition of knowledge, instead of regarding the possession of knowledge as subordinate to the cultivation of our faculties; and, in consequence of this error, those sciences which afford a greater number of more certain facts, have been deemed superior in utility to those which bestow a higher cultivation on the higher faculties of the mind.

As to the first of these errors, the fallacy is so palpable, that we may well wonder at its prevalence. It is manifest, indeed, that man, in so far as he is a mean for the glory of God, must be an end unto himself, for it is only in the accomplishment of his own perfection, that, as a creature, he can manifest the glory of his Creator. Though therefore man, by relation to God, be but a mean,—for that very reason, in relation to all else, is he an end. Wherefore, now speaking of him exclusively in his natural capacity and temporal relations, I say it is manifest that man is by nature necessarily an end to himself,—that his perfection and happiness constitute the goal of his activity, to which he tends, and ought to tend, when not diverted from this, his general and native destination, by peculiar and accidental circumstances. But it is equally evident, that, under the condition of society, individual men are, for the most part, to a greater or less degree, actually so diverted. To

live, the individual must have the means of living;
and these means, (unless he already possess them), he
must procure,—he must purchase. But purchase
with what? With his services,—*i.e.* he must reduce
himself to an instrument,—an instrument of utility to
others, and the services of this instrument he must
barter for those means of subsistence of which he is
in want. In other words, he must exercise some
trade, calling, or profession.

Thus, in the actualities of social life, each man,
instead of being solely an end to himself,—instead
of being able to make everything subordinate to that
full and harmonious development of his individual
faculties, in which his full perfection and his true
happiness consist,—is, in general, compelled to degrade
himself into the mean or instrument towards the
accomplishment of some end, external to himself, and
for the benefit of others.

Liberal and
profes-
sional edu-
cation.
Now the perfection of man as an end, and the perfection of man as a mean or instrument, are not only
not the same, they are, in reality, generally opposed.
And as these two perfections are different, so the train-
ing requisite for their acquisition is not identical, and
has, accordingly, been distinguished by different names.
The one is styled Liberal, the other Professional education,—the branches of knowledge cultivated for these
purposes being called respectively liberal and professional, or liberal and lucrative sciences. By the
Germans, the latter are usually distinguished as the
Brodwissenschaften, which we may translate, *The Bread
and Butter sciences.*[a] A few of the professions, indeed,
as requiring a higher development of the higher facul-

a Schelling, *Vorlesungen über die* p. 67.—ED.
Methode des Academischen Studium,

ties, and involving, therefore, a greater or less amount
of liberal education, have obtained the name of liberal
professions. We must, however, recollect that this
is only an accidental and a very partial exception.
But though the full and harmonious development of
our faculties be the high and natural destination of all,
while the cultivation of any professional dexterity is
only a contingency, though a contingency incumbent
upon most, it has, however, happened that the paramount and universal end of man,—of man absolutely,
—has been often ignorantly lost sight of, and the term
useful appropriated exclusively to those acquirements
which have a value only to man considered in his
relative, lower, and accidental character of an instrument. But, because some have thus been led to
appropriate the name of useful to those studies and
objects of knowledge, which are conducive to the
inferior end, it assuredly does not follow that those
conducive to the higher have not a far preferable title
to the name thus curiously denied to them. Even
admitting, therefore, that the study of mind is of
no immediate advantage in preparing the student
for many of the subordinate parts in the mechanism of society, its utility cannot, on that account, be
called in question, unless it be asserted that man
" liveth by bread alone," and has no higher destination than that of the calling by which he earns his
subsistence.

Misapplica-
tion of the
term use-
ful.

The second error to which I have adverted, reverses
the relative subordination of knowledge and of intellectual cultivation. In refutation of this, I shall
attempt briefly to show, firstly, that knowledge and
intellectual cultivation are not identical; secondly,
that knowledge is itself principally valuable as a

Knowledge
and intel-
lectual cul-
tivation.

mean of intellectual cultivation; and, lastly, that intellectual cultivation is more directly and effectually accomplished by the study of mind than by any other of our rational pursuits.

But to prevent misapprehension, I may premise what I mean by knowledge, and what by intellectual cultivation. By knowledge is understood the mere possession of truths; by intellectual cultivation, or intellectual development, the power acquired through exercise by the higher faculties, of a more varied, vigorous, and protracted activity.

Not identical.
In the first place, then, it will be requisite, I conceive, to say but little to show that knowledge and intellectual development are not only not the same, but stand in no necessary proportion to each other. This is manifest if we consider the very different conditions under which these two qualities are acquired. The one condition under which all powers, and consequently the intellectual faculties, are developed, is exercise. The more intense and continous the exercise, the more vigorously developed will be the power.

But a certain quantity of knowledge,—in other words, a certain amount of possessed truths,—does not suppose, as its condition, a corresponding sum of intellectual exercise. One truth requires much, another truth requires little, effort in acquisition; and, while the original discovery of a truth evolves perhaps a maximum of the highest quality of energy, the subsequent learning of that truth elicits probably but a minimum of the very lowest.

Is truth or mental exercise the superior end?
But, as it is evident that the possession of truths, and the development of the mind in which they are deposited, are not identical, I proceed, in the second place, to show that, considered as ends, and in relation

to each other, the knowledge of truths is not supreme, LECT.
I. but subordinate to the cultivation of the knowing mind. The question—Is Truth, or is the Mental Exercise in the pursuit of truth, the superior end?—this is perhaps the most curious theoretical, and certainly the most important practical, problem in the whole compass of philosophy. For, according to the solution at which we arrive, must we accord the higher or the lower rank to certain great departments of study; and, what is of more importance, the character of its solution, as it determines the aim, regulates from first to last the method, which an enlightened science of education must adopt.

But, however curious and important, this question *Popular solution of this question.* has never, in so far as I am aware, been regularly discussed. Nay, what is still more remarkable, the erroneous alternative has been very generally assumed as true. The consequence of this has been, that sciences of far inferior, have been elevated above sciences of far superior, utility; while education has been systematically distorted,—though truth and nature have occasionally burst the shackles which a perverse theory had imposed. The reason of this is sufficiently obvious. At first sight, it seems even absurd to doubt that truth is more valuable than its pursuit; for is this not to say that the end is less important than the mean?—and on this superficial view is the prevalent misapprehension founded. A slight consideration will, however, expose the fallacy.

Knowledge is either practical or speculative. In *Practical knowledge: its end.* practical knowledge it is evident that truth is not the ultimate end; for, in that case, knowledge is, *ex hypothesi*, for the sake of application. The knowledge of a moral, of a political, of a religious truth, is of value only as it affords the preliminary or condition of its exercise.

The end of speculative knowledge. In speculative knowledge, on the other hand, there may indeed, at first sight, seem greater difficulty; but further reflection will prove that speculative truth is only pursued, and is only held of value, for the sake of intellectual activity. "Sordet cognita veritas" is a shrewd aphorism of Seneca. A truth, once known, falls into comparative insignificance. It is now prized, less on its own account than as opening up new ways to new activity, new suspense, new hopes, new discoveries, new self-gratulation. Every votary of science is wilfully ignorant of a thousand established facts,—of a thousand which he might make his own more easily than he could attempt the discovery of even one. But it is not knowledge,—it is not truth, —that he principally seeks; he seeks the exercise of his faculties and feelings; and, as in following after the one he exerts a greater amount of pleasurable energy than in taking formal possession of the thousand, he disdains the certainty of the many, and prefers the chances of the one. Accordingly, the sciences always studied with keenest interest are those in a state of progress and uncertainty: absolute certainty and absolute completion would be the paralysis of any study; and the last worst calamity that could befall man, as he is at present constituted, would be that full and final possession of speculative truth, which he now vainly anticipates as the consummation of his intellectual happiness.

"Quæsivit cœlo lucem, ingemuitque reperta." [a]

a Virgil, Æn., iv. 692.—ED.

But what is true of science is true, indeed, of all human activity. "In life," as the great Pascal observes, "we always believe that we are seeking repose, while, in reality, all that we ever seek is agitation."[a] When Pyrrhus proposed to subdue a part of the world, and then to enjoy rest among his friends, he believed that what he sought was possession, not pursuit; and Alexander assuredly did not foresee that the conquest of one world would only leave him to weep for another world to conquer. It is ever the contest that pleases us, and not the victory. Thus it is in play; thus it is in hunting; thus it is in the search of truth;[b] thus it is in life. The past does not interest, the present does not satisfy, the future alone is the object which engages us.

"[Nullo votorum fine beati]
Victuros agimus semper, nec vivimus unquam."[y]
"Man never is, but always to be blessed."[δ]

The question, I said, has never been regularly discussed,—probably because it lay in too narrow a compass; but no philosopher appears to have ever seriously proposed it to himself, who did not resolve it in contradiction to the ordinary opinion. A contradiction of this opinion is even involved in the very term Philosophy; and the man who first declared that he was not a σοφὸς, or possessor, but a φιλόσοφος,[e]

a *Pensées*, partie i. art. vii. § 1, (vol. ii. p. 34, ed. Faugère): "Ils croient chercher sincèrement le repos, et ne cherchent en effet que l'agitation." "Le conseil qu'on donnait à Pyrrhus, de prendre le repos qu'il allait chercher par tant de fatigues, recevait bien des difficultés."—ED.

β "Rien ne nous plait que le combat, mais non pas la victoire. . . . Ainsi dans le jeu, ainsi dans la recherche de la vérité. On aime à voir dans les disputes le combat des opinions; mais de contempler la vérité trouvée, point du tout. . . . Nous ne cherchons jamais les choses, mais la recherche des choses."—Pascal, *Pensées*, vol. i. p. 205, ed. Faugère.—ED.

γ Manilius, *Astronomicon*, lib. iv. 4.—ED.

δ Pope, *Essay on Man*, i. 96.—ED.

e Pythagoras, according to the ordinary account; see Cicero, *Tusc. Quæst.*, v. 3. Sir W. Hamilton, however, probably meant Socrates. See Lecture III., p. 47.—ED.

or seeker of truth, at once enounced the true end of human speculation, and embodied it in a significant name. Under the same conviction Plato defines man "the hunter of truth," [a] for science is a chase, and in a chase the pursuit is always of greater value than the game.

> "Our hopes, like towering falcons, aim
> At objects in an airy height ;
> But all the pleasure of the game
> Is afar off to view the flight." [β]

"The intellect," says Aristotle, in one passage, "is perfected, not by knowledge but by activity ; [γ] and in another, "The arts and sciences are powers, but every power exists only for the sake of action ; the end of philosophy, therefore, is not knowledge, but the energy conversant about knowledge." [δ] Descending to the schoolmen : "The intellect," says Aquinas, "commences in operation, and in operation it ends;"[e] and Scotus even declares that a man's knowledge is measured by the amount of his mental activity— "tantum scit homo, quantum operatur." [ζ] The profoundest thinkers of modern times have emphatically

a This definition is not to be found in the Platonic Dialogues ; a passage something like it occurs in the *Euthydemus*, p. 290. Cf. Diog. Laert., lib. viii. *Pythagoras*, § 8 : 'Εν τῷ βίῳ, οἱ μὲν ἀνδραποδώδεις φύονται, δόξης καὶ πλεονεξίας θηραταί· οἱ δὲ φιλόσοφοι, τῆς ἀληθείας.—ED.

β Prior, *Lines to the Hon. C. Montague ; British Poets*, vol. vii. p. 393, (Anderson's ed.)—ED.

γ Said of moral knowledge, *Eth. Nic.*, i. 3: Τέλος οὐ γνῶσις, ἀλλὰ πρᾶξις. Cf. *ibid.*, i. 7, 13 ; i. 8, 9 ; ix. 7, 4 ; xi. 9, 7 ; x. 7, 1. *Met.*, xi. 7: 'Η νοῦ ἐνέργεια ζωή.—ED.

δ This sentence seems to be made up from two separate passages in the *Metaphysics*. Lib. viii. c. 2 : Πᾶσαι

αἱ τέχναι καὶ αἱ ποιητικαὶ καὶ ἐπιστῆμαι δυνάμεις εἰσίν. Lib. viii. c. 8 : Τέλος δ' ἡ ἐνέργεια, καὶ τούτου χάριν ἡ δύναμις λαμβάνεται· . . . καὶ τὴν θεωρητικὴν (ἔχουσιν) ἵνα θεωρῶσιν ἀλλ' οὐ θεωροῦσιν ἵνα θεωρητικὴν ἔχωσιν.—ED.

e This is perhaps the substance of *Summa*, Pars i., Q. lxxix., art. ii. and iii.—ED.

ζ These words contain the substance of the doctrine of Scotus regarding science, given in his *Quæstiones in Aristotelis Logicam*, p. 318— *Super Lib. Post.*, Q. i. "Scire in *actu*," says the subtle doctor, " est quum aliquis cognoscit majorem et minorem, et, simul cum hoc, applicat præmissas ad conclusionem. Sic igitur patet quod actualitas scientiæ est ex appli-

testified to the same great principle. "If," says Malebranche, "I held truth captive in my hand, I should open my hand and let it fly, in order that I might again pursue and capture it."ᵃ "Did the Almighty," says Lessing, "holding in his right hand *Truth*, and in his left *Search after Truth*, deign to tender me the one I might prefer,—in all humility, but without hesitation, I should request *Search after Truth*."β "Truth," says Von Müller, "is the property of God, the pursuit of truth is what belongs to man;"ᵞ and Jean Paul Richter : "It is not the goal, but the course, which makes us happy."ᵟ But there would be no end of similar quotations.ᵉ

But if speculative truth itself be only valuable as a mean of intellectual activity, those studies which determine the faculties to a more vigorous exertion, will, in every liberal sense, be better entitled, absolutely, to the name of useful, than those which, with a greater complement of more certain facts, awaken them to a less intense, and consequently to a less improving exercise. On this ground I would rest one of the pre-eminent utilities of mental philosophy. That it comprehends all the sublimest objects of our theoretical and moral interest ; that every (natural) conclusion concerning God, the soul, the present worth and the future destiny of man, is exclusively deduced from the philosophy of mind, will be at once admitted.

Philosophy best entitled to the appellation useful.

ratione causæ ad affectum." Compare Quæst. ii., —"An acquisitio scientiæ sit nobis per doctrinam ?"—for his view of the end and means of education.—Ed.

a ["Malebranche disait avec une ingénieuse exagération, 'Si je tenais la vérité captive dans ma main, j'ouvrirais la main afin de poursuivre en- core la vérité.'"—Mazure, *Cours de Philosophie*, tom. i. p. 20.]

β *Eine Duplik*, § 1 ; *Schriften*, edit. Lachmann, x. p. 49.—Ed.

γ ["Die Wahrheit ist in Gott, uns bleibt das Forschen."]

δ *Leben*, drittes Heft, § 257. See Scheidler's *Psychologie*, p. 45.—Ed.

ε Compare *Discussions*, p. 40.

LECT. I.

But I do not at present found the importance on the paramount dignity of the pursuit. It is as the best gymnastic of the mind,—as a mean, principally, and almost exclusively, conducive to the highest education of our noblest powers,—that I would vindicate to these speculations the necessity which has too frequently been denied them. By no other intellectual application is the mind thus reflected on itself, and its faculties aroused to such independent, vigorous, unwonted, and continued energy;—by none, therefore, are its best capacities so variously and intensely evolved. " By turning," says Burke, " the soul inward on itself, its forces are concentred, and are fitted for greater and stronger flights of science; and in this pursuit, whether we take or whether we lose our game, the chase is certainly of service." [a]

Application of the foregoing principles to the conduct of a class of philosophy.

These principles being established, I have only now to offer a few observations in regard to their application, that is, in regard to the mode in which I conceive that this class ought to be conducted. From what has already been said, my views on this subject may be easily anticipated. Holding that the paramount end of liberal study is the development of the student's mind, and that knowledge is principally useful as a mean of determining the faculties to that exercise, through which this development is accomplished,—it follows, that I must regard the main duty of a Professor to consist not simply in communicating information, but in doing this in such a manner, and with such an accompaniment of subsidiary means, that the information he conveys may be the occasion of awakening his pupils to a vigorous and varied exertion of their faculties. Self-activity is the indis-

a *On the Sublime and Beautiful,* Preface, p. 8.—ED.

pensable condition of improvement; and education is
education,—that is, accomplishes its purpose, only by
affording objects and supplying incitements to this
spontaneous exertion.   Strictly speaking, every one
must educate himself.

But as the end of education is thus something more
than the mere communication of knowledge, the communication of knowledge ought not to be all that
academical education should attempt.   Before printing
was invented, Universities were of primary importance
as organs of publication, and as centres of literary con-
fluence: but since that invention, their utility as media
of communication is superseded; consequently, to justify the continuance of their existence and privileges,
they must accomplish something that cannot be accomplished by books.   But it is a remarkable circumstance that, before the invention of printing, universities viewed the activity of the pupil as the great
mean of cultivation, and the communication of knowledge as only of subordinate importance ; whereas,
since that invention, universities, in general, have
gradually allowed to fall into disuse the powerful
means which they possess of rousing the pupil to exertion, and have been too often content to act as mere
oral instruments of information, forgetful, it would
almost seem, that Fust and Coster ever lived.   It is
acknowledged, indeed, that this is neither the principal nor the proper purpose of a university.   Every
writer on academical education from every corner of
Europe proclaims the abuse, and, in this and other
universities, much has been done by individual effort
to correct it.[a]

But though the common duty of all academical

a Compare *Discussions*, p. 772.—ED.

LECT.
I.

The true
end of
liberal
education.

The condi-
tions of in-
struction in
intellectual
philosophy.

instructors be the cultivation of the student, through
the awakened exercise of his faculties, this is more
especially incumbent on those to whom is intrusted
the department of liberal education; for, in this
department, the pupil is trained, not to any mere
professional knowledge, but to the command and
employment of his faculties in general. But, moreover, the same obligation is specially imposed upon a
professor of intellectual philosophy, by the peculiar
nature of his subject, and the conditions under which
alone it can be taught. The phænomena of the external world are so palpable and so easily described,
that the experience of one observer suffices to render
the facts he has witnessed intelligible and probable to
all. The phænomena of the internal world, on the
contrary, are not capable of being thus described: all
that the prior observer can do, is to enable others to
repeat his experience. In the science of mind, we
can neither understand nor be convinced of anything
at second hand. Here testimony can impose no belief; and instruction is only instruction as it enables
us to teach ourselves. A fact of consciousness, however accurately observed, however clearly described,
and however great may be our confidence in the
observer, is for us as zero, until we have observed and
recognised it ourselves. Till that be done, we cannot
realise its possibility, far less admit its truth. Thus
it is that, in the philosophy of mind, instruction can
do little more than point out the position in which
the pupil ought to place himself, in order to verify,
by his own experience, the facts which his instructor
proposes to him as true. The instructor, therefore,
proclaims, οὐ φιλοσοφία, ἀλλὰ φιλοσοφεῖν; he does
not profess to teach *philosophy, but to philosophise.*

It is this condition imposed upon the student of
doing everything himself, that renders the study of ——
the mental sciences the most improving exercise of Use and
intellect.   But everything depends upon the condition importance
of examinations in
being fulfilled; and, therefore, the primary duty of a a class of
Philosophy.
teacher of philosophy is to take care that the student
does actually perform for himself the necessary process.   In the first place, he must discover, by examination, whether his instructions have been effective,—
whether they have enabled the pupil to go through
the intellectual operation; and, if not, it behoves him
to supply what is wanting,—to clear up what has
been misunderstood.   In this view, examinations are
of high importance to a Professor; for without such a
medium between the teacher and the taught, he can
never adequately accommodate the character of his
instruction to the capacity of his pupils.

But, in the second place, besides placing his pupil in The intellectual in-
a condition to perform the necessary process, the in- structor
must seek
structor ought to do what in him lies to determine the to influence
the will of
pupil's *will* to the performance.   But how is this to be his pupils.
effected?   Only by rendering the effort more pleasurable than its omission.   But every effort is at first
difficult,—consequently irksome.   The ultimate benefit
it promises is dim and remote, while the pupil is often
of an age at which present pleasure is more persuasive
than future good.   The pain of the exertion must,
therefore, be overcome by associating with it a still
higher pleasure.   This can only be effected by enlist-
ing some passion in the cause of improvement.   We
must awaken emulation, and allow its gratification only
through a course of vigorous exertion.   Some rigorists,
I am aware, would proscribe, on moral and religious
grounds, the employment of the passions in education;

but such a view is at once false and dangerous. The affections are the work of God; they are not radically evil; they are given us for useful purposes, and are, therefore, not superfluous. It is their abuse that is alone reprehensible. In truth, however, there is no alternative. In youth, passion is preponderant. There is then a redundant amount of energy which must be expended; and this, if it find not an outlet through one affection, is sure to find it through another. The aim of education is thus to employ for good those impulses which would otherwise be turned to evil. The passions are never neutral; they are either the best allies, or the worst opponents, of improvement. "Man's nature," says Bacon, "runs either to herbs or weeds; therefore let him seasonably water the one and destroy the other."[a] Without the stimulus of emulation, what can education accomplish? The love of abstract knowledge, and the habit of application, are still unformed, and if emulation intervene not, the course by which these are acquired is, from a strenuous and cheerful energy, reduced to an inanimate and dreary effort; and this too at an age when pleasure is all-powerful, and impulse predominant over reason. The result is manifest.

These views have determined my plan of practical instruction. Regarding the communication of knowledge as a high, but not the highest, aim of academical instruction, I shall not content myself with the delivery of Lectures. By all the means in my power I shall endeavour to rouse you, Gentlemen, to the free and vigorous exercise of your faculties; and shall deem my task accomplished, not by teaching Logic and Philosophy, but by teaching to reason and philosophise.[β]

a Essay xxxviii.—"Of Nature in Men,"—*Works*, ed. Montagu, vol. i. p. 133.—ED.

β For Fragment containing the Author's views on the subject of Academical Honours, see Appendix I.—ED.

LECTURE II.[a]

PHILOSOPHY—ITS ABSOLUTE UTILITY.

(B) OBJECTIVE.

IN the perverse estimate which is often made of the ends and objects of education, it is impossible that the Science of Mind,—Philosophy Proper,—the Queen of Sciences, as it was denominated of old, should not be degraded in common opinion from its pre-eminence, as the highest branch of general education; and, therefore, before attempting to point out to you what constitutes the value of Philosophy, it becomes necessary to clear the way by establishing a correct notion of what the value of a study is.

Some things are valuable, finally, or for themselves, —these are ends; other things are valuable, not on their own account, but as conducive towards certain ulterior ends,—these are means. The value of ends is absolute,—the value of means is relative. Absolute value is properly called a *good*,—relative value is properly called a *utility*.[β] Of goods, or absolute ends,

LECT.
II.

The value of a study.

Ends and means.

---

a It is to be observed, that the Lectures here printed as First and Second, were not uniformly delivered by the Author in that order. The one or other was, however, usually given as the Introductory Lecture of the Course. This circumstance accounts for the repetition of the principal doctrines of Lecture I. in the opening of Lecture II.—ED.

β [Cf. Aristotle, *Eth. Nic.*, lib. i. c. 7, § 1.]

there are for man but two,—perfection and happiness. By perfection is meant the full and harmonious development of all our faculties, corporeal and mental, intellectual and moral ; by happiness, the complement of all the pleasures of which we are susceptible.

Human perfection and happiness coincide.
Now, I may state, though I cannot at present attempt to prove, and I am afraid many will not even understand the statement, that human perfection and human happiness coincide, and thus constitute, in reality, but a single end. For as, on the one hand, the perfection or full development of a power is in proportion to its capacity of free, vigorous, and continued action, so, on the other, all pleasure is the concomitant of activity ; its degree being in proportion as that activity is spontaneously intense, its prolongation in proportion as that activity is spontaneously continued ; whereas, pain arises either from a faculty being restrained in its spontaneous tendency to action, or from being urged to a degree, or to a continuance, of energy beyond the limit to which it of itself freely tends.

To promote our perfection is thus to promote our happiness ; for to cultivate fully and harmoniously our various faculties, is simply to enable them by exercise to energise longer and stronger without painful effort ; that is, to afford us a larger amount of a higher quality of enjoyment.

Criterion of the utility of a study.
Perfection (comprising happiness) being thus the one end of our existence, in so far as man is considered either as an end unto himself, or as a mean to the glory of his Creator ; it is evident that, absolutely speaking, that is, without reference to special circumstances and relations, studies and sciences must, in common with all other pursuits, be judged useful as they contribute, and only as they contribute, to the

perfection of our humanity,—that is, to our perfection simply as men. It is manifest that in this relation alone can anything distinctively, emphatically, and without qualification, be denominated useful; for as our perfection as men is the paramount and universal end proposed to the species, whatever we may style useful in any other relation, ought, as conducive only to a subordinate and special end, to be so called, not simply, but with qualifying limitation. Propriety has, however, in this case been reversed in common usage. For the term Useful has been exclusively bestowed, in ordinary language, on those branches of instruction which, without reference to his general cultivation as a man or a gentleman, qualify an individual to earn his livelihood by a special knowledge or dexterity in some lucrative calling or profession; and it is easy to see how, after the word had been thus appropriated to what, following the Germans, we may call the *Bread and Butter* sciences, those which more proximately and obtrusively contribute to the intellectual and moral dignity of man, should, as not having been styled the useful, come, in popular opinion, to be regarded as the useless branches of instruction.

As it is proper to have different names for different things, we may call the higher utility, or that conducive to the perfection of a man viewed as an end in himself, by the name of Absolute or General; the inferior utility, or that conducive to the skill of an individual viewed as an instrument for some end out of himself, by the name of Special or Particular.

Now, it is evident, that in estimating the utility of any branch of education, we ought to measure it both by the one kind of utility and by the other; but it is also evident that a neglect of the former standard will

Margin notes:
LECT. II.

General and Particular Utility.

lead us further wrong in appreciating the value of
any branch of common or general instruction, than a
neglect of the latter.

It has been the tendency of different ages, of different countries, of different ranks and conditions of
society, to measure the utility of studies rather by one
of these standards than by both.   Thus it was the bias
of antiquity, when the moral and intellectual cultivation of the citizen was viewed as the great end of all
political institutions, to appreciate all knowledge principally by the higher standard ; on the contrary, it is
unfortunately the bias of our modern civilisation, since
the accumulation, (and not to the distribution), of
riches in a country, has become the grand problem of
the statesman, to appreciate it rather by the lower.

In considering, therefore, the utility of philosophy,
we have, first, to determine its Absolute, and, in the
second place, its Special utility—I say its special utility,
for, though not itself one of the professional studies, it
is mediately more or less conducive to them all.

In the present Lecture I must, of course, limit myself to one branch of this division; and even a part of
the first or Absolute utility will more than occupy our
hour.

Philoso-
phy: its
Absolute
utility.
Limiting myself, therefore, to the utility of philosophy as estimated by the higher standard alone, it is
further to be observed that, on this standard, a science
or study is useful in two different ways, and, as these
are not identical,—this pursuit being more useful in
the one way, that pursuit more useful in the other,—
these in reality constitute two several standards of
utility, by which each branch of knowledge ought to
be separately measured.

The cultivation, the intellectual perfection, of a man,

may be estimated by the amount of two different elements; it may be estimated by the mere sum of truths which he has learned, or it may be estimated by the greater development of his faculties, as determined by their greater exercise in the pursuit and contemplation of truth. For, though this may appear a paradox, these elements are not merely not convertible, but are, in fact, very loosely connected with each other; and as an individual may possess an ample magazine of knowledge, and still be little better than an intellectual barbarian, so the utility of one science may be principally seen in affording a greater number of higher and more indisputable truths,—the utility of another in determining the faculties to a higher energy, and consequently to a higher cultivation. The former of these utilities we may call the Objective, as it regards the object matter about which our cognitive faculties are occupied; the other the Subjective, inasmuch as it regards our cognitive faculties themselves as the subject in which knowledge is inherent.

I shall not at present enter on the discussion which of these utilities is the higher. In the opening Lecture of last year, I endeavoured to show that all knowledge is only for the sake of energy, and that even merely speculative truth is valuable only as it determines a greater quantity of higher power into activity. In that Lecture, I also endeavoured to show that, on the standard of subjective utility, philosophy is of all our studies the most useful; inasmuch as more than any other it exercises, and consequently develops, to a higher degree and in a more varied manner, our noblest faculties. At present, on the contrary, I shall confine myself to certain views of the importance of philosophy, estimated by the standard of its Objective

LECT.
II.

Absolute utility of a science of two kinds —Objective and Subjective.

Philosophy: its Objective utility.

utility. The discussion, I am aware, will be found
somewhat disproportioned to the age and average
capacity of my hearers; but, on this occasion, and
before this audience, I hope to be excused if I venture
for once on matters which, to be adequately understood,
require development and illustration from the matured
intelligence of those to whom they are presented.

The human
mind the
noblest
object of
speculation.
Considered in itself, a knowledge of the human mind,
whether we regard its speculative or its practical importance, is confessedly of all studies the highest and the
most interesting. "On earth," says an ancient philosopher, "there is nothing great but man; in man, there
is nothing great but mind." [a] No other study fills and
satisfies the soul like the study of itself. No other
science presents an object to be compared in dignity, in
absolute or in relative value, to that which human consciousness furnishes to its own contemplation. What
is of all things the best? asked Chilon of the Oracle.
"To know thyself," was the response. This is, in fact,
the only science in which all are always interested, for,
while each individual may have his favourite occupation, it still remains true of the species that

"The proper study of mankind is man." [B]

Sir Thomas
Browne
quoted.
"Now for my life," says Sir Thomas Browne, "it is
a miracle of thirty years, which to relate were not a
history, but a piece of poetry, and would sound to
common ears like a fable.

"For the world, I count it not an inn, but an hospital; and a place not to live but to die in. The

a [Phavorinus, quoted by Joannes
Picus Mirandulanus, In Astrologiam,
lib. iii. p. 351, Basil. ed.] For notice
of Phavorinus, see Vossius, De Hist.
Græc., lib. ii. c. 10.—ED.

B Pope, Essay on Man, ii. 2.—ED.
[Cf. Charron, De la Sagesse, liv. i.
chap. i. "Le vray estude de l'homme
est l'homme."]

world that I regard is myself; it is the microcosm of
my own frame that I cast mine eye on : for the other,
I use it but like my globe, and turn it round sometimes for my recreation.   Men that look upon my out-
side, perusing only my condition and fortunes, do err
in my altitude ; for I am above Atlas his shoulders.
The earth is a point not only in respect of the heavens
above us, but of that heavenly and celestial part within
us.   That mass of flesh that circumscribes me, limits
not my mind.   That surface that tells the heavens it
hath an end, cannot persuade me I have any.   I take
my circle to be above three hundred and sixty.   Though
the number of the ark do measure my body, it compre-
hendeth not my mind.   Whilst I study to find how I
am a microcosm, or little world, I find myself something
more than the great.   There is surely a piece of divinity
in us ; something that was before the elements, and
owes no homage unto the sun.   Nature tells me, I am
the image of God, as well as Scripture.   He that understands not thus much hath not his introduction or
first lesson, and is yet to begin the alphabet of man."[a]

But, though mind, considered in itself, be the noblest
object of speculation which the created universe presents to the curiosity of man, it is under a certain relation that I would now attempt to illustrate its utility ; for mind rises to its highest dignity when viewed
as the object through which, and through which alone,
our unassisted reason can ascend to the knowledge of
a God.   The Deity is not an object of immediate contemplation ; as existing and in himself, he is beyond
our reach ; we can know him only mediately through
his works, and are only warranted in assuming his existence as a certain kind of cause necessary to account

Relation of
Psychology
to Theo-
logy.

a Browne's *Religio Medici*, part ii. § 11.   *Discussions*, p. 311.—ED.

Existence
of Deity an
inference
from a spe-
cial class of
effects.

for a certain state of things, of whose reality our faculties are supposed to inform us. The affirmation of a
God being thus a regressive inference, from the existence of a special class of effects to the existence of a
special character of cause, it is evident, that the whole
argument hinges on the fact,—Does a state of things
really exist such as is only possible through the agency
of a Divine Cause? For if it can be shown that such
a state of things does not really exist, then, our inference to the kind of cause requisite to account for it, is
necessarily null.

These af-
forded ex-
clusively
by the
phæno-
mena of
mind.

This being understood, I now proceed to show you
that the class of phænomena which requires that kind
of cause we denominate a Deity, is exclusively given
in the phænomena of mind,—that the phænomena of
matter, taken by themselves, (you will observe the
qualification, taken by themselves), so far from war-
ranting any inference to the existence of a God, would,
on the contrary, ground even an argument to his
negation,—that the study of the external world taken
with, and in subordination to, that of the internal,
not only loses its atheistic tendency, but, under such
subservience, may be rendered conducive to the great
conclusion, from which, if left to itself, it would dissuade us.

We must first of all then consider what kind of
cause it is which constitutes a Deity, and what kind
of effects they are which allow us to infer that a
Deity must be.

The notion of a God is not contained in the notion
of a mere First Cause; for in the admission of a first
cause, Atheist and Theist are at one. Neither is this
notion completed by adding to a first cause the attribute of Omnipotence, for the atheist who holds mat-

ter or necessity to be the original principle of all that is, does not convert his blind force into a God, by merely affirming it to be all-powerful.   It is not until the two great attributes of Intelligence and Virtue (and be it observed that virtue involves Liberty)—I say, it is not until the two attributes of intelligence and virtue or holiness, are brought in, that the belief in a primary and omnipotent cause becomes the belief in a veritable Divinity.   But these latter attributes are not more essential to the divine nature than are the former.   For as original and infinite power does not of itself constitute a God, neither is a God constituted by intelligence and virtue, unless intelligence and goodness be themselves conjoined with this original and infinite power.   For even a creator, intelligent and good and powerful, would be no God, were he dependent for his intelligence and goodness and power on any higher principle.   On this supposition, the perfections of the creator are viewed as limited and derived.   He is himself, therefore, only a dependency,—only a creature ; and if a God there be, he must be sought for in that higher principle, from which this subordinate principle derives its attributes. Now is this highest principle, (*ex hypothesi* all-powerful), also intelligent and moral,—then it is itself alone the veritable Deity ; on the other hand is it, though the author of intelligence and goodness in another, itself unintelligent,—then is a blind Fate constituted the first and universal cause, and atheism is asserted.

The peculiar attributes which distinguish a Deity from the original omnipotence or blind fate of the atheist, being thus those of intelligence and holiness of will,—and the assertion of theism being only the assertion that the universe is created by intelligence and governed not only by physical but by moral laws, we have next to consider how we are warranted in these two affirmations, 1°, That intelligence stands first in the absolute order of existence,—in other words, that final preceded efficient causes; and, 2°, That the universe is governed by moral laws.

The proof of these two propositions is the proof of a God; and it establishes its foundation exclusively on the phænomena of mind. I shall endeavour, Gentlemen, to show you this, in regard to both these propositions; but, before considering how far the phænomena of mind and of matter do and do not allow us to infer the one position or the other, I must solicit your attention to the characteristic contrasts which these two classes of phænomena in themselves exhibit.

In the compass of our experience, we distinguish two series of facts,—the facts of the external or material world, and the facts of the internal world or world of intelligence. These concomitant series of phænomena are not like streams which merely run parallel to each other; they do not, like the Alpheus and Arethusa, flow on side by side without a commingling of their waters. They cross, they combine, they are interlaced; but notwithstanding their intimate connection, their mutual action and reaction, we are able to discriminate them without difficulty, because they are marked out by characteristic differences.

The phænomena of the material world are subjected to immutable laws, are produced and reproduced in the same invariable succession, and manifest only the blind force of a mechanical necessity.

The phænomena of man are, in part, subjected to the laws of the external universe. As dependent

upon a bodily organisation, as actuated by sensual
propensities and animal wants, he belongs to matter,
and in this respect he is the slave of necessity. But
what man holds of matter does not make up his personality. They are his, not he; man is not an organism,—he is an intelligence served by organs.[a] For
in man there are tendencies,—there is a law,—which
continually urge him to prove that he is more powerful than the nature by which he is surrounded and
penetrated. He is conscious to himself of faculties
not comprised in the chain of physical necessity, his
intelligence reveals prescriptive principles of action,
absolute and universal, in the Law of Duty, and a
liberty capable of carrying that law into effect, in
opposition to the solicitations, the impulsions of his
material nature. From the coexistence of these opposing forces in man there results a ceaseless struggle
between physical necessity and moral liberty; in the
language of Revelation, between the Flesh and the
Spirit; and this struggle constitutes at once the distinctive character of humanity, and the essential condition of human development and virtue.

In the facts of intelligence we thus become aware
of an order of existence diametrically in contrast to
that displayed to us in the facts of the material universe. There is made known to us an order of things,
in which intelligence, by recognising the unconditional
law of duty and an absolute obligation to fulfil it,
recognises its own possession of a liberty incompatible
with a dependence upon fate, and of a power capable
of resisting and conquering the counteraction of our
animal nature.

a ["Mens cujusque, is est quis- Scipionis, c. 8—after Plato.] Cf.
que; non ea figura, quæ digito de- Plato, Alc. Prim., p. 130, and infra,
monstrari potest."—Cicero, Somnium p. 164.—ED.

LECT.
II.

Conscious-
ness of
freedom,
and of a
law of
duty, the
conditions
of Theo-
logy.

Now, it is only as man is a free intelligence, a moral power, that he is created after the image of God, and it is only as a spark of divinity glows as the life of our life in us, that we can rationally believe in an Intelligent Creator and Moral Governor of the universe. For, let us suppose, that in man intelligence is the product of organisation, that our consciousness of moral liberty is itself only an illusion, in short, that acts of volition are results of the same iron necessity which determines the phænomena of matter ;—on this supposition, I say, the foundations of all religion, natural and revealed, are subverted.[a]

The truth of this will be best seen by applying the supposition of the two positions of theism previously stated—viz. that the notion of God necessarily supposes, 1°, That in the absolute order of existence intelligence should be first, that is, not itself the product of an unintelligent antecedent; and, 2°, That the universe should be governed not only by physical but by moral laws.

First con-
dition of
the proof
of a Deity,
drawn from
Psycho-
logy.

Now, in regard to the former, how can we attempt to prove that the universe is the creation of a free original intelligence, against the counter-position of the atheist, that liberty is an illusion, and intelligence, or the adaptation of means to ends, only the product of a blind fate? As we know nothing of the absolute

Analogy
between
our experi-
ence and
the abso-
lute order
of exist-
ence.

order of existence in itself, we can only attempt to infer its character from that of the particular order within the sphere of our experience, and as we can affirm naught of intelligence and its conditions, except what we may discover from the observation of our own minds, it is evident that we can only analogically carry out into the order of the universe, the relation

a See *Discussions*, p. 623.—ED.

in which we find intelligence to stand in the order of the human constitution. If in man, intelligence be a free power,—in so far as its liberty extends, intelligence must be independent of necessity and matter; and a power independent of matter necessarily implies the existence of an immaterial subject,—that is, a spirit. If then the original independence of intelligence on matter in the human constitution, in other words, if the spirituality of mind in man be supposed a datum of observation, in this datum is also given both the condition and the proof of a God. For we have only to infer, what analogy entitles us to do, that intelligence holds the same relative supremacy in the universe which it holds in us, and the first positive condition of a Deity is established, in the establishment of the absolute priority of a free creative intelligence. On the other hand, let us suppose the result of our study of man to be, that intelligence is only a product of matter, only a reflex of organisation, such a doctrine would not only afford no basis on which to rest any argument for a God, but, on the contrary, would positively warrant the atheist in denying his existence. For if, as the materialist maintains, the only intelligence of which we have any experience be a consequent of matter,— on this hypothesis, he not only cannot assume this order to be reversed in the relations of an intelligence beyond his observation, but, if he argue logically, he must positively conclude, that, as in man, so in the universe, the phænomena of intelligence or design are only in their last analysis the products of a brute necessity. Psychological materialism, if carried out fully and fairly to its conclusions, thus inevitably results in theological atheism; as it has been well expressed by Dr Henry More,

LECT.
II.

Psychological Materialism: its issue.

LECT.
II.

Nullus in microcosmo spiritus, nullus in macrocosmo Deus.[a] I do not of course mean to assert that all materialists deny, or actually disbelieve, a God. For, in very many cases, this would be at once an un-merited compliment to their reasoning, and an un-merited reproach to their faith.

Second condition of the proof of a Deity, drawn from Psychology.

Such is the manifest dependence of our theology on our psychology in reference to the first condition of a Deity,—the absolute priority of a free intelligence. But this is perhaps even more conspicuous in relation to the second, that the universe is governed not merely by physical but by moral laws, for God is only God in-asmuch as he is the Moral Governor of a Moral World.

Our interest also in its establishment is incompar-ably greater, for while a proof that the universe is the work of an omnipotent intelligence, gratifies only our speculative curiosity,—a proof that there is a holy legislator by whom goodness and felicity will be ulti-mately brought into accordance, is necessary to satisfy both our intellect and our heart. A God is, indeed, to us only of practical interest, inasmuch as he is the condition of our immortality.

Now, it is self-evident, in the first place, that if there be no moral world, there can be no moral gover-nor of such a world; and, in the second, that we have, and can have, no ground on which to believe in the reality of a moral world, except in so far as we our-selves are moral agents. This being undeniable, it is further evident, that, should we ever be convinced that we are not moral agents, we should likewise be convinced that there exists no moral order in the uni-verse, and no supreme intelligence by which that moral order is established, sustained, and regulated.

a Cf. *Antidotus adversus Atheis-mum,* lib. iii. c. 16, (*Opera Omnia,* vol. ii. p. 143, Londini, 1679); and the Author's *Discussions,* p. 788.—ED.

Theology is thus again wholly dependent on Psycho- logy ; for, with the proof of the moral nature of man, stands or falls the proof of the existence of a Deity.

But in what does the character of man as a moral agent consist? Man is a moral agent only as he is accountable for his actions,—in other words, as he is the object of praise or blame ; and this he is, only inasmuch as he has prescribed to him a rule of duty, and as he is able to act, or not to act, in conformity with its precepts. The possibility of morality thus depends on the possibility of liberty ; for if man be not a free agent, he is not the author of his actions, and has, therefore, no responsibility,—no moral personality at all.

Now the study of Philosophy, or mental science, operates in three ways to establish that assurance of human liberty, which is necessary for a rational belief in our own moral nature, in a moral world, and in a moral ruler of that world.

In the first place, an attentive consideration of the phænomena of mind is requisite in order to a luminous and distinct apprehension of liberty as a fact or datum of intelligence. For though, without philosophy, a natural conviction of free agency lives and works in the recesses of every human mind, it requires a process of philosophical thought to bring this conviction to clear consciousness and scientific certainty.

In the second place, a profound philosophy is necessary to obviate the difficulties which meet us when we attempt to explain the possibility of this fact, and to prove that the datum of liberty is not a mere illusion. For though an unconquerable feeling compels us to recognise ourselves as accountable, and therefore free, agents, still, when we attempt to realise in thought how the fact of our liberty can be, we soon find that this altogether transcends our understanding, and that every effort to bring the fact of liberty within the compass of our conceptions, only results in the substitution in its place of some more or less disguised form of necessity. For,—if I may be allowed to use expressions which many of you cannot be supposed at present to understand,—we are only able to conceive a thing, inasmuch as we conceive it under conditions; while the possibility of a free act supposes it to be an act which is not conditioned or determined. The tendency of a superficial philosophy is, therefore, to deny the fact of liberty, on the principle that what cannot be conceived is impossible. A deeper and more comprehensive study of the facts of mind, overturns this conclusion, and disproves its foundation. It shows that, —so far from the principle being true, that what is inconceivable is impossible,—on the contrary, all that is conceivable is a mean between two contradictory extremes, both of which are inconceivable, but of which, as mutually repugnant, the one or the other must be true. Thus philosophy, in demonstrating that the limits of thought are not to be assumed as the limits of possibility, while it admits the weakness of our discursive intellect, re-establishes the authority of consciousness, and vindicates the veracity of our primitive convictions. It proves to us, from the very laws of mind, that while we can never understand *how* any original datum of intelligence is possible, we have no reason from this inability to doubt *that* it is true. A learned ignorance is thus the end of philosophy, as it is the beginning of theology.[a]

a See *Discussions*, p. 634.—ED.

In the third place, the study of mind is necessary to counterbalance and correct the influence of the study of matter; and this utility of Metaphysics rises in proportion to the progress of the natural sciences, and to the greater attention which they engross.

An exclusive devotion to physical pursuits, exerts an evil influence in two ways. In the first place, it diverts from all notice of the phænomena of moral liberty, which are revealed to us in the recesses of the human mind alone; and it disqualifies from appreciating the import of these phænomena, even if presented, by leaving uncultivated the finer power of psychological reflection, in the exclusive exercise of the faculties employed in the easier and more amusing observation of the external world. In the second place, by exhibiting merely the phænomena of matter and extension, it habituates us only to the contemplation of an order in which everything is determined by the laws of a blind or mechanical necessity. Now, what is the inevitable tendency of this one-sided and exclusive study? That the student becomes a materialist, if he speculate at all. For, in the first place, he is familiar with the obtrusive facts of necessity, and is unaccustomed to develop into consciousness the more recondite facts of liberty : he is, therefore, disposed to disbelieve in the existence of phænomena whose reality he may deny, and whose possibility he cannot understand. At the same time, the love of unity, and the philosophical presumption against the multiplication of essences, determine him to reject the assumption of a second, and that an hypothetical, substance,—ignorant as he is of the reasons by which that assumption

Twofold evils of exclusive physical study

LECT.
II.

Physical
study in
its infancy
not mate-
rialising.

is legitimated. In the infancy of science, this tendency of physical study was not experienced. When
men first turned their attention on the phænomena
of nature, every event was viewed as a miracle, for
every effect was considered as the operation of an intelligence. God was not exiled from the universe of
matter; on the contrary, he was multiplied in proportion to its phænomena. As science advanced, the
deities were gradually driven out; and long after the
sublunary world had been disenchanted, they were
left for a season in possession of the starry heavens.
The movement of the celestial bodies, in which Kepler
still saw the agency of a free intelligence, was at
length by Newton resolved into a few mechanical
principles: and at last even the irregularities which
Newton was compelled to leave for the miraculous
correction of the Deity, have been proved to require
no supernatural interposition; for La Place has shown
that all contingencies, past and future, in the heavens,
find their explanation in the one fundamental law of
gravitation.

But the very contemplation of an order and adaptation so astonishing, joined to the knowledge that
this order and adaptation are the necessary results of
a brute mechanism,—when acting upon minds which
have not looked into themselves for the light of which
the world without can only afford them the reflection,—far from elevating them more than any other
aspect of external creation to that inscrutable Being
who reigns beyond and above the universe of nature,
tends, on the contrary, to impress on them, with peculiar force, the conviction, that as the mechanism of
nature can explain so much, the mechanism of nature
can explain all.

"Wonder," says Aristotle, "is the first cause of philosophy:"[a] but in the discovery that all existence is but mechanism, the consummation of science would be an extinction of the very interest from which it originally sprang. "Even the gorgeous majesty of the heavens," says a great religious philosopher, "the object of a kneeling adoration to an infant world, subdues no more the mind of him who comprehends the one mechanical law by which the planetary systems move, maintain their motion, and even originally form themselves. He no longer wonders at the object, infinite as it always is, but at the human intellect alone which in a Copernicus, Kepler, Gassendi, Newton, and La Place, was able to transcend the object, by science to terminate the miracle, to reave the heaven of its divinities, and to exorcise the universe. But even this, the only admiration of which our intelligent faculties are now capable, would vanish, were a future Hartley, Darwin, Condillac, or Bonnet, to succeed in displaying to us a mechanical system of the human mind as comprehensive, intelligible, and satisfactory as the Newtonian mechanism of the heavens."[b]

To this testimony I may add that, should Physiology ever succeed in reducing the facts of intelligence to Phænomena of matter, Philosophy would be subverted in the subversion of its three great objects, —God, Free-Will, and Immortality. True wisdom would then consist, not in speculation, but in repressing thought during our brief transit from nothingness to nothingness. For why? Philosophy would have become a meditation, not merely of death but of an-

LECT. II.

If all existence be but mechanism, philosophical interest extinguished.

a *Metaph.*, i. 2, 9. Compare Plato, *Theætetus*, p. 155.—ED.

β Jacobi, *Werke*, vol. ii. p. 52-54. Quoted in *Discussions*, p. 812.—ED.

LECT.
II.

nihilation; the precept, *Know thyself,* would have been replaced by the terrific oracle to Œdipus—

> " May'st thou ne'er know the truth of what thou art ;"

and the final recompense of our scientific curiosity would be wailing deeper than Cassandra's, for the ignorance that saved us from despair.

Coincidence of the views here given with those of previous philosophers.

The views which I have now taken of the respective influence of the sciences of mind and of matter in relation to our religious belief, are those which have been deliberately adopted by the profoundest thinkers, ancient and modern. Were I to quote to you the testimonies that crowd on my recollection to the effect that ignorance of Self is ignorance of God, I should make no end, for this is a truth proclaimed by Jew and Gentile, Christian and Mahommedan.[a] I shall content myself with adducing three passages from three philosophers, which I select, both as articulately confirming all that I have now advanced, and because there are not, in the whole history of speculation, three authorities on the point in question more entitled to respect.

Plato.

The first quotation is from Plato, and it corroborates the doctrine I have maintained in regard to the conditions of a God, and of our knowledge of his existence. " The cause," he says, " of all impiety and irreligion among men is, that reversing in themselves the relative subordination of mind and body, they have, in like manner, in the universe, made that to be first which is second, and that to be second which is first ; for while, in the generation of all things, intelligence and final causes precede matter and efficient causes, they, on the contrary, have viewed matter and

---

a On Self-Knowledge, as the con- *sions,* pp. 787, 788, and the authorities
dition of knowing God, see *Discus-*   there cited.—ED.

material things as absolutely prior, in the order of LECT.
existence, to intelligence and design ; and thus depart- II.
ing from an original error in relation to themselves,
they have ended in the subversion of the Godhead." [a]

The second quotation is from Kant ; it finely illus- Kant.
trates the influences of material and mental studies by
contrasting them in reference to the very noblest object
of either, and the passage is worthy of your attention.
not only for the soundness of its doctrine, but for the
natural and unsought-for sublimity of its expression.
" Two things there are, which, the oftener and the more
steadfastly we consider them, fill the mind with an
ever new, an ever rising admiration and reverence ;—
the STARRY HEAVEN *above, the* MORAL LAW *within.*
Of neither am I compelled to seek out the reality, as
veiled in darkness, or only to conjecture the possibility,
as beyond the hemisphere of my knowledge.　Both I
contemplate lying clear before me, and connect both
immediately with my consciousness of existence.　The
one departs from the place I occupy in the outer world
of sense ; expands, beyond the bounds of imagination,
this connection of my body with worlds rising beyond
worlds, and systems blending into systems ; and pro-
tends it also into the illimitable times of their periodic
movement — to its commencement and perpetuity.
The other departs from my invisible self, from my personality ; and represents me in a world, truly infinite
indeed, but whose infinity can be tracked out only by
the intellect, with which also my connection, unlike
the fortuitous relation I stand in to all worlds of sense,
I am compelled to recognise as universal and neces-

a *De Legibus,* lib. x. pp. 888, 889.　sect. iv. (p. 435 *et seq.* of vol. iii. Lond.
Quoted in *Discussions,* p. 312.　Com-　ed. 1845), and *Eternal and Immut. Mo-*
pare Cudworth, *Intell. System,* c. v.　*rality,* book iv. c. vi. § 6, *seq.*—Ed.

sary. In the former, the first view of a countless multitude of worlds annihilates, as it were, my importance as an *animal product*, which, after a brief and that incomprehensible endowment with the powers of life, is compelled to refund its constituent matter to the planet—itself an atom in the universe—on which it grew. The other, on the contrary, elevates my worth as an *intelligence* even without limit; and this through my personality, in which the moral law reveals a faculty of life independent of my animal nature, nay, of the whole material world :—at least if it be permitted to infer as much from the regulation of my being, which a conformity with that law exacts ; proposing, as it does, my moral worth for the absolute end of my activity, conceding no compromise of its imperative to a necessitation of nature, and spurning, in its infinity, the conditions and boundaries of my present transitory life." [a]

Jacobi.

The third quotation is from the pious and profound Jacobi, and it states the truth boldly and without disguise in regard to the relation of Physics and Metaphysics to Religion. "But is it unreasonable to confess, that we believe in God, not by reason of the nature [β] which conceals him, but by reason of the supernatural in man, which alone reveals and proves him to exist ?

"*Nature conceals God:* for through her whole domain Nature reveals only fate, only an indissoluble chain of mere efficient causes without beginning and without end, excluding, with equal necessity, both

---

a *Kritik der praktischen Vernunft,* Beschluss. Quoted in *Discussions,* p. 310.—Ed.

β [In the philosophy of Germany, *Natur,* and its correlatives, whether of Greek or Latin derivation, are, in general, expressive of the world of Matter, in contrast to the world of Intelligence.]—*Oral Interpolation,* supplied from *Reid's Works,* p. 216.—Ed.

providence and chance. An independent agency, a
free original commencement within her sphere and
proceeding from her powers, is absolutely impossible.
Working without will, she takes counsel neither of the
good nor of the beautiful ; creating nothing, she casts
up from her dark abyss only eternal transformations
of herself, unconsciously and without an end ; further-
ing with the same ceaseless industry decline and increase, death and life,—never producing what alone
is of God and what supposes liberty,—the virtuous,
the immortal.

" *Man reveals God :* for Man by his intelligence
rises above nature, and in virtue of this intelligence is
conscious of himself as a power not only independent
of, but opposed to, nature, and capable of resisting,
conquering, and controlling her. As man has a living
faith in this power, superior to nature, which dwells
in him ; so has he a belief in God, a feeling, an experience of his existence. As he does not believe in
this power, so does he not believe in God ; he sees,
he experiences naught in existence but nature,—necessity,—fate." [a]

Such is the comparative importance of the sciences These uses of Psycho-
of mind and of matter in relation to the interests of logy not
religion. But it may be said, how great soever be the superseded by the
value of philosophy in this respect, were man left Christian revelation.
to rise to the divinity by the unaided exercise of his
faculties, this value is superseded under the Christian
dispensation, the Gospel now assuring us of all and
more than all philosophy could ever warrant us in
surmising. It is true, indeed, that in Revelation there
is contained a great complement of truths of which
natural reason could afford us no knowledge or assur-

[a] *Von den Göttlichen Dingen. Werke,* iii. p. 424-6.—ED.

ance, but still the importance of mental science to
theology has not become superfluous in Christianity;
for whereas anterior to Revelation, religion rises out
of psychology as a result, subsequently to revelation,
it supposes a genuine philosophy of mind as the condition of its truth. This is at once manifest. Revelation is a revelation to man and concerning man;
and man is only the object of revelation, inasmuch
as he is a moral, a free, a responsible being. The
Scriptures are replete with testimonies to our natural
liberty; and it is the doctrine of every Christian
church that man was originally created with a will
capable equally of good as of evil, though this will,
subsequently to the Fall, has lost much of its primitive
liberty. Christianity thus, by universal confession,
supposes as a condition the moral nature of its object;
and if some individual theologians be found who have
denied to man a higher liberty than a machine, this is
only another example of the truth, that there is no
opinion which has been unable to find not only its
champions but its martyrs. The differences which
divide the Christian churches on this question, regard
only the liberty of man in certain particular relations,
for fatalism, or a negation of human responsibility in
general, is equally hostile to the tenets of the Calvinist
and Arminian.

In these circumstances it is evident, that he who
disbelieves the moral agency of man must, in consistency with that opinion, disbelieve Christianity. And
therefore inasmuch as Philosophy,—the Philosophy of
Mind,—scientifically establishes the proof of human
liberty, philosophy, in this, as in many other relations
not now to be considered, is the true preparative and
best aid of an enlightened Christian Theology.

LECTURE III.

THE NATURE AND COMPREHENSION OF PHILOSOPHY.

I HAVE been in the custom of delivering sometimes together, more frequently in alternate years, two systematic courses of lectures,—the one on PSYCHOLOGY, that is, the science which is conversant about the phænomena of mind in general,—the other on LOGIC, that is, the science of the laws regulating the manifestation and legitimacy of the highest faculty of Cognition, —Thought, strictly so denominated—the faculty of Relations,—the Understanding proper. As first, or initiative, courses of philosophy,—each has its peculiar advantages; and I know not, in truth, which I should recommend a student to commence with. What, however, I find it expedient to premise to each is an *Introduction*, in which the nature and general relations of philosophy are explained, and a summary view taken of the faculties, (particularly the Cognitive faculties), of mind.

In the ensuing course, we shall be occupied with the General Philosophy of Mind.

You are, then, about to commence a course of philosophical discipline,—for Psychology is pre-eminently a philosophical science. It is therefore proper, before proceeding to a consideration of the special objects of our course, that you should obtain at least a general notion of what philosophy is. But in affording you this

LECT.
III.

information, it is evident that there lie considerable difficulties in the way. For the definition and the divisions of philosophy are the results of a lofty generalisation from particulars, of which particulars you are, or must be presumed to be, still ignorant. You cannot, therefore, it is manifest, be made adequately to comprehend, in the commencement of your philosophical studies, notions which these studies themselves are intended to enable you to understand. But although you cannot at once obtain a full knowledge of the nature of philosophy, it is desirable that you should be enabled to form at least some vague conception of the road you are about to travel, and of the point to which it will conduct you. I must, therefore, beg that you will, for the present, hypothetically believe,—believe upon authority,—what you may not now adequately understand; but this only to the end that you may not hereafter be under the necessity of taking any conclusion upon trust. Nor is this temporary exaction of credit peculiar to philosophical education. In the order of nature, belief always precedes knowledge,—it is the condition of instruction. The child (as observed by Aristotle) must believe, in order that he may learn;[a] and even the primary facts of intelligence,—the facts which precede, as they afford the conditions of, all knowledge,—would not be original were they revealed to us under any other form than that of natural or necessary beliefs. Without further preamble, therefore, I shall now endeavour to afford you some general notion of what philosophy is.[β]

In doing this, there are two questions to be an-

---

a *Soph. Elench.*, c. 2.—ED.   *inter Antiquos*, see Brandis, *Geschichte*
β On comprehension of Philosophy   *der Philosophie*, &c., vol. i. § 6, p. 7, *seq.*

swered :—1st, What is the meaning of the *name?*
and, 2d, What is the meaning of the *thing?*   An answer to the former question is afforded in a nominal
definition of the term *philosophy,* and in a history of
its employment and application.

In regard to the etymological signification of the
word, you are of course aware that Philosophy is a
term of Greek origin—that it is a compound of φίλος,
a *lover* or *friend,* and σοφία,[a] *wisdom*—speculative
wisdom.   Philosophy is thus, literally, *a love of wisdom.*   But if the grammatical meaning of the word
be unambiguous, the history of its application is, I
think, involved in considerable doubt.   According to
the commonly received account, the designation of
philosopher (*lover* or *suitor of wisdom*) was first
assumed and applied by Pythagoras ; whilst the
occasion and circumstances of its assumption, we
have a story by Cicero,[β] on the authority of Heraclides
Ponticus ; [γ] and by Diogenes Laertius, in one place,[δ]
on the authority of Heraclides, and in another,[ε] on
that of Sosicrates,—although it be doubtful whether
the word Sosicrates be not in the second passage a
corrupted lection for Heraclides ;[ζ] in which case the

---

a Σοφία in Greek, though sometimes used in a wide sense, like the
term *wise* applied to skill in handi-
craft, yet properly denoted speculative, not practical wisdom or prudence.   See Aristotle, *Eth. Nic.,* lib.
vi. c. 7, with the commentary of
Eustratius. Διὸ Ἀναξαγόραν, καὶ Θαλῆν
καὶ τοὺς τοιούτους, σοφοὺς μὲν, φρονίμους
δ' οὔ φασιν εἶναι, ὅταν ἴδωσιν ἀγνοοῦντας
τὰ συμφέροντ' ἑαυτοῖς· καὶ περιττὰ μὲν,
καὶ θαυμαστὰ, καὶ χαλεπὰ, καὶ δαιμόνια
εἰδέναι αὐτοὺς φασιν, ἀχρηστὰ δ', ὅτι οὐ
τὰ ἀνθρώπινα ἀγαθὰ ζητοῦσιν. Ἡ δὲ
φρόνησις περὶ τὰ ἀνθρώπινα, καὶ περὶ ὧν

ἐστι βουλεύσασθαι.   From the long
commentary of Eustratius, the following extract will be sufficient : Ἀλλὰ τὸ
τέλος τοῦ σοφοῦ, ἡ θεωρία τῆς ἀληθείας
ἐστι, καὶ ἡ τοῦ ὄντος κατάληψις· οὐχι
δέ τι πρακτὸν ἀγαθόν.   Πρακτὸν γάρ
ἐστιν ἀγαθὸν τὸ διὰ πράξεως κατορθού-
μενον, θεωρία δὲ πράξεως ἑτέρα.—ED.

β *Tusc. Quæst.,* lib. v. c. 3.

γ Heraclides Ponticus — scholar
both of Plato and of Aristotle.

δ Lib. i. 12.

ε Lib. viii. 8.

ζ See Menage, *Commentary on
Laertius,* viii. 8.

The inter-
view of
Pythagoras
and Leon.

whole probability of the story will depend upon the trustworthiness of Heraclides alone, for the comparatively recent testimony of Iamblichus, in his Life of Pythagoras, must go for nothing. As told by Cicero, it is as follows :—Pythagoras, once upon a time (says the Roman orator), having come to Phlius, a city of Peloponnesus, displayed in a conversation which he had with Leon, who then governed that city, a range of knowledge so extensive, that the prince, admiring his eloquence and ability, inquired to what art he had principally devoted himself. Pythagoras answered, that he professed no art, and was simply a *philosopher*. Leon, struck by the novelty of the name, again inquired who were the philosophers, and in what they differed from other men. Pythagoras replied, that human life seemed to resemble the great fair, held on occasion of those solemn games which all Greece met to celebrate. For some, exercised in athletic contests, resorted thither in quest of glory and the crown of victory; while a greater number flocked to them in order to buy and sell, attracted by the love of gain. There were a few, however,—and they were those distinguished by their liberality and intelligence,—who came from no motive of glory or of gain, but simply to look about them, and to take note of what was done, and in what manner. So likewise, continued Pythagoras, we men all make our entrance into this life on our departure from another. Some are here occupied in the pursuit of honours, others in the search of riches; a few there are who, indifferent to all else, devote themselves to an inquiry into the nature of things. These, then, are they whom I call students of wisdom, for such is meant by philosopher.

Pythagoras was a native of Samos, and flourished

about 560 years before the advent of Christ,[a]—about 130 years before the birth of Plato.   Heraclides and Sosicrates, the two vouchers of this story,—if Sosicrates be indeed a voucher,—lived long subsequently to the age of Pythagoras; and the former is, moreover, confessed to have been an egregious fabulist.   From the principal circumstances of his life, mentioned by Laertius after older authors, and from the fragments we possess of the works of Heraclides,—in short, from all opinions, ancient and modern, we learn that he was at once credulous and deceitful,—a dupe and an impostor.[β]   The anecdote, therefore, rests on very slender authority.   It is probable, I think, that Socrates was the first who adopted, or, at least, the first who familiarised, the expression.[γ]   It was natural that he should be anxious to contradistinguish himself from the Sophists, (οἱ σοφοὶ, οἱ σοφισταὶ, sophistæ), literally, the *wise* men;[δ] and no term could more appropriately ridicule the arrogance of these pretenders, or afford a happier contrast to their haughty designation, than that of philosopher (*i.e.* the *lover of wisdom*); and, at the same time, it is certain that the substantives φιλοσοφία and φιλόσοφος, first ap-

---

a The exact dates of the birth and death of Pythagoras are uncertain. Nearly all authorities, however, are agreed that he "flourished" B.C. 540-510, in the times of Polycrates and Tarquinius Superbus.(Clinton, *F. H.*, 510).   His birth is usually placed in the 49th Olympiad (B.C. 584).   See Brandis, *Gesch. der Phil.*, vol. i. p. 422; Zeller, *Phil. der Griechen.*, vol. i. p. 217, 2d ed.—Ed.

β Compare Meiners, *Geschichte der Wissenschaften in Griechenland und Rom*, vol. i. p. 118; and Krug, *Lexikon*, vol. iii. p. 211.—Ed.

γ There is, however, the ἰητρὸς φιλόσοφος ἰσόθεος of Hippocrates. But this occurs in one of the Hippocratic writings which is manifestly spurious, and of date subsequent to the father of medicine.   Hippocrates was an early contemporary of Socrates. [The expression occurs in the Περὶ Εὐσχημοσύνης, *Opera—Quarta Classis*, p. 41, ed. Venice, 1588.—Ed.]

δ Perhaps rather, "the Professors of Wisdom." See an able paper by Mr Cope in the *Journal of Classical and Sacred Philology*, vol. i. p. 182. —Ed.

pear in the writings of the Socratic school.[a]   It is
true, indeed, that the verb φιλοσοφεῖν is found in
Herodotus, in the address by Crœsus to Solon ;[β] and
that too in a participal form, to designate the latter
as a man who had travelled abroad for the purpose of
acquiring knowledge, (ὡς φιλοσοφέων γῆν πολλὴν θεω-
ρίης εἵνεκεν ἐπελήλυθας).   It is, therefore, not impossible that, before the time of Socrates, those who devoted themselves to the pursuit of the higher branches
of knowledge, were occasionally designated philosophers : but it is far more probable that Socrates and
his school first appropriated the term as a distinctive
appellation ; and that the word *philosophy*, in consequence of this appropriation, came to be employed for
the complement of all higher knowledge, and, more
especially, to denote the science conversant about the
principles or causes of existence.   The term *philosophy*,
I may notice, which was originally assumed in modesty, soon lost its Socratic and etymological signification, and returned to the meaning of σοφία, or
wisdom.   Quintillian[γ] calls it *nomen insolentissimum*;
Seneca,[δ] *nomen invidiosum*; Epictetus[e] counsels his
scholars not to call themselves " Philosophers ;" and
proud is one of the most ordinary epithets with which
philosophy is now associated.   Thus Campbell, in his
Address to the Rainbow, says :

> " I ask not *proud* philosophy
> To tell me what thou art."

So much for the name signifying ; we proceed now
to the thing signified.   Were I to detail to you the

---

a See especially Plato, *Phædrus*,
p. 278 : Τὸ μέν σοφόν, ὦ Φαῖδρε, καλεῖν
ἔμοιγε μέγα εἶναι δοκεῖ καὶ θεῷ μόνῳ
πρέπειν. τὸ δὲ ἢ φιλόσοφον ἢ τοιοῦτόν
τι μᾶλλόν τε ἂν αὐτῷ ἁρμόττοι καὶ ἐμ-
μελεστέρως ἔχοι.   Compare also the
description of the philosopher in the

Symposium, p. 204, as μεταξὺ σοφοῦ
καὶ ἀμαθοῦς.—ED.
β Lib. i. 30.
γ *Inst. Orat.*, Proœm.
δ *Epist.*, v.
e *Ench.*, c. 68, ed. Wolf ; 46, ed.
Schweigh.

various definitions [a] of philosophy which philosophers have promulgated—far more, were I to explain the grounds on which the author of each maintains the exclusive adequacy of his peculiar definition—I should, in the present stage of your progress, only perplex and confuse you. Philosophy, for example,—and I select only a few specimens of the more illustrious definitions,—philosophy has been defined :—The science of things divine and human, and of the causes in which they are contained ; [β]—The science of effects by their causes ; [γ]—The science of sufficient reasons ; [δ]—The science of things possible, inasmuch as they are possible ; [ε]—The science of things, evidently deduced from first principles [ζ]—The science of truths, sensible and abstract ; [η]—The application of reason to its legitimate objects ; [θ]—The science of the relations of all knowledge to the necessary ends of human reason ; [ι]—

a Vide Gassendi, i. p. 1, seq.; Denzinger, Instit. Log., i. p. 40; Scheidler's Encyclop., pp. 56, 75; Weiss, Log., p. 8; Scheiblerus, Op. Log., i. p. 1, seq.

β Cicero, De Officiis, ii. 2 : " Nec quidquam aliud est philosophia, si interpretari velis, quam studium sapientiæ. Sapientia autem est, (ut a veteribus philosophis definitum est), rerum divinarum et humanarum, causarumque quibus hæ res continentur, scientia." Cf. Tusc. Quæst., iv. 26, v. 3. De Fin., ii. 12 ; Seneca, Epist. 89 ; Pseudo-Plutarch, De Plac. Philos., Procem.: Οἱ μὲν οὖν Στωϊκοὶ ἔφασαν τὴν μὲν σοφίαν εἶναι θείων τε καὶ ἀνθρωπίνων ἐπιστήμην· τὴν δὲ φιλοσοφίαν, ἄσκησιν τέχνης ἐπιτηδείου. Cf. Plato, Phædrus, p. 259 ; Rep., vi. p. 486.—Ed.

γ Hobbes, Computatio sive Logica, c. 1 : " Philosophia est effectuum sive Phænomenon ex conceptis eorum causis seu generationibus, et rursus generationum quæ esse possunt, ex cognitis effectibus per rectam ratiocinationem acquisita cognitio." Cf. Arist. Metaph., i. 1: Τὴν ὀνομαζομένην σοφίαν περὶ τὰ πρῶτα αἴτια καὶ τὰς ἀρχὰς ὑπολαμβάνουσι πάντες.—Ed.

δ Leibnitz, quoted by Mazure, Cours de Philosophie, tom. i. p. 2 ; see also Wenzel, Elementa Philosophiæ, tom. I. § 7. Cf. Leibnitz, Lettres entre Leibnitz et Clarke,—Opera, p. 778, (ed. Erd.)—Ed.

ε Wolf, Philosophia Rationalis, § 29.—Ed.

ζ Descartes, Principia, Epistola Authoris. Cf. Wolf, Phil. Rat., § 33.—Ed.

η Condillac, L'Art de Raisonner, Cours, tom. iii. p. 3, (ed. 1780). Cf. Clemens Alex., Strom., viii. 8, p. 782: Ἡ δὲ τῶν φιλοσόφων πραγματεία περὶ τε τὰ νοήματα καὶ τὰ ὑποκείμενα καταγίνεται.—Ed.

θ Compare Tennemann, Geschichte der Philosophie, Einleitung, § 13.—Ed.

ι Kant, Kritik der reinen Vernunft, Methodenlehre, c. 3 ; Krug, Philosophisches Lexikon, iii. p. 213.—Ed.

The science of the original form of the ego or mental
self; [a]—The science of science ;[β]—The science of the
absolute ; [γ]—The science of the absolute indifference
of the ideal and real [δ]—or, The identity of identity
and non-identity, &c. &c.[ε] All such definitions are
(if not positively erroneous), either so vague that they
afford no precise knowledge of their object; or they
are so partial, that they exclude what they ought to
comprehend ; or they are of such a nature that they
supply no preliminary information, and are only to
be understood, (if ever), after a knowledge has been
acquired of that which they profess to explain. It is,
indeed, perhaps impossible, adequately to define philosophy. For what is to be defined comprises what
cannot be included in a single definition. For philosophy is not regarded from a single point of view,—
it is sometimes considered as theoretical,—that is, in
relation to man as a thinking and cognitive intelligence ; sometimes as practical,—that is, in relation
to man as a moral agent ; and sometimes, as comprehending both theory and practice. Again, philosophy
may either be regarded objectively,—that is, as a complement of truths known ; or subjectively,—that is, as
a habit or quality of the mind knowing. In these circumstances, I shall not attempt a definition of philosophy, but shall endeavour to accomplish the end which
every definition proposes,—make you understand, as
precisely as the unprecise nature of the object-matter

a Krug, *Philosophisches Lexikon*,
iii. p. 213. The definition is substantially Fichte's. See his *Grundlage
der Gesammten Wissenschaftslehre*,
(*Werke*, i. p. 283) ; and his *Zweite
Einleitung in die Wissenschaftslehre*,
(*Werke*, i. p. 515.)—ED.

β Fichte, *Über den Begriff der Wissenschaftslehre*, § 1 (*Werke*, i. 45.)—ED.

γ Schelling, *Vom Ich als Princip
der Philosophie*, §§ 6, 9 ; Krug, *Lexikon*, iii. p. 213.—ED.

δ Schelling, *Bruno*, p. 205 (2d ed.)
Cf. *Philosophie der Natur*, Einleitung,
p. 64, and Zusatz zur Einleitung, p.
65-88 (2d ed.)—ED.

ε Hegel, *Logik*, (*Werke*, iii. p. 64.)
— ED.

permits, what is meant by philosophy, and what are the sciences it properly comprehends within its sphere.

As a matter of history I may here, however, paren- thetically mention, that in Greek antiquity there were in all six definitions of philosophy which obtained celebrity. On these collectively there are extant various treatises. Among the commentators of Aristotle, that of Ammonius Hermiæ[a] is the oldest; and the fullest is one by an anonymous author, lately published by Dr Cramer in the fourth volume of his *Anecdota Græca Parisiensia.*[β] Of the six, the first and second define philosophy from its object-matter,—that which it is about; the third and fourth, from its end,—that for the sake of which it is; the fifth, from its relative pre-eminence; and the sixth, from its etymology.

The first of these definitions of philosophy is,—"the knowledge of things existent, as existent," (γνῶσις τῶν ὄντων ᾗ ὄντα).[γ]

The second is—"the knowledge of things divine and human," (γνῶσις θείων καὶ ἀνθρωπίνων πραγμάτων).[δ] These are both from the object-matter; and both were referred to Pythagoras.

The third and fourth, the two definitions of philosophy from its end, are, again, both taken from Plato. Of these the third is,—" philosophy is a meditation of death," (μελέτη θανάτου) ;[ε] the fourth,—"philosophy

a *Ammonii in quinque voces Porphyrii Commentarius*, p. 1 (ed. Ald.) Given in part by Brandis, *Scholia in Aristotelem*, p. 9.—ED.

β P. 389. Extracted also in part by Brandis, *Scholia in Aristotelem*, p. 8. This commentary is conjectured by Val. Rose (*De Aristotelis Librorum Ordine et Auctoritate*, p. 243) to be the work of Olympiodorus. The definitions quoted in the text are given by Tzetzes, *Chiliads*, x. 600.—ED.

γ Cf. Arist. *Metaph.*, iii. 1.—ED.

δ See *ante*, p. 49, note β.—ED.

ε *Phædo*, p. 80: Τοῦτο δὲ οὐδὲν ἄλλο ἐστὶν ἢ ὀρθῶς φιλοσοφοῦσα καὶ τῷ ὄντι τεθνάναι μελετῶσα ῥᾳδίως· ἢ οὐ τοῦτ᾽ ἂν εἴη μελέτη θανάτου: Cf. Cicero, *Tusc. Quæst.*, i. 30, with the relative commentary by Davis; Macrobius, *In Som. Scipionis*, i. 13; Damascenus, *Dialectica*, c. 3.—ED.

LECT.
III.

is a resembling of the Deity in so far as that is competent to man," (ὁμοίωσις θεῷ κατὰ τὸ δυνατὸν ἀνθρώπῳ).ᵃ

The fifth, that from its pre-eminence, was borrowed from Aristotle, and defined philosophy "the art of arts, and science of sciences," (τέχνη τεχνῶν καὶ ἐπιστήμη ἐπιστημῶν).ᵝ

Finally, the sixth, that from the etymology, was, like the first and second, carried up to Pythagoras—it defined philosophy "the love of wisdom," (φιλία σοφίας).ᵞ

To these a seventh and even an eighth were sometimes added,—but the seventh was that by the physicians, who defined medicine the philosophy of bodies, (ἰατρικὴ ἐστι φιλοσοφία σωμάτων); and philosophy, the medicine of souls, (φιλοσοφία ἐστὶν ἰατρικὴ ψυχῶν).ᵟ This was derided by the philosophers; as, to speak with Homer, being an exchange of brass for gold, and of gold for brass, (χρύσεα χαλκείων); and as defining the more known by the less known.

The eighth is from an expression of Plato, who, in the Theætetus,ᵉ calls philosophy "the greatest music," (μεγίστη μουσικὴ), meaning thereby the harmony of

---

a *Theætetus*, p. 176: Διὸ καὶ πειρᾶσ-θαι χρὴ ἐνθένδε ἐκεῖσε φεύγειν ὅτι τάχιστα· φυγὴ δὲ ὁμοίωσις θεῷ κατὰ τὸ δυνατόν.—ED.

β The anonymous commentator quotes this as a passage from the *Metaphysics*. It does not occur literally, but the sense is substantially that expressed in Book i. c. 2: Ἀκριβέστα-ται δὲ τῶν ἐπιστημῶν αἱ μάλιστα τῶν πρώτων εἰσὶν . . . Ἀλλὰ μὴν καὶ διδασ-καλικὴ γε ἡ τῶν αἰτιῶν θεωρητικὴ μᾶλ-λον· . . . οὕτε τῆς τοιαύτης ἄλλην χρὴ νομίζειν τιμιωτέραν· ἡ γὰρ θειοτάτη καὶ τιμιωτάτη. Cf. *Eth. Nic.*, vi. 7: Δῆλον

ὅτι ἡ ἀκριβεστάτη ἂν τῶν ἐπιστημῶν εἴη ἡ σοφία. The nearest approach to a definition of Philosophy in the *Metaphysics* is in A minor, c. 1: Ὀρθῶς δ' ἔχει καὶ τὸ καλεῖσθαι τὴν φιλοσοφίαν ἐπιστήμην τῆς ἀληθείας.—ED.

γ See *ante*, p. 45.—ED.

δ Anon. apud Cramer, *Anecdota*, iv. p. 398; Brandis, *Scholia*, p. 7.—ED.

e So quoted by the commentator; but the passage occurs in the *Phædo*, p. 61: Καὶ ἐμοὶ οὕτω τὸ ἐνύπνιον ὅπερ ἔπραττον, τοῦτο ἐπικελεύειν, μουσικὴν ποιεῖν, ὡς φιλοσοφίας μὲν οὔσης μεγίστης μουσικῆς.—ED.

the rational, irascible, and appetent, parts of the soul, LECT.
III.
(λόγος, θυμός, ἐπιθυμία).

But to return : All philosophy is knowledge, but all Philosophi-
cal and em-
knowledge is not philosophy. Philosophy is, therefore, pirical
knowledge.
a kind of knowledge. What, then, is philosophical
knowledge, and how is it discriminated from knowledge in general ? We are endowed by our Creator
with certain faculties of observation, which enable us
to become aware of certain appearances or phænomena.
These faculties may be stated as two,—Sense, or External Perception, and Self-Consciousness, or Internal
Perception; and these faculties severally afford us
the knowledge of a different series of phænomena.
Through our senses, we apprehend what exists, or what
occurs, in the external or material world ; by our
self-consciousness,[a] we apprehend what is, or what
occurs, in the internal world, or world of thought.
What is the extent, and what the certainty, of the
knowledge acquired through sense and self-consciousness, we do not at present consider. It is now sufficient that the simple fact be admitted, that we do
actually thus know ; and that fact is so manifest, that
it requires, I presume, at my hands, neither proof nor
illustration.

The information which we thus receive,—that cer- Empirical
knowledge
tain phænomena are, or have been, is called Historical, —what.
or Empirical knowledge.[β] It is called historical, because, in this knowledge, we know only the fact, only
that the phænomenon is; for history is properly only the
narration of a consecutive series of phænomena in time,
or the description of a coexistent series of phænomena

---

a On the place and sphere of Con-
sciousness, see *Discussions*, p. 47.—
ED. β Brandis, *Geschichte der Philoso-*
phie, vol. i. p. 2. [Cf. Wolf, *Phil.*
Rat., § 3.—ED.]

in space. Civil history is an example of the one ; natural history of the other. It is called empirical or experiential, if we might use that term, because it is given us by experience or observation, and not obtained as the result of inference or reasoning. I may notice, by parenthesis, that you must discharge from your minds the by-meaning accidentally associated with the word *empiric* or *empirical*, in common English. This term is with us more familiarly used in reference to medicine, and from its fortuitous employment in that science, in a certain sense, the word empirical has unfortunately acquired, in our language, a one-sided and an unfavourable meaning. Of the origin of this meaning many of you may not be aware. You are aware, however, that ἐμπειρία is the Greek term for experience, and ἐμπειρικὸς an epithet applied to one who uses experience. Now, among the Greek physicians, there arose a sect who, professing to employ experience alone to the exclusion of generalisation, analogy, and reasoning, denominated themselves οἱ ἐμπειρικοί—the Empirics. The opposite extreme was adopted by another sect, who, rejecting observation, founded their doctrine exclusively on reasoning and theory ;—and these called themselves οἱ μεθοδικοί—or Methodists. A third school, of whom Galen was the head, opposed equally to the two extreme sects of the Empirics and of the Methodists, and availing themselves both of experience and reasoning, were styled οἱ δογματικοί — the Dogmatists, or rational physicians.[a] A keen controversy arose ; the Empirics

a See Galen, *De Sectis*, c. i., and the *Definitiones Medicæ* and *Introductio seu Medicus*, ascribed to the same author; Celsus, *De Re Medica*, Præf.; Dan. Le Clerc, *Histoire de la Médecine*, part ii., liv. ii., ch. 1—liv. iv., ch. 1.—ED.

were defeated; they gradually died out; and their doctrine, of which nothing is known to us, except through the writings of their adversaries,[a] has probably been painted in blacker colours than it deserved. Be this, however, as it may, the word was first naturalised in English, at a time when the Galenic works were of paramount authority in medicine, as a term of medical import—of medical reproach; and the collateral meaning, which it had accidentally obtained in that science, was associated with an unfavourable signification, so that an Empiric, in common English, has been long a synonym for a charlatan or quack-doctor, and, by a very natural extension, in general, for any ignorant pretender in science. In philosophical language, the term *empirical* means simply what belongs to, or is the product of, experience or observation, and, in contrast to another term afterwards to be explained, is now technically in general use through every other country of Europe. Were there any other word to be found of a corresponding signification in English, it would perhaps, in consequence of the by-meaning attached to empirical, be expedient not to employ this latter. But there is not. *Experiential* is not in common use, and *experimental* only designates a certain kind of experience—viz. that in which the fact observed has been brought about by a certain intentional pre-arrangement of its co-efficients. But this by the way.

Returning, then, from our digression : Historical or empirical knowledge is simply the knowledge that something is. Were we to use the expression, *the knowledge that*, it would sound awkward and unusual in our modern languages. In Greek, the most philosophical of all tongues, its parallel, however, was fami-

a Le Clerc, *Histoire de la Médecine*, part ii., liv. ii., ch. 1.—ED.

liarly employed, more especially in the Aristotelic philosophy,[a] in contrast to another knowledge of which
we are about to speak. It was called τὸ ὅτι, that is,
ἡ γνῶσις ὅτι ἔστιν.[β] I should notice, that with us
the knowledge that, is commonly called the knowledge
of the *fact*.[γ] As examples of empirical knowledge,
take the facts, whether known on our own experience
or on the testified experience of others,—that a stone
falls,—that smoke ascends,—that the leaves bud in
spring and fall in autumn,—that such a book contains
such a passage,—that such a passage contains such an
opinion,—that Cæsar, that Charlemagne, that Napoleon, existed.[δ]

Philosophi-
cal know-
ledge—
what.

But things do not exist, events do not occur, isolated,
—apart—by themselves; they exist, they occur, and
are by us conceived, only in connection. Our observation affords us no example of a phænomenon which
is not an effect; nay, our thought cannot even realise
to itself the possibility of a phænomenon without a
cause. We do not at present inquire into the nature

a See *Anal. Post.*, ii. 1: Τὰ ζητού-
μενά ἐστιν ἴσα τὸν ἀριθμὸν ὅσαπερ ἐπι-
στάμεθα. Ζητοῦμεν δὲ τέτταρα, τὸ ὅτι,
τὸ διότι, εἰ ἔστι, τί ἐστιν. These were
distinguished by the Latin logicians
as the *quæstiones scibiles*, and were
usually rendered *quod sit, cur sit, an
sit, quid sit.*—ED.

β This expression in Latin, at least
in Latin not absolutely barbarous,
can only be translated vaguely by
an accusative and an infinitive, for
you are probably aware that the con-
junctive *quod*, by which the Greek
ὅτι is often translated, has always a
causal signification in genuine Latinity. Thus, we cannot say, *scio quod
res sit, credo quod tu sis doctus:*—this
is barbarous. We must say, *scio
rem esse, credo te esse doctum.*

γ [Empirical is also used in contrast with Necessary knowledge;
the former signifying the knowledge
simply of what is, the latter of what
must be.]—*Oral Interpolation.*

δ The terms historical and empirical are used as synonymous by Aristotle, as both denoting a knowledge
of the ὅτι. (Compare the *De Incessu
Animalium*, c. 1; *Metaph.*, i. 1.)
Aristotle, therefore, calls his empirical work on animals, *History of Animals;*—Theophrastus, his empirical
work on plants, *History of Plants;*—
Pliny, his empirical book on nature
in general, *Natural History.* Pliny
says: "Nobis propositum est *natu-
ras rerum indicare manifestas, non
causas indagare dubias.*" See Brandis, *Geschichte der Philosophie*, i. p. 2.

of the connection of effect and cause,[a]—either in reality or in thought. It is sufficient for our present purpose to observe that, while, by the constitution of our nature, we are unable to conceive anything to begin to be, without referring it to some cause,—still the knowledge of its particular cause is not involved in the knowledge of any particular effect. By this necessity which we are under of thinking some cause for every phænomenon; and by our original ignorance of what particular causes belong to what particular effects,—it is rendered impossible for us to acquiesce in the mere knowledge of the fact of a phænomenon: on the contrary, we are determined,—we are necessitated, to regard each phænomenon as only partially known until we discover the causes on which it depends for its existence. For example, we are struck with the appearance in the heavens called the rainbow. Think we cannot that this phænomenon has no cause, though we may be wholly ignorant of what that cause is. Now, our knowledge of the phænomenon as a mere fact,—as a mere isolated event,—does not content us; we therefore set about an inquiry into the cause,—which the constitution of our mind compels us to suppose,—and at length discover that the rainbow is the effect of the refraction of the solar rays by the watery particles of a cloud. Having ascertained the cause, but not till then, we are satisfied that we fully know the effect.

Now, this knowledge of the cause of a phænomenon is different from, is something more than, the knowledge of that phænomenon simply as a fact; and these two cognitions or knowledges[b] have, accordingly, re-

---

a See on this point the Author's *Discussions*, p. 609.—ED.

β [*Knowledges* is a term in frequent use by Bacon, and though now obso-

ceived different names. The latter, we have seen, is called *historical*, or *empirical* knowledge; the former is called *philosophical*, or *scientific*, or *rational* knowledge.[a] Historical, is the knowledge that a thing is—philosophical, is the knowledge why or how it is. And as the Greek language, with peculiar felicity, expresses historical knowledge by the ὅτι—the γνῶσις ὅτι ἔστι: so, it well expresses philosophical knowledge by the διότι[β]—the γνῶσις διότι ἔστι, though here its relative superiority is not the same. To recapitulate what has now been stated :—There are two kinds or degrees of knowledge. The first is the knowledge that a thing is—ὅτι χρῆμα ἔστι, *rem esse;*—and it is called the knowledge of the fact, historical, or empirical knowledge. The second is the knowledge why or how a thing is, διότι χρῆμα ἔστι, *cur res sit;*—and is termed the knowledge of the cause, philosophical, scientific, rational knowledge.

Philosophy implies a search after first causes.

Philosophical knowledge, in the widest acceptation of the term, and as synonymous with science, is thus the knowledge of effects as dependent on their causes. Now, what does this imply? In the first place, as every cause to which we can ascend is itself also an effect,—it follows that it is the scope, that is, the aim of philosophy, to trace up the series of effects and causes, until we arrive at causes which are not also themselves effects. These first causes do not indeed lie within the reach of philosophy, nor even within the sphere of our comprehension; nor, consequently, on

lete, should be revived, as, without it, we are compelled to borrow *cognitions* to express its import.}—*Oral Interpolation.* [See Bacon's *Advancement of Learning,* p. 176, (*Works,* vol. ii., ed. Mont.); and Sergeant's *Method to Science,* Preface, p. xxv., p. 166, *et alibi passim.*—ED.]

a Wolf, *Philosophia Rationalis,* § 6; Kant, *Kritik der reinen Vernunft,* Methodenlehre, c. 3.—ED.

β Arist. *Anal. Post.,* ii. 1.—ED.

the actual reaching them does the existence of philosophy depend.   But as philosophy is the knowledge
of effects in their causes, the tendency of philosophy
is ever upwards; and philosophy can, in thought, in
theory, only be viewed as accomplished,—which in
reality it never can be,—when the ultimate causes,—
the causes on which all other causes depend,—have
been attained and understood.[a]

But, in the second place, as every effect is only produced by the concurrence of at least two causes, (and
by cause, be it observed, I mean everything without
which the effect could not be realised), and as these
concurring or co-efficient causes, in fact, constitute
the effect, it follows, that the lower we descend in the
series of causes, the more complex will be the product;
and that the higher we ascend, it will be the more
simple.   Let us take, for example, a neutral salt. This,
as you probably know, is the product—the combination of an alkali and an acid.   Now, considering the
salt as an effect, what are the concurrent causes,—the
co-efficients,—which constitute it what it is?   These
are, *first*, the acid, with its affinity to the alkali;
secondly, the alkali, with its affinity to the acid; and
thirdly, the translating force (perhaps the human hand)
which made their affinities available, by bringing the
two bodies within the sphere of mutual attraction.
Each of these three concurrents must be considered as
a partial cause, for, abstract any one, and the effect
is not produced.   Now, these three partial causes are
each of them again effects; but effects evidently less
complex than the effect which they, by their concur-

a Arist. *Anal. Post.*, i. 24: Ἔτι μέ-
χρι τούτου ζητοῦμεν τὸ διὰ τί, καὶ τότε
οἰόμεθα εἰδέναι, ὅταν μὴ ᾖ ὅτι τι ἄλλο
τοῦτο ἢ γινόμενον ἢ ὄν· τέλος γὰρ καὶ
πέρας τὸ ἔσχατον ἤδη οὕτως ἐστίν. Cf.
Metaph., i. 2: Δεῖ γὰρ ταύτην τῶν
πρώτων ἀρχῶν καὶ αἰτίων εἶναι θεωρητι-
κήν.—ED.

rence, constituted. But each of these three constituents is an effect, and therefore to be analysed into
its causes; and these causes again into others, until
the procedure is checked by our inability to resolve
the last constituent into simpler elements. But, though
thus unable to carry our analysis beyond a limited extent, we neither conceive, nor are we able to conceive,
the constituent in which our analysis is arrested, as
itself anything but an effect. We therefore carry on
the analysis in imagination; and as each step in the
procedure carries us from the more complex to the more
simple, and, consequently, nearer to unity, we at last
arrive at that unity itself,—at that ultimate cause
which, as ultimate, cannot again be conceived as an
effect.[a]

Philosophy
necessarily
tends to-
wards a
first cause.

Philosophy thus, as the knowledge of effects in their
causes, necessarily tends, not towards a plurality of
ultimate or first causes, but towards one alone. This
first cause,—the Creator,—it can indeed never reach,
as an object of immediate knowledge; but, as the convergence towards unity in the ascending series is manifest, in so far as that series is within our view, and as
it is even impossible for the mind to suppose the convergence not continuous and complete, it follows,—
unless all analogy be rejected,—unless our intelligence
be declared a lie,—that we must, philosophically, believe in that ultimate or primary unity which, in our
present existence, we are not destined in itself to
apprehend.

a I may notice that an ultimate
cause, and a first cause, are the same,
but viewed in different relations.
What is called the ultimate cause in
ascending from effects to causes,—
that is, in the regressive order, is
called the first cause in descending

from causes to effects,—that is, in
the progressive order. This synonymous meaning of the terms ultimate
and primary it is important to recollect, for these words are in very common use in philosophy.

Such is philosophical knowledge in its most extensive signification ; and, in this signification, all the
sciences, occupied in the research of causes, may be
viewed as so many branches of philosophy.

There is, however, one section of these sciences Sciences
which is denominated philosophical by pre-eminence; nated phi-
—sciences which the term philosophy exclusively de- by pre-
notes, when employed in propriety and vigour. What eminence.
these sciences are, and why the term philosophy has
been specially limited to them, I shall now endeavour
to make you understand.

"Man," says Protagoras, "is the measure of the Man's
universe;"[a] and, in so far as the universe is an object relative.
of human knowledge, the paradox is a truth. What-
ever we know, or endeavour to know, God or the
world,—mind or matter,—the distant or the near,—
we know, and can know, only in so far as we possess
a faculty of knowing in general ; and we can only
exercise that faculty under the laws which control
and limit its operations. However great, and infinite, and various, therefore, may be the universe
and its contents,—these are known to us, not as they
exist, but as our mind is capable of knowing them.
Hence the brocard—"Quicquid recipitur, recipitur ad
modum recipientis."[β]

In the first place, therefore, as philosophy is a

---

a See Plato, *Theætetus,* p. 152 ;
Arist., *Metaph.,* x. 6.—Ed.

β Boethius, *De Consol. Phil.* v.
Prosa iv.: "Omne enim quod cognoscitur, non secundum sui vim, sed secundum cognoscentium potius comprehenditur facultatem." Proclus, *In
Plat. Parm.,* p. 748, ed. Stallbaum: Τὸ
γιγνῶσκον κατὰ τὴν ἑαυτοῦ γιγνώσκει
φύσιν. Aquinas, *Summa,* Pars i. Q.
70, art. 3 : "Similitudo agentis reci-

pitur in patientem secundum modum
patientis." *Ibid.,* Pars i. Q. 14, art.
1 : "Scientia est secundum modum
cognoscentis. Scitum enim est in
sciente secundum modum scientis."
Chauvin gives the words of the text.
See *Lexicon Philosophicum,* art. *Fi-
nitas.* See also other authorities to
the same effect quoted in the Author's
Discussions, p. 644.—Ed.

LECT.
III.

The primary problem of philosophy.

knowledge, and as all knowledge is only possible under the conditions to which our faculties are subjected,—the grand,—the primary problem of philosophy must be to investigate and determine these conditions, as the necessary conditions of its own possibility.

The study of mind the philosophical study.

In the second place, as philosophy is not merely a knowledge, but a knowledge of causes, and as the mind itself is the universal and principal concurrent cause in every act of knowledge; philosophy is, consequently, bound to make the mind its first and paramount object of consideration. The study of mind is thus the philosophical study by pre-eminence. There is no branch of philosophy which does not suppose this as its preliminary, which does not borrow from this its light. A considerable number, indeed, are only the science of mind viewed in particular aspects, or considered in certain special applications. Logic, for example, or the science of the laws of thought, is only a fragment of the general science of mind, and presupposes a certain knowledge of the operations which are regulated by these laws. Ethics is the science of the laws which govern our actions as moral agents; and a knowledge of these laws is only possible through a knowledge of the moral agent himself. Political science, in like manner, supposes a knowledge of man in his natural constitution, in order to appreciate the modifications which he receives, and of which he is susceptible, in social and civil life. The Fine Arts have all their foundation in the theory of the beautiful; and this theory is afforded by that part of the philosophy of mind, which is conversant with the phænomena of feeling. Religion, Theology, in fine, is not independent of the same philosophy.

Branches of this study.

Logic.

Ethics.

Politics.

The Fine Arts.

Theology dependent on study of mind.

For as God only exists for us as we have faculties LECT.
capable of apprehending his existence, and of fulfilling III.
his behests, nay, as the phænomena from which we
are warranted to infer his being are wholly mental,
the examination of these faculties and of these phænomena is, consequently, the primary condition of every
sound theology. In short, the science of mind, whether considered in itself, or in relation to the other
branches of our knowledge, constitutes the principal
and most important object of philosophy,—constitutes
in propriety, with its suite of dependent sciences,
philosophy itself.[a]

This limitation of the term philosophy to the sciences Misappli-
of mind, when not expressly extended to the other cation of
branches of science, has been always that generally Philosophy
prevalent;—yet it must be confessed that, in this in this
country, the word is applied to subjects with which, country.
on the continent of Europe, it is rarely, if ever, associated. With us the word philosophy, taken by itself,
does not call up the precise and limited notion which
it does to a German, a Hollander, a Dane, an Italian,
or a Frenchman ; and we are obliged to say the philosophy of mind, if we do not wish it to be vaguely
extended to the sciences conversant with the phænomena of matter. We not only call Physics by the
name of Natural Philosophy, but every mechanical
process has with us its philosophy. We have books
on the philosophy of Manufactures, the philosophy of
Agriculture, the philosophy of Cookery, &c. In all
this we are the ridicule of other nations. Socrates, it
is said, brought down philosophy from the clouds,—
the English have degraded her to the kitchen ; and

a Cf. Cousin, *Cours de l'Histoire de*   Programme de la Première Partie du
la Phil. Mod., Prem. Sér. tom. ii. ;   Cours.—ED.

this, our prostitution of the term, is, by foreigners, alleged as a significant indication of the low state of the mental sciences in Britain.[*]

From what has been said, you will, without a definition, be able to form at least a general notion of what is meant by philosophy. In its more extensive signification, it is equivalent to a knowledge of things by their causes,—and this is, in fact, Aristotle's definition;[β] while, in its stricter meaning, it is confined to the sciences which constitute, or hold immediately of, the science of mind.

a See Hegel, *Werke*, vi. 13; xiii. 72; Scheidler, *Encyclop. de Philosophie*, i. p. 27.—Ed.

β *Metaph.*, v. 1: Πᾶσα ἐπιστήμη διανοητικὴ περὶ αἰτίας καὶ ἀρχάς ἐστιν ἢ ἀκριβεστέρας ἢ ἁπλουστέρας. *Ibid.*, i. 1: Τὴν ὀνομαζομένην σοφίαν περὶ τὰ πρῶτα αἴτια καὶ τὰς ἀρχὰς ὑπολαμβάνουσι πάντες . . . ὅτι μὲν οὖν ἡ σοφία περὶ τινας αἰτίας καὶ ἀρχάς ἐστιν ἐπιστήμη, δῆλον. *Eth. Nic.*, vi. 7: Δεῖ ἄρα τὸν σοφὸν μὴ μόνον τὰ ἐκ τῶν ἀρχῶν εἰδέναι, ἀλλὰ καὶ περὶ τὰς ἀρχὰς ἀληθεύειν.— Ed.

LECTURE IV.

THE CAUSES OF PHILOSOPHY.

HAVING thus endeavoured to make you vaguely appre- LECT IV.
hend what cannot be precisely understood,—the Nature
and Comprehension of Philosophy,—I now proceed to The causes of philoso-
another question,—What are the Causes of Philosophy? phy in the elements of
The causes of philosophy lie in the original elements our con- stitution.
of our constitution.   We are created with the faculty
of knowledge, and, consequently, created with the tendency to exert it.   Man philosophises as he lives.   He
may philosophise well or ill, but philosophise he must.
Philosophy can, indeed, only be assailed through philosophy itself.   " If," says Aristotle, in a passage preserved to us by Olympiodorus,[a] " we must philosophise, we must philosophise ; if we must not philosophise, we must philosophise ;—in any case, therefore,
we must philosophise."   " Were philosophy," says
Clement of Alexandria,[β] " an evil, still philosophy is
to be studied, in order that it may be scientifically These causes either
contemned."   And Averroes,[γ]—" Philosophi solum est essential or comple-
spernere philosophiam."   Of the causes of philosophy mentary.

a Olympiodori in Platonis Alcibia- dem Priorem Commentarii, ed. Creu- zer, p. 144: Καὶ Ἀριστοτέλης ἐν τῷ Προτρεπτικῷ ἔλεγεν ὅτι εἴτε φιλοσοφη- τέον, φιλοσοφητέον· εἴτε μὴ φιλοσοφη- τέον, φιλοσοφητέον· πάντως δὲ φιλο- σοφητέον.   Quoted also by the anon- ymous commentator in Cramer's Anecdota, iv. p. 391.—ED.

β Εἰ καὶ ἄχρηστος εἴη φιλοσοφία, εἰ εὔχρηστος ἡ τῆς ἀχρηστίας βεβαίωσις, εὔχρηστος.   Stromata, i. 2.—ED.

γ See Discussions, p. 786.—ED. [" Se moquer de la philosophie c'est vraiment philosopher." Pascal, Pensées, part i. art. xi. § 36. Com- pare Montaigne, Essais, lib. ii. c. xii. —tom. ii. p. 216, ed. 1725.]

LECT.
IV.
some are therefore contained in man's very capacity
for knowledge; these are essential and necessary. But
there are others, again, which lie in certain feelings
with which he is endowed; these are complementary
and assistant.

The first
class appa-
rently two-
fold.
1. The prin-
ciple of
Cause and
Effect.
Of the former class,—that is, of the essential causes,
—there are in all two: the one is, the necessity we feel
to connect Causes with Effects; the other, to carry up
our knowledge into Unity. These tendencies, however,
if not identical in their origin, coincide in their result;
for, as I have previously explained to you, in ascending from cause to cause, we necessarily, (could we
carry our analysis to its issue), arrive at absolute
unity. Indeed, were it not a discussion for which
you are not as yet prepared, it might be shown, that
both principles originate in the same condition,—that
both emanate, not from any original power, but from
the same original powerlessness of mind.[a] Of the
former,—namely, the tendency, or rather the necessity, which we feel to connect the objects of our experience with others which afford the reasons of their
existence,—it is needful to say but little. The nature
of this tendency is not a matter on which we can
at present enter; and the fact of its existence is
too notorious to require either proof or illustration.
It is sufficient to say, or rather to repeat what we
have already stated, that the mind is unable to realise
in thought the possibility of any absolute commencement; it cannot conceive that anything which begins
to be is anything more than a new modification
of pre-existent elements; it is unable to view any
individual thing as other than a link in the mighty
chain of being; and every isolated object is viewed

a This is partially argued in the *Discussions*, p. 609.—ED.

by it only as a fragment which, to be known, must be known in connection with the whole of which it constitutes a part. It is thus that we are unable to rest satisfied with a mere historical knowledge of existence; and that even our happiness is interested in discovering causes, hypothetical at least, if not real, for the various phænomena of the existence of which our experience informs us.

"Felix qui potuit rerum cognoscere causas." a

The second tendency of our nature, of which philo- 2. The love of Unity. sophy is the result, is the desire of Unity. On this, which indeed involves the other, it is necessary to be somewhat more explicit. This tendency is one of the most prominent characteristics of the human mind. It, in part, originates in the imbecility of our faculties. We are lost in the multitude of the objects presented to our observation, and it is only by assorting them in classes that we can reduce the infinity of nature to the finitude of mind. The conscious Ego, the conscious Self, by its nature one, seems also constrained to require that unity by which it is distinguished, in everything which it receives, and in everything which it produces. I regret that I can illustrate this only by examples which cannot, I am aware, as yet be fully intelligible to all. We are conscious of a scene presented to our senses only by uniting its parts into a perceived whole. Perception is thus a unifying act. The imagination cannot represent an object without uniting, in a single combination, the various elements of which it is composed. Generalisation is only the apprehension of the one in the many, and language little else than a registry of the factitious

a Virgil, Georgics, ii. 490.

unities of thought. The judgment cannot affirm or deny one notion of another, except by uniting the two in one indivisible act of comparison. Syllogism is simply the union of two judgments in a third. Reason, Intellect, νοῦς, in fine, concatenating thoughts and objects into system, and tending always upwards from particular facts to general laws, from general laws to universal principles, is never satisfied in its ascent till it comprehend, (what, however, it can never do), all laws in a single formula, and consummate all conditional knowledge in the unity of unconditional existence. Nor is it only in science that the mind desiderates the one. We seek it equally in works of art. A work of art is only deserving of the name, inasmuch as an idea of the work has preceded its execution, and inasmuch as it is itself a realisation of the ideal model in sensible forms. All languages express the mental operations by words which denote a reduction of the many to the one. Σύνεσις, περίληψις συναίσθησις, συνεπίγνωσις, &c., in Greek ;—in Latin, cogere (co-agere), cogitare (co-agitare), concipere, cognoscere, comprehendere, conscire, with their derivatives, may serve for examples.

The history of philosophy is only the history of this tendency ; and philosophers have amply testified to its reality. "The mind," says Anaxagoras,[a] "only knows when it subdues its objects, when it reduces the many to the one." "All knowledge," says the Platonists,[β] "is the gathering up into one, and the

a Arist., De Anima, iii. 4 : Ἀνάγκη ἄρα, ἐπεὶ πάντα νοεῖ, ἀμιγῆ εἶναι, ὥσπερ φησὶν Ἀναξαγόρας, ἵνα κρατῇ, τοῦτο δ' ἐστὶν ἵνα γνωρίζῃ. The passage of Anaxagoras is given at length in the Commentary of Simplicius, and quot-

ed in part by Trendelenburg on the De Anima, p. 466.—ED.

β Priscianus Lydus : Κατὰ τὴν εἰς ἓν συναίρεσιν, καὶ τὴν ἀμέριστον τοῦ γνωστοῦ παντὸς περίληψιν, ἀπάσης ἱσταμένης γνώσεως. (Μετάφρασις τῶν

indivisible apprehension of this unity by the knowing LECT.
IV. mind." Leibnitz [a] and Kant [β] have, in like manner, defined knowledge by the representation of multitude in unity. "The end of philosophy," says Plato,[γ] "is the intuition of unity;" and Plotinus, among many others,[δ] observes that our knowledge is perfect as it is one. The love of unity is by Aristotle applied to solve a multitude of psychological phænomena.[ε] St Augustin even analyses pain into a feeling of the frustration of unity. "Quid est enim aliud dolor, nisi quidam sensus divisionis vel corruptionis impatiens? Unde luce clarius apparet, quam sit illa anima in sui corporis universitate avida unitatis et tenax."[ζ]

This love of unity, this tendency of mind to gene- Love of unity a guiding principle in philosophy. ralise its knowledge, leads us to anticipate in nature a corresponding uniformity; and as this anticipation is found in harmony with experience, it not only affords the efficient cause of philosophy, but the guiding principle to its discoveries. "Thus, for instance,

Θεοφράστου Περὶ Αἰσθήσεως — Opera Theoph. ed. Basil., p. 273.) Thus rendered in the Latin version of Ficinus: "Cognitio omnis constat secundum quandam in unum congregationem, atque secundum impartibilem cognoscibilis totius comprehensionem."—ED.

a Monadologie, § 14.—ED.

β Kritik der reinen Vernunft, p. 359, ed. 1799.—ED.

γ Cf. Philebus, sub. init., especially p. 16: Δεῖν ἡμᾶς ἀεὶ μίαν ἰδέαν περὶ παντὸς ἑκάστοτε θεμένους ζητεῖν: and Republic, v. p. 475 et seq. —ED.

δ Enn., iii. lib. viii. c. 2, on which Ficinus says: "Cognoscendi potentia in ipso actu cognitionis unum quodammodo fit cum objecto, et quo magis fit unum, eo perfectior est cognitio, atque vicissim." Enn., vi. lib. ix. c. 1: Ἀρετὴ δὲ ψυχῆς ὅταν εἰς ἓν, καὶ εἰς μίαν ὁμολογίαν ἐλθῇ .... Ἐπειδὴ τὰ

πάντα εἰς ἓν ἄγει, δημιουργοῦσα καὶ πλάττουσα καὶ μορφοῦσα καὶ συντάττουσα. Proclus: Γνῶσις οὐδενὸς ἔσται τῶν ὄντων, εἴπερ μή ἐστι τὸ ἓν .... Οὐδὲ λόγος ἔσται· καὶ γὰρ ὁ λόγος ἐκ πολλῶν εἷς, εἴπερ τέλειος· καὶ ἡ γνῶσις, ὅταν τὸ γινώσκον ἓν γίνηται πρὸς τὸ γνωστόν. In Platonis Theologiam, p. 76 (ed. 1618).—ED.

ε See De Memoria, § 5, for application of this principle to the problem of Reminiscence. Cf. Reid's Works, p. 900. See also Problems, xviii. 9, where it is used to explain the higher pleasure we derive from those narratives that relate to a single subject.—ED.

ζ De Libero Arbitrio, lib. iii. 23. [St Augustin applied the principle of Unity to solve the theory of the Beautiful: "Omnis pulchritudinis forma unitas est." Epist. xviii.]—Oral Interpolation.

when it is observed that solid bodies are compressible, we are induced to expect that liquids will be found to be so likewise; we subject them, consequently, to a series of experiments; nor do we rest satisfied until it be proved that this quality is common to both classes of substances. Compressibility is then proclaimed a physical law,—a law of nature in general; and we experience a vivid gratification in this recognition of unconditioned universality. Another example: Kant,[a] reflecting on the differences among the planets, or rather among the stars revolving round the sun, and having discovered that these differences betrayed a uniform progress and proportion,—a proportion which was no longer to be found between Saturn and the first of the comets,—the law of unity and the analogy of nature, led him to conjecture that, in the intervening space, there existed a star, the discovery of which would vindicate the universality of the law. This anticipation was verified. Uranus was discovered by Herschel, and our dissatisfaction at the anomaly appeased. Franklin, in like manner, surmised that lightning and the electric spark were identical; and when he succeeded in verifying this conjecture, our love of unity was gratified. From the moment an isolated fact is discovered, we endeavour to refer it to other facts which it resembles. Until this be accomplished, we do not view it as understood. This is the case, for example, with sulphur, which, in a certain degree of temperature, melts like other bodies, but at a higher degree of heat, instead of evaporating, again

a *Allgemeine Naturgeschichte und Theorie des Himmels,* 1755; *Werke,* vol. vi. p. 88. Kant's conjecture was founded on a supposed progressive increase in the eccentricities of the planetary orbits. This progression, however, is only true of Venus, the Earth, Jupiter, and Saturn. The eccentricity diminishes again in Uranus, and still more in Neptune. Subsequent discoveries have thus rather weakened than confirmed the theory.—Ed.

consolidates.   When a fact is generalised, our discontent is quieted, and we consider the generality itself as tantamount to an explanation.   Why does this apple fall to the ground?   Because all bodies gravitate towards each other.   Arrived at this general fact, we inquire no more, although ignorant now as previously of the cause of gravitation; for gravitation is nothing more than a name for a general fact, the *why* of which we know not.   A mystery, if recognised as universal, would no longer appear mysterious.

"But this thirst of unity,—this tendency of mind to generalise its knowledge, and our concomitant belief in the uniformity of natural phænomena, is not only an effective mean of discovery, but likewise an abundant source of error.   Hardly is there a similarity detected between two or three facts, than men hasten to extend it to all others; and if, perchance, the similarity has been detected by ourselves, self-love closes our eyes to the contradictions which our theory may encounter from experience."[a]   "I have heard," says Condillac, "of a philosopher who had the happiness of thinking that he had discovered a principle which was to explain all the wonderful phænomena of chemistry, and who, in the ardour of his self-gratulation, hastened to communicate his discovery to a skilful chemist.   The chemist had the kindness to listen to him, and then calmly told him that there was but one unfortunate circumstance for his discovery,—that the chemical facts were precisely the converse of what he had supposed them to be.   'Well, then,' said the philosopher, 'have the goodness to tell me what they are, that I may explain them on my system.'"[b]   We are

a Garnier, *Cours de Psychologie*, p. 192-94. [Cf. Ancillon, *Nouv. Mé-langes*, i. p. 1 *et seq.*]

β *Traité des Systèmes*, chap. xii. *Œuvres Philos.*, tom. iv. p. 146 (ed. 1795).

naturally disposed to refer everything we do not know
to principles with which we are familiar. As Aristotle
observes,[a] the early Pythagoreans, who first studied
arithmetic, were induced, by their scientific predilec-
tions, to explain the problem of the universe by the
properties of number ; and he notices also that a certain musical philosopher was, in like manner, led to
suppose that the soul was but a kind of harmony.[β]
The musician suggests to my recollection a passage of
Dr Reid. " Mr Locke," says he, " mentions an eminent
musician who believed that God created the world
in six days, and rested the seventh, because there are
but seven notes in music. I myself," he continues,
" knew one of that profession who thought that there
could be only three parts in harmony—to wit, bass,
tenor, and treble ; because there are but three persons
in the Trinity."[γ]   The alchemists would see in nature
only a single metal, clothed with the different appearances which we denominate gold, silver, copper, iron,
mercury, &c., and they confidently explained the mysteries not only of nature, but of religion, by salt,
sulphur, and mercury.[δ] Some of our modern zoolo-
gists recoil from the possibility of nature working on
two different plans, and rather than renounce the
unity which delights them, they insist on recognising
the wings of insects in the gills of fishes, and the
sternum of quadrupeds in the antennæ of butterflies,
—and all this that they may prove that man is only
the evolution of a molluscum. Descartes saw in the
physical world only matter and motion ;[ε] and, more
recently, it has been maintained that thought itself

a *Metaph.*, i. 5.—ED.
β *De Anima*, i. 4 ; Plato, *Phædo*,
p. 86. The same theory was afterwards adopted by Aristotle's own
pupil, Aristoxenus. See Cicero, *Tusc.
Quæst.*, i. 10.—ED.

γ *Intellectual Powers*, Ess. vi. chap.
viii.; *Coll. Works*, p. 473.
δ See Brucker, *Hist. Philosophiæ*,
vol. iv. p. 677 *et seq.*—ED.
ε *Principia*, pars ii. 23.—ED.

is only a movement of matter.[a] Of all the faculties
of the mind, Condillac recognised only one, which
transformed itself like the Protean metal of the alchemists; and he maintains that our belief in the rising
of to-morrow's sun is a sensation.[β] It is this tendency, indeed, which has principally determined philosophers, as we shall hereafter see, to neglect or violate the original duality of consciousness; in which,
as an ultimate fact,—a self and not-self,—mind knowing and matter known,—are given in counterpoise
and mutual opposition; and hence the three Unitarian
schemes of Materialism, Idealism, and absolute Identity.[γ] In fine, Pantheism, or the doctrine which identifies mind and matter,—the Creator and the creature,
—God and the universe,—how are we to explain the
prevalence of this modification of atheism in the most
ancient and in the most recent times? Simply because it carries our love of unity to its highest fruition.
To sum up what has just been said in the words of
Sir John Davies, a highly philosophic poet of the
Elizabethan age :—

> " Musicians think our souls are harmonies ;
>   Physicians hold that they complexions be ; .
> Epicures make them swarms of atomies :
>   Which do by chance into our bodies flee.
>
> One thinks the soul is air ; another fire ;
>   Another blood, diffused about the heart ;
> Another saith the elements conspire,
>   And to her essence each doth yield a part.
>
> Some think one gen'ral soul fills every brain,
>   As the bright sun sheds light in every star ;
> And others think the name of soul is vain,
>   And that we only well-mix'd bodies are.

---

a Priestley, *Disquisitions relating
to Matter and Spirit*, sect. iii. p. 24
et seq.; *Free Discussion of Materialism and Necessity*, pp. 258, 267 et
seq.—ED.

β The preceding illustrations are
borrowed from Garnier, *Psychologie*,
p. 194.—ED.

γ See the Author's Supplementary
Dissertations to Reid, Note C.—ED.

> Thus these great clerks their little wisdom show,
>   While with their doctrines they at hazard play ;
> Tossing their light opinions to and fro,
>   To mock the lewd,[a] as learn'd in this as they ;
>
> For no craz'd brain could ever yet propound,
>   Touching the soul so vain and fond a thought ;
> But some among these masters have been found,
>   Which, in their schools, the self-same thing have taught." [β]

Influence of preconceived opinion reducible to love of unity.

To this love of unity—to this desire of reducing the objects of our knowledge to harmony and system —a source of truth and discovery if subservient to observation, but of error and delusion if allowed to dictate to observation what phænomena are to be perceived ; to this principle, I say, we may refer the influence which preconceived opinions exercise upon our perceptions and our judgments, by inducing us to see and require only what is in unison with them. What we wish, says Demosthenes, that we believe ; [γ] what we expect, says Aristotle, that we find [δ]—truths which have been re-echoed by a thousand confessors, and confirmed by ten thousand examples. Opinions once adopted become part of the intellectual system of their holders. If opposed to prevalent doctrines, self-love defends them as a point of honour, exaggerates whatever may confirm, overlooks or extenuates whatever may contradict. Again, if accepted as a general doctrine, they are too often recognised, in consequence of their prevalence, as indisputable truths, and all counter-appearances peremptorily overruled as manifest illusions. Thus it is that men will not see

a *Lewd*, according to Tooke, from Anglo-Saxon, *Lœwed*, past participle of *Lœwan, to mislead.* It was formerly applied to the (lay) people in contradistinction from the clergy. See Richardson, *Eng. Dict.*, v. *Lewd.*—Ed.

β *On the Immortality of the Soul*, stanza 9 *et seq.*

γ Βούλεται τοῦθ' ἕκαστος καὶ οἴεται. Demosth. *Olynth.*, iii. p. 68.—Ed.

δ *Rhet.*, ii. 1: Τῷ μὲν ἐπιθυμοῦντι καὶ εὐέλπιδι ὄντι, ἐὰν ᾖ τὸ ἐσόμενον ἡδύ, καὶ ἔσεσθαι καὶ ἀγαθὸν ἔσεσθαι φαίνεται, τῷ δ' ἀπαθεῖ, καὶ δυσχεραίνοντι, τοὐναντίον.— Ed.

in the phænomena what alone is to be seen ; in their observations, they interpolate and they expunge ; and this mutilated and adulterated product they call a fact. And why? Because the real phænomena, if admitted, would spoil the pleasant music of their thoughts, and convert its factitious harmony into discord. "Quæ volunt sapiunt, et nolunt sapere quæ vera sunt."[a] In consequence of this, many a system, professing to be reared exclusively on observation and fact, rests in reality mainly upon hypothesis and fiction. A pretended experience is, indeed, the screen behind which every illusive doctrine regularly retires. "There are more false facts," says Cullen,[β] "current in the world, than false theories ;"—and the livery of Lord Bacon has been most ostentatiously paraded by many who were no members of his household. Fact,— observation,—induction, have always been the watch-words of those who have dealt most extensively in fancy. It is now above three centuries since Agrippa, in his *Vanity of the Sciences*, observed of Astrology, Physiognomy, and Metoposcopy, (the Phrenology of those days), that experience was always professedly their only foundation and their only defence : "Solent omnes illæ divinationum prodigiosæ artes non, nisi experientiæ titulo, si defendere et se objectionum vinculis extricare."[γ] It was on this ground, too, that at a later period, the great Kepler vindicated the first of these arts, Astrology. For, said he, how could the principle of a science be false, where experience showed that its predictions were uniformly fulfilled?[δ] Now,

a [St Hilarii, *De Trinitate*, lib. viii., sub. init.]

β For Cullen's illustrations of the influence of a pretended experience in Medicine, see his *Materia Medica*, vol. i. c. ii. art. iv., second edition.—ED.

γ *Opera*, vol. ii. c. 32, p. 64.

δ *De Stella Nova*, cc. 8, 10 ; *Harmonice Mundi*, lib. iv. c. 7.—ED.

truth was with Kepler even as a passion; and his, too, was one of the most powerful intellects that ever cultivated and promoted a science. To him astronomy, indeed, owes perhaps even more than to Newton. And yet, even his great mind, preoccupied with a certain prevalent belief, could observe and judge only in conformity with that belief. This tendency to look at realities only through the spectacles of an hypothesis, is perhaps seen most conspicuously in the fortunes of medicine. The history of that science is, in truth, little else than an incredible narrative of the substitution of fictions for facts; the converts to an hypothesis, (and every, the most contradictory, doctrine has had its day), regularly seeing and reporting only in conformity with its dictates.[a] The same is also true of the philosophy of mind; and the variations and alternations in this science, which are perhaps only surpassed by those in medicine, are to be traced to a refusal of the real phænomenon revealed in consciousness, and to the substitution of another, more in unison with preconceived opinions of what it ought to be. Nor, in this commutation of fact with fiction, should we suspect that there is any *mala fides*. Prejudice, imagination, and passion, sufficiently explain the illusion. "Fingunt simul creduntque."[b] "When," says Kant, "we have once heard a bad report of this or that individual, we incontinently think that we read the rogue in his countenance; fancy here mingles with observation, which is still farther vitiated when affection or passion interferes."

"The passions," says Helvetius,[y] "not only concentrate our attention on certain exclusive aspects of the

a See the Author's Article, "On the Revolutions of Medicine," *Discussions*, p. 242.—ED.

β Tacitus, *Hist.*, lib. ii. c. 8.—ED.
y De l'Esprit, Discours i. chap. ii.

objects which they present, but they likewise often deceive us in showing these same objects where they do not exist. The story is well known of a parson and a gay lady. They had both heard that the moon was peopled,—believed it,—and, telescope in hand, were attempting to discover the inhabitants. If I am not mistaken, says the lady, who looked first, I perceive two shadows; they bend toward each other, and, I have no doubt, are two happy lovers. Lovers, madam, says the divine, who looked second; oh, fie ! the two shadows you saw are the two steeples of a cathedral. This story is the history of man. In general, we perceive only in things what we are desirous of finding: on the earth, as in the moon, various prepossessions make us always recognise either lovers or cathedrals."

Such are the two intellectual necessities which afford the two principal sources of philosophy :—the intellectual necessity of refunding effects into their causes ; [a] —and the intellectual necessity of carrying up our knowledge into unity or system. But, besides these intellectual necessities, which are involved in the very existence of our faculties of knowledge, there is another powerful subsidiary to the same effect,—in a certain affection of our capacities of feeling. This feeling, according to circumstances, is denominated *surprise*, *astonishment, admiration, wonder*, and, when blended with the intellectual tendencies we have considered, it obtains the name of *curiosity*. This feeling, though it cannot, as some have held, be allowed to be the principal, far less the only, cause of philosophy, is, however, a powerful auxiliary to speculation ; and, though

Auxiliary cause of philosophy —Wonder.

a [This expression is employed by Sergeant. See *Method to Science*, p. 222. Cf. pp. 144, 145.]

inadequate to account for the existence of philosophy absolutely, it adequately explains the preference with which certain parts of philosophy have been cultivated, and the order in which philosophy in general has been developed. We may err both in exaggerating, and in extenuating, its influence. Wonder has been contemptuously called the daughter of ignorance; true, but wonder, we should add, is the mother of knowledge. Among others, Plato, Aristotle, Plutarch, and Bacon, have all concurred in testifying to the influence of this principle. "Admiration," says the Platonic Socrates in the *Theætetus*,*—"admiration is a highly philosophical affection; indeed, there is no other principle of philosophy but this."—"That philosophy," says Aristotle, "was not originally studied for any practical end, is manifest from those who first began to philosophise. It was, in fact, wonder which then, as now, determined men to philosophical researches. Among the phænomena presented to them, their admiration was first directed to those more proximate and more on a level with their powers, and then rising by degrees, they came at length to demand an explanation of the higher phænomena,—as the different states of the moon, sun, and stars, and the origin of the universe. Now, to doubt and to be astonished, is to recognise our ignorance. Hence it is that the lover of wisdom is in a certain sort a lover of mythi, (φιλόμυθός πως), for the subject of mythi is the astonishing and marvellous. If, then, men philosophise to escape ignorance, it is clear that they pursue knowledge on its own account, and not for the sake of any foreign utility. This is proved by the fact; for it was only after all that pertained to the wants, welfare,

* P. 155.—Ed.

and conveniences of life had been discovered, that men commenced their philosophical researches. It is, therefore, manifest that we do not study philosophy for the sake of anything ulterior; and, as we call him a free man who belongs to himself and not to another, so philosophy is of all sciences the only free or liberal study, for it alone is unto itself an end." [a]—"It is the business of philosophy," says Plutarch, "to investigate, to admire, and to doubt." [β] You will find in the first book of the *De Augmentis* of Bacon,[γ] a recognition of the principle " admiratio est semen sapientiæ," and copious illustrations of its truth,—illustrations which I shall not quote, but they deserve your private study.

No one, however, has so fully illustrated the play and effect of this motive as a distinguished philosopher of this country, Adam Smith; although he has attributed too little to the principle, too much to the subsidiary, momenta. He seems not to have been aware of what had been, previously to him, observed in regard to this principle by others. You will find the discussion among his posthumous essays, in that entitled *The Principles which lead and direct Philosophical Inquiries illustrated by the History of Astronomy;*—to this I must simply refer you.

We have already remarked, that the principle of wonder affords an explanation of the order in which the different objects of philosophy engaged the attention of mankind. The aim of all philosophy is the discovery of principles, that is, of higher causes; but, in the procedure to this end, men first endeavoured to explain those phænomena which attracted their

Affords an explanation of the order in which objects studied.

a *Metaph.*, lib. i. c. 2. See also for a passage to a similar effect, *Rhetoric*, lib. i. c. 11.

β Plutarch, Περὶ τοῦ Εἰ τοῦ ἐν Δελ- φοῖς, vol. ii. p. 385 (ed. 1599): 'Έπεὶ δὲ τοῦ φιλοσοφεῖν, ἔφη, τὸ ζητεῖν, τὸ θαυμάζειν, καὶ ἀπορεῖν.—Ed.

γ Vol. viii. p. 8, (Montagu's ed.)

attention by arousing their wonder. The child is wholly absorbed in the observation of the world without; the world within first engages the contemplation of the man. As it is with the individual, so was it with the species. Philosophy, before attempting the problem of intelligence, endeavoured to resolve the problem of nature. The spectacle of the external universe was too imposing not first to solicit curiosity, and to direct upon itself the prelusive efforts of philosophy. Thales and Pythagoras, in whom philosophy finds its earliest representatives, endeavoured to explain the organisation of the universe, and to substitute a scientific for a religious cosmogony. For a season their successors toiled in the same course; and it was only after philosophy had tried, and tired, its forces on external nature, that the human mind recoiled upon itself, and sought in the study of its own nature the object and end of philosophy. The mind now became to itself its point of departure, and its principal object; and its progress, if less ambitious, was more secure. Socrates was he who first decided this new destination of philosophy. From his epoch man sought in himself the solution of the great problem of existence, and the history of philosophy was henceforward only a development, more or less successful, more or less complete, of the inscription on the Delphic temple—Γνῶθι σεαυτόν—Know thyself.[a]

a Plato, *Protagoras*, p. 343.—ED. *Philosophie*, p. 1.]
[See Géruzez, *Nouveau Cours de*

LECTURE V.

THE DISPOSITIONS WITH WHICH PHILOSOPHY OUGHT TO BE STUDIED.

HAVING, in the previous Lectures, informed you,—1°, What Philosophy is, and 2°, What are its causes, I would now, in the third place, say a few words to you on the Dispositions with which Philosophy ought to be studied, for, without certain practical conditions, a speculative knowledge of the most perfect Method of procedure, (our next following question), remains barren and unapplied.

" To attain to a knowledge of ourselves," says Socrates, " we must banish prejudice, passion, and sloth ; "[a] and no one who neglects this precept can hope to make any progress in the philosophy of the human mind, which is only another term for the knowledge of ourselves.

In the first place, then, all prejudices,—that is, all opinions formed on irrational grounds,—ought to be removed. A preliminary doubt is thus the fundamental condition of philosophy ; and the necessity of such a doubt is no less apparent than is its difficulty. We do not approach the study of philosophy ignorant, but perverted. "There is no one who has not grown up under a load of beliefs—beliefs which he owes to the accidents of country and family, to the

First condition of the study of Philosophy,—renunciation of prejudice.

a [See Gatien-Arnoult, *Doctrine Philosophique*, p. 39.]

books he has read, to the society he has frequented, to
the education he has received, and, in general, to the
circumstances which have concurred in the formation
of his intellectual and moral habits. These beliefs
may be true or they may be false, or, what is more
probable, they may be a medley of truths and errors.
It is, however, under their influence that he studies,
and through them, as through a prism, that he views
and judges the objects of knowledge. Everything is
therefore seen by him in false colours, and in distorted
relations. And this is the reason why philosophy, as
the science of truth, requires a renunciation of prejudices, (præ-judicia, opiniones præ-judicatæ),—that is,
conclusions formed without a previous examination of
their grounds."ᵃ   In this, if I may without irreverence
compare things human with things divine, Christianity
and Philosophy coincide,—for truth is equally the end
of both. What is the primary condition which our
Saviour requires of his disciples? That they throw off
their old prejudices, and come with hearts willing to
receive knowledge, and understandings open to conviction. " Unless," He says, " ye become as little children, ye shall not enter into the kingdom of heaven."
Such is true religion ; such also is true philosophy.
Philosophy requires an emancipation from the yoke of
foreign authority, a renunciation of all blind adhesion
to the opinions of our age and country, and a puri-
fication of the intellect from all assumptive beliefs.
Unless we can cast off the prejudices of the man, and
become as children, docile and unperverted, we need
never hope to enter the temple of philosophy. It is
the neglect of this primary condition which has mainly
occasioned men to wander from the unity of truth, and

ᵃ [Gatien-Arnoult, *Doct. Phil.*, pp. 39, 40.]

caused the endless variety of religious and philosophical sects. Men would not submit to approach the
word of God in order to receive from that alone their
doctrine and their faith; but they came in general
with preconceived opinions, and, accordingly, each
found in revelation only what he was predetermined
to find. So, in like manner, is it in philosophy. Consciousness is to the philosopher what the Bible is to
the theologian. Both are revelations of the truth,—
and both afford the truth to those who are content
to receive it, as it ought to be received, with reverence and submission. But as it has, too frequently,
fared with the one revelation, so has it with the other.
Men turned, indeed, to consciousness, and professed to
regard its authority as paramount, but they were not
content humbly to accept the facts which consciousness revealed, and to establish these without retrenchment or distortion, as the only principles of their philosophy; on the contrary, they came with opinions
already formed, with systems already constructed, and
while they eagerly appealed to consciousness when
its data supported their conclusions, they made no
scruple to overlook, or to misinterpret, its facts when
these were not in harmony with their speculations.
Thus religion and philosophy, as they both terminate
in the same end, so they both depart from the same
fundamental condition. "Aditus ad regnum hominis,
quod fundatur in scientiis, quam ad regnum cœlorum,
in quod, nisi sub persona infantis, intrare non datur."[a]

But the influence of early prejudice is the more
dangerous, inasmuch as this influence is unobtrusive.
Few of us are, perhaps, fully aware of how little we
owe to ourselves,—how much to the influence of

a Bacon, *Nov. Org.*, lib. i., aph. lxviii.

others. "Non licet," says Seneca, "ire recta via; trahunt in pravum parentes; trahunt servi; nemo errat uni sibi, sed dementiam spargit in proximos accipitque invicem. Et ideo, in singulis vitia populorum sunt, quia illa populus dedit; dum facit quisque pejorem, factus est. Didicit deteriora, deinde docuit: effectaque est ingens illa nequitia, congesto in unum, quod cuique pessimum scitur. Sit ergo aliquis custos, et aurem subinde pervellat, abigatque rumores et reclamet populis laudantibus." [a]

Source of
the power
of custom.
Man a so-
cial animal.
Man is by nature a social animal. "He is more political," says Aristotle, "than any bee or ant." [β] But the existence of society, from a family to a state, supposes a certain harmony of sentiment among its members; and nature has, accordingly, wisely implanted in us a tendency to assimilate in opinions and habits of thought to those with whom we live and act. There is thus, in every society great or small, a certain gravitation of opinions towards a common centre. As, in our natural body, every part has a necessary sympathy with every other, and all together form, by their harmonious conspiration, a healthy whole; so, in the social body, there is always a strong predisposition in each of its members to act and think in unison with the rest. This universal sympathy, or fellow-feeling, of our social nature, is the principle of the different spirit dominant in different ages, countries, ranks, sexes, and periods of life. It is the cause why fashions, why political and religious enthusiasm, why moral example, either for good or evil, spread so rapidly, and exert so powerful an influence. As men are naturally prone to imitate others, they consequently regard, as important or insignificant, as honourable or disgraceful, as true

a *Epist.* xciv.　　　　β *Polit.*, i. 2.—ED.

or false, as good or bad, what those around them consider in the same light. They love and hate what they see others desire and eschew. This is not to be regretted ; it is natural, and, consequently, it is right. Indeed, were it otherwise, society could not subsist, for nothing can be more apparent than that mankind in general, destined as they are to occupations incompatible with intellectual cultivation, are wholly incapable of forming opinions for themselves on many of the most important objects of human consideration. If such, however, be the intentions of nature with respect to the unenlightened classes, it is manifest that a heavier obligation is thereby laid on those who enjoy the advantages of intellectual cultivation, to examine with diligence and impartiality the foundations of those opinions which have any connection with the welfare of mankind. If the multitude must be led, it is of consequence that it he led by enlightened conductors.[a] That the great multitude of mankind are, by natural disposition, only what others are, is a fact at all times so obtrusive, that it could not escape observation from the moment a reflective eye was first turned upon man. "The whole conduct of Cambyses," says Herodotus,[b] the father of history, "towards the Egyptian gods, sanctuaries, and priests, convinces me that this king was in the highest degree insane, for otherwise he would not have insulted the worship and holy things of the Egyptians. If any one should accord to all men the permission to make free choice of the best among all customs, undoubtedly each would choose his own. That this would certainly happen can be shown by many examples, and, among others, by the

a See Stewart, *Elements,* Introd. β Lib. iii. cc. 37, 38.
Part ii. § 1; *Works,* vol. ii. p. 67.—ED.

following. The King Darius once asked the Greeks who were resident at his court, at what price they could be induced to devour their dead parents. The Greeks answered, that to this no price could bribe them. Thereupon the king asked some Indians who were in the habit of eating their dead parents, what they would take not to eat but to burn them; and the Indians answered even as the Greeks had done." Herodotus concludes this narrative with the observation, that "Pindar had justly entitled Custom—the Queen of the World."

The ancient sceptics, from the conformity of men in every country, their habits of thinking, feeling, and acting, and from the diversity of different nations in these habits, inferred that nothing was by nature beautiful or deformed, true or false, good or bad, but that these distinctions originated solely in custom. The modern scepticism of Montaigne terminates in the same assertion; and the sublime misanthropy of Pascal has almost carried him to a similar exaggeration. "In the just and the unjust," says the latter, "we find hardly anything which does not change its character in changing its climate. Three degrees of an elevation of the pole reverses the whole of jurisprudence. A meridian is decisive of truth, and a few years of possession. Fundamental laws change. Right has its epochs. A pleasant justice which a river or a mountain limits. Truth, on this side the Pyrenees, error on the other!" [a] This doctrine is exaggerated, but it has a foundation in truth; and the most zealous champions of the immutability of moral distinctions are unanimous in acknowledging the powerful influence which the opinions, tastes, manners, affections, and actions of the society

a *Pensées*, partie i. art. vi. § 8 (vol. ii. p. 126, ed. Faugère).

in which we live, exert upon all and each of its members. [a]

LECT.
V.

Nor is this influence of man on man less unambiguous in times of social tranquillity, than in crises of social convulsion. In seasons of political and religious revolution, there arises a struggle between the resisting force of ancient habits and the contagious sympathy of new modes of feeling and thought. In one portion of society, the inveterate influence of custom prevails over the contagion of example ; in others, the contagion of example prevails over the conservative force of antiquity and habit. In either case, however, we think and act always in sympathy with others. "We remain," says an illustrious philosopher, "submissive so long as the world continues to set the example. As we follow the herd in forming our conceptions of what is respectable, so we are ready to follow the multitude also, when such conceptions come to be questioned or rejected ; and are no less vehement reformers, when the current of opinion has turned against former establishments, than we were zealous abettors while that current continued to set in a different direction." [β]

This influ-
ence of man
on man in
times both
of tranquil-
lity and
convulsion.

Thus it is that no revolution in public opinion is the work of an individual, of a single cause, or of a day. When the crisis has arrived, the catastrophe must ensue ; but the agents through whom it is apparently accomplished, though they may accelerate, cannot originate its occurrence. Who believes that but for Luther or Zwingli the Reformation would not have been ? Their individual, their personal energy and zeal, perhaps, hastened by a year or two the event ;

Relation
of the in-
dividual to
social
crises.

a See Meiners, *Untersuchungen über die Denkkräfte und Willenskräfte des Menschen*, ii. 325 *et seq.* (ed. 1806); from whom most of the preceding ob-

servations in the text are borrowed.

β Ferguson's *Moral and Political Science*, vol. i. part i. chap. ii. § 11, p. 135.

but had the public mind not been already ripe for their
revolt, the fate of Luther and Zwingli, in the sixteenth
century, would have been that of Huss and Jerome of
Prague in the fifteenth.   Woe to the revolutionist who
is not himself a creature of the revolution !   If he anticipate, he is lost ; for it requires, what no individual
can supply, a long and powerful counter-sympathy in
a nation to untwine the ties of custom which bind a
people to the established and the old.   This is finely
expressed by Schiller, in a soliloquy from the mouth
of the revolutionary Wallenstein :—

Schiller.

> " What is thy purpose ?   Hast thou fairly weighed it ?
> Thou seekest even from its broad base to shake
> The calm enthroned majesty of power,
> By ages of possession consecrate—
> Firm rooted in the rugged soil of custom—
> And with the people's first and fondest faith,
> As with a thousand stubborn tendrils twined.
> That were no strife where strength contends with strength.
> It is not strength I fear—I fear no foe
> Whom with my bodily eye I see and scan;
> Who, brave himself, inflames my courage too.
> It is an unseen enemy I dread,
> Who, in the hearts of mankind, fights against me—
> Fearful to me but from his own weak fear.
> Not that which proudly towers in life and strength
> Is truly dreadful ; but the mean and common,
> The memory of the eternal yesterday,
> Which, ever warning, ever still returns,
> And weighs to-morrow, for it weighed to-day ;
> Out of the common is man's nature framed,
> And custom is the nurse to whom he cleaves.
> Woe then to him whose daring hand profanes
> The honoured heir-looms of his ancestors !
> There is a consecrating power in time ;
> And what is grey with years to man is godlike.
> Be in possession, and thou art in right ;
> The crowd will lend thee aid to keep it sacred." [a]

This may enable you to understand how seductive

[a] *The Death of Wallenstein*, (translated by Mr George Moir,) Act i. scene 4.

is the influence of example; and I should have no
end were I to quote to you all that philosophers have
said of the prevalence and evil influence of prejudice
and opinion.

We have seen that custom is called, by Pindar and
Herodotus, the Queen of the world—and the same
thing is expressed by the adage—"Mundus regitur
opinionibus." "Opinion," says the great Pascal, "dis-
poses of all things. It constitutes beauty, justice, happiness; and these are the all in all of the world. I
would with all my heart see the Italian book of which
I know only the title,—a title, however, which is itself
worth many books—*Della opinione regina del mondo.*
I subscribe to it implicitly."[a] "Coutume," says Regnier,

> "Coutume, opinion, reines de notre sort,
>   Vous réglez des mortels, et la vie, et la mort!"

Almost every opinion we have," says the pious Charron, "we have but by authority; we believe, judge,
act, live, and die on trust, as common custom teaches
us; and rightly, for we are too weak to decide and
choose of ourselves. But the wise do not act thus."[β]
"Every opinion," says Montaigne, "is strong enough
to have had its martyrs;"[γ] and Sir W. Raleigh—
"It is opinion, not truth, that travelleth the world
without passport."[δ] "Opinion," says Heraclitus, "is a
falling sickness;"[ε] and Luther—"O doxa! doxa! quam
es communis noxa." In a word, as Hommel has it,
"An ounce of custom outweighs a ton of reason."[ζ]

Such being the recognised universality and evil ef-

Pensées, partie i. art. vi. §3. [Vol.
ii. p. 52, ed. Faugère. M. Faugère has
restored the original text of Pascal—
"*L'imagination* dispose de tout." The
ordinary reading is *L'opinion.*—Ed.]

β *De la Sagesse*, liv. i. chap. xvi.

γ *Essais*, liv. i. chap. xl.

δ Preface to his *History of the
World.*

ε Diog. Laert., lib. ix. §7.

ζ [Alex. v. Joch (Hommel), *Über
Belohnung und Strafe*, p. 111. See
Krug, *Philosophisches Lexikon*, vol. v.
p. 467, art. *Gewohnheit.*]

Testi-
monies of
philoso-
phers to the
power of
received
opinion.

LECT.
V.

Philoso-
phers una-
nimous in
making
doubt the
first step
to philoso-
phy.

Bacon.

Descartes.

fect of prejudice, philosophers have, consequently, been unanimous in making doubt the first step towards philosophy. Aristotle has a fine chapter in his *Metaphysics*[a] on the utility of doubt, and on the things which we ought first to doubt of; and he concludes by establishing that the success of philosophy depends on the art of doubting well. This is even enjoined on us by the Apostle. For in saying " Prove " (which may be more correctly translated *test*)—" Test all things," he implicitly commands us to doubt all things.

" He," says Bacon, " who would become philosopher, must commence by repudiating belief; "[β] and he concludes one of the most remarkable passages of his writings with the observation, that " were there a single man to be found with a firmness sufficient to efface from his mind the theories and notions vulgarly received, and to apply his intellect free and without prevention, the best hopes might be entertained of his success."[γ] " To philosophise," says Descartes, " seriously, and to good effect, it is necessary for a man to renounce all prejudices; in other words, to apply the greatest care to doubt of all his previous opinions so long as these have not been subjected to a new examination, and been recognised as true."[δ] But it is needless to multiply authorities in support of so

a Lib. ii. c. 1.—Ed.

β This saying is attributed by Ga-tien-Arnoult to Diderot. See *Doct. Phil.*, p. 39.—Ed.

γ " Nemo adhuc tanta mentis constantia inventus est, ut decreverit, et sibi imposuerit, theorias et notiones communes penitus abolere, et intellectum abrasum et æquum ad particularia, de integro, applicare. Itaque illa ratio humana quam habemus, ex multa fide, et multo etiam casu, nec non ex puerilibus, quas primo hausi-mus, notionibus, farrago quædam est, et congeries. Quod siquis ætate matura, et sensibus integris, et mente repurgata, se ad experientiam, et ad particularia de integro applicet, de eo melius sperandum est."— *Nov. Org.*, i. aph. xcvii.; *Works*, vol. ix. p. 252, (Montagu's ed.) See also *omnino Nov. Org.*, i. aph. lxviii.

δ *Prin. Phil.* pars i. § 75. [Cf. Clauberg, *De Dubitatione Cartesiana*, cc. i. ii. *Opera*, p. 1131.—Ed.]

obvious a truth. The ancient philosophers refused to
admit slaves to their instruction. Prejudice makes
men slaves ; it disqualifies them for the pursuit of
truth ; and their emancipation from prejudice is what
philosophy first inculcates on, what it first requires
of, its disciples.[a] Let us, however, beware that we
act not the part of revolted slaves ; that in asserting
our liberty we do not run into licence. Philosophical Philosophi-
doubt is not an end but a mean. We doubt in cal doubt.
order that we may believe; we begin that we may
not end with doubt. We doubt once that we may
believe always; we renounce authority that we may
follow reason ; we surrender opinion that we may
obtain knowledge. We must be protestants, not in-
fidels, in philosophy. "There is a great difference," Male-
says Malebranche, " between doubting and doubting. branche.
We doubt through passion and brutality ; through
blindness and malice, and finally through fancy and
from the very wish to doubt; but we doubt also from
prudence and through distrust, from wisdom and
through penetration of mind. The former doubt is a
doubt of darkness, which never issues to the light, but
leads us always further from it; the latter is a doubt
which is born of the light, and which aids in a certain
sort to produce light in its turn."[b] Indeed, were the
effect of philosophy the establishment of doubt, the
remedy would be worse than the disease. Doubt, as
a permanent state of mind, would be, in fact, little
better than an intellectual death. The mind lives as
it believes,—it lives in the affirmation of itself, of
nature, and of God; a doubt upon any one of these
would be a diminution of its life,—a doubt upon the

a [Cf. Gatien-Arnoult, *Doct. Phil.*,   β *Recherche de la Vérité*, liv. i.
p. 41.]   chap. xx. § 3.

three, were it possible, would be tantamount to a mental annihilation. It is well observed, by Mr Stewart, "that it is not merely in order to free the mind from the influence of error, that it is useful to examine the foundation of established opinions. It is such an examination alone, that, in an inquisitive age like the present, can secure a philosopher from the danger of unlimited scepticism. To this extreme, indeed, the complexion of the times is more likely to give him a tendency, than to implicit credulity. In the former ages of ignorance and superstition, the intimate association which had been formed, in the prevailing systems of education, between truth and error, had given to the latter an ascendant over the minds of men, which it could never have acquired if divested of such an alliance. The case has, of late years, been most remarkably reversed : the common-sense of mankind, in consequence of the growth of a more liberal spirit of inquiry, has revolted against many of those absurdities which had so long held human reason in captivity ; and it was, perhaps, more than could have been reasonably expected, that, in the first moments of their emancipation, philosophers should have stopped short at the precise boundary which cooler reflection and more moderate views would have prescribed. The fact is, that they have passed far beyond it ; and that, in their zeal to destroy prejudices, they have attempted to tear up by the roots many of the best and happiest and most essential principles of our nature. That implicit credulity is a mark of a feeble mind, will not be disputed ; but it may not, perhaps, be as generally acknowledged, that the case is the same with unlimited scepticism : on the contrary, we are sometimes apt to ascribe this disposition to a more than ordinary vigour of intellect. Such a prejudice was by no means unnatural, at that period in the history of modern Europe, when reason first began to throw off the yoke of authority, and when it unquestionably required a superiority of understanding, as well as of intrepidity, for an individual to resist the contagion of prevailing superstition. But, in the present age, in which the tendency of fashionable opinions is directly opposite to those of the vulgar, the philosophical creed, or the philosophical scepticism, of by far the greater number of those who value themselves on an emancipation from popular errors, arises from the very same weakness with the credulity of the multitude; nor is it going too far to say, with Rousseau, that ' he who, in the end of the eighteenth century, has brought himself to abandon all his early principles without discrimination, would probably have been a bigot in the days of the League.' In the midst of these contrary impulses of fashionable and vulgar prejudices, he alone evinces the superiority and the strength of his mind, who is able to disentangle truth from error; and to oppose the clear conclusions of his own unbiassed faculties to the united clamours of superstition and of false philosophy. Such are the men whom nature marks out to be the lights of the world; to fix the wavering opinions of the multitude, and to impress their own characters on that of their age." [a]

In a word, philosophy is, as Aristotle has justly expressed it, not the art of doubting, but the art of doubting well.[β]

Aristotle.

---

[a] *Elements*, vol. i. book ii. § 1; *Coll. Works*, vol. ii. p. 68 *et seq.*—ED.

[β] *Metaph.*, ii. 1 : Ἔστι δὲ τοῖς εὐπορῆσαι βουλομένοις προὔργου τὸ δι-απορῆσαι καλῶς· ἡ γὰρ ὕστερον εὐπορία λύσις τῶν πρότερον ἀπορουμένων ἐστί, λύειν δ' οὐκ ἔστιν ἀγνοοῦντας τὸν δεσ-μόν.—ED.

Second
practical
condition—
subjuga-
tion of the
passions.

Sloth.

Pride.

In the second place, in obedience to the precept of
Socrates, the passions, under which we shall include
sloth, ought to be subjugated.

These ruffle the tranquillity of the mind, and consequently deprive it of the power of carefully considering all that the solution of a question requires should
be examined. A man under the agitation of any
lively emotion, is hardly aware of aught but what has
immediate relation to the passion which agitates and
engrosses him. Among the affections which influence
the will, and induce it to adhere to scepticism or error,
there is none more dangerous than sloth. The greater
proportion of mankind are inclined to spare themselves
the trouble of a long and laborious inquiry ; or they
fancy that a superficial examination is enough ; and
the slightest agreement between a few objects, in a
few petty points, they at once assume as evincing the
correspondence of the whole throughout. Others apply
themselves exclusively to the matters which it is
absolutely necessary for them to know, and take no
account of any opinion but that which they have
stumbled on,—for no other reason than that they have
embraced it, and are unwilling to recommence the
labour of learning. They receive their opinion on the
authority of those who have had suggested to them
their own ; and they are always facile scholars, for
the slightest probability is, for them, all the evidence
that they require.

Pride is a powerful impediment to a progress in
knowledge. Under the influence of this passion, men
seek honour but not truth. They do not cultivate
what is most valuable in reality, but what is most
valuable in opinion. They disdain, perhaps, what can
be easily accomplished, and apply themselves to the

obscure and recondite ; but as the vulgar and easy is the foundation on which the rare and arduous is built, they fail even in attaining the object of their ambition, and remain with only a farrago of confused and ill-assorted notions. In all its phases, self-love is an enemy to philosophical progress; and the history of philosophy is filled with the illusions of which it has been the source. On the one side, it has led men to close their eyes against the most evident truths which were not in harmony with their adopted opinions. It is said that there was not a physician in Europe, above the age of forty, who would admit Harvey's discovery of the circulation of the blood. On the other hand, it is finely observed by Bacon, that "the eye of human intellect is not dry, but receives a suffusion from the will and from the affections, so that it may almost be said to engender any sciences it pleases. For what a man wishes to be true, that he prefers believing." [a] And, in another place, "if the human intellect hath once taken a liking to any doctrine, either because received and credited, or because otherwise pleasing,—it draws everything else into harmony with that doctrine, and to its support; and albeit there may be found a more powerful array of contradictory instances, these, however, it either does not observe, or it contemns, or by distinction extenuates and rejects."[β]

LECT.
V.

a *Nov. Org.*, lib. i. aph. xlix.   β *Ibid.*, aph. xlvi.

LECTURE VI.

THE METHOD OF PHILOSOPHY.

THE next question we proceed to consider is,—What is the true Method or Methods of Philosophy?

There is only one possible method in philosophy; and what have been called the different methods of different philosophers, vary from each other only as more or less perfect applications of this one Method to the objects of knowledge.

Method a progress towards an end.

All method[a] is a rational progress,—a progress towards an end; and the method of philosophy is the procedure conducive to the end which philosophy proposes. The ends,—the final causes,—of philosophy, as we have seen, are two;—first, the discovery of efficient causes, secondly, the generalisation of our knowledge into unity;—two ends, however, which fall together into one, inasmuch as the higher we proceed in the discovery of causes, we necessarily approximate more and more to unity. The detection of the one in the many might, therefore, be laid down as the end to which philosophy, though it can never reach it, tends continually to approximate. But, considering philo-

Philosophy has but one possible method.

a [On the difference between Order and Method, see Facciolati, *Rudimenta Logica*, pars iv. c. 1, note: "Methodus differt ab Ordine; quia ordo facit ut rem unam discamus post aliam; Methodus ut unam per aliam." Cf. Zabarella, *Op. Log.*, pp. 139, 149, 223, 225; Molinæus, *Log.*, p. 234 *et seq.*, p. 244 *et seq.*, ed. 1613.]

sophy in relation to both these ends, I shall endeavour to show you that it has only one possible method.

Considering philosophy, in the first place, in relation to its first end,—the discovery of causes,—we have seen that causes, (taking that term as synonymous for all without which the effect would not be), are only the coefficients of the effect; an effect being nothing more than the sum or complement of all the partial causes, the concurrence of which constitute its existence. This being the case,—and as it is only by experience that we discover what particular causes must conspire to produce such or such an effect,— it follows, that nothing can become known to us as a cause except in and through its effect; in other words, that we can only attain to the knowledge of a cause by extracting it out of its effect. To take the example we formerly employed, of a neutral salt. This, as I observed, is made up by the conjunction of three proximate causes,—viz., an acid, — an alkali, — and the force which brought the alkali and the acid into the requisite approximation. This last, as a transitory condition, and not always the same, we shall throw out of account. Now, though we might know the acid and the alkali in themselves as distinct phænomena, we could never know them as the concurrent causes of the salt, unless we had known the salt as their effect. And though, in this example, it happens that we are able to compose the effect by the union of its causes, and to decompose it by their separation, — this is only an accidental circumstance; for the far greater number of the objects presented to our observation, can only be decomposed, but not actually recomposed, and in those which can be recomposed, this possibility is itself only the result of a knowledge of the causes previously obtained by an original decomposition of the effect.

Analysis.   In so far, therefore, as philosophy is the research of causes, the one necessary condition of its possibility is the decomposition of effects into their constituted causes. This is the fundamental procedure of philosophy, and is called by a Greek term *Analysis*. But though analysis be the fundamental procedure, it is still only a mean towards an end. We analyse only that we may comprehend; and we comprehend only inasmuch as we are able to reconstruct in thought the complex effects which we have analysed into their elements. This mental reconstruction is, therefore, the final, the consummative procedure of philosophy, and Synthesis. it is familiarily known by the Greek term *Synthesis*. Analysis and synthesis, though commonly treated as two different methods, are, if properly understood, only the two necessary parts of the same method. Each is the relative and the correlative of the other. Analysis, without a subsequent synthesis, is incomplete; it is a mean cut off from its end. Synthesis, without a previous analysis, is baseless; for synthesis receives from analysis the elements which it recomposes. And, as synthesis supposes analysis as the prerequisite of its possibility, so it is also dependent on analysis for the qualities of its existence. The value of every synthesis depends upon the value of the foregoing analysis. If the precedent analysis afford false elements, the subsequent synthesis of these elements will necessarily afford a false result. If the elements furnished by analysis are assumed, and not really discovered,—in other words, if they be hypothetical,—the synthesis of these hypothetical elements will con-

stitute only a conjectural theory. The legitimacy of
every synthesis is thus necessarily dependent on the
legitimacy of the analysis which it presupposes, and
on which it founds.

These two relative procedures are thus equally necessary to each other. On the one hand, analysis
without synthesis affords only a commenced, only an
incomplete, knowledge. On the other, synthesis without analysis is a false knowledge,—that is, no knowledge at all.   Both, therefore, are absolutely necessary
to philosophy, and both are, in philosophy, as much
parts of the same method as, in the animal body, inspiration and expiration are of the same vital function.   But though these operations are each requisite
to the other, yet were we to distinguish and compare
what ought only to be considered as conjoined, it is
to analysis that the preference must be accorded. An
analysis is always valuable ; for though now without
a synthesis, this synthesis may at any time be added ;
whereas a synthesis without a previous analysis is
radically and *ab initio* null.

So far, therefore, as regards the first end of philosophy, or the discovery of causes, it appears that there
is only one possible method,—that method of which
analysis is the foundation, synthesis the completion.
In the second place, considering philosophy in relation
to its second end,—the carrying up our knowledge
into unity,—the same is equally apparent.

Everything presented to our observation, whether
external or internal, whether through sense or self-
consciousness, is presented in complexity. Through
sense the objects crowd upon the mind in multitudes,
and each separate individual of these multitudes is
itself a congeries of many various qualities. The same

Only one
possible
method—
shown in
relation to
the second
end of Phi-
losophy.

is the case with the phænomena of self-consciousness. Every modification of mind is a complex state; and the different elements of each state manifest themselves only in and through each other. Thus, nothing but multiplicity is ever presented to our observation; and yet our faculties are so limited that they are able to comprehend at once only the very simplest conjunctions. There seems, therefore, a singular disproportion between our powers of knowledge and the objects to be known. How is the equilibrium to be restored? This is the great problem proposed by nature, and which analysis and synthesis, in combination, enable us to solve. For example, I perceive a tree, among other objects of an extensive landscape, and I wish to obtain a full and distinct conception of that tree. What ought I to do? *Divide et impera:* I must attend to it by itself, that is, to the exclusion of the other constituents of the scene before me. I thus analyse that scene; I separate a petty portion of it from the rest, in order to consider that portion apart. But this is not enough, the tree itself is not a unity, but, on the contrary, a complex assemblage of elements, far beyond what my powers can master at once. I must carry my analysis still farther. Accordingly, I consider successively its height, its breadth, its shape; I then proceed to its trunk, rise from that to its branches, and follow out its different ramifications; I now fix my attention on the leaves, and severally examine their form, colour, &c. It is only after having thus, by analysis, detached all these parts, in order to deal with them one by one, that I am able, by reversing the process, fully to comprehend them again in a series of synthetic acts. By synthesis, rising from the ultimate analysis step by step, I view the parts in relation to each other, and, finally, to the whole of which they are the constituents; I reconstruct them; and it is only through these two counter-processes of analysis and synthesis that I am able to convert the confused perception of the tree, which I obtained at first sight, into a clear, and distinct, and comprehensive knowledge.[a]

But if analysis and synthesis be required to afford us a perfect knowledge even of one individual object of sense, still more are they required to enable the mind to reduce an indefinite multitude of objects,—the infinitude we may say of nature,—to the limits of its own finite comprehension. To accomplish this, it is requisite to extract the one out of the many, and thus to recall multitude to unity,—confusion to order. And how is this performed? The one in the many being that in which a plurality of objects agree,—that is, may be considered as the same; and the agreement of objects in any common quality being discoverable only by an observation and comparison of the objects themselves: it follows that a knowledge of the one can only be evolved out of a foregoing knowledge of the many. But this evolution can only be accomplished by an analysis and a synthesis. By analysis, from the infinity of objects presented to our observation, we select some. These we consider apart, and, further, only in certain points of view,—and we compare these objects with others also considered in the same points of view. So far the procedure is analytic. Having discovered, however, by this observation and comparison, that certain objects agree in certain respects, we generalise the qualities in which they coincide,—that is, from a certain number of individual instances we infer a general law; we perform what is called an act of induction. This induction is erroneously viewed **Induction**.

a [On the subject of analysis and synthesis, compare Condillac, *Logique*, cc. i. ii.]

as analytic; it is purely a synthetic process.[a] For example, from our experience,—and all experience, be it that of the individual or of mankind, is only finite, —from our limited experience, I say, that bodies, as observed by us, attract each other, we infer by induction the unlimited conclusion that all bodies gravitate towards each other. Now, here the consequent contains much more than was contained in the antecedent. Experience, the antecedent only says, and only can say—this, that, and the other body gravitate, (that is, *some* bodies gravitate); the consequent educed from that antecedent says,—*all* bodies gravitate. The antecedent is limited, the consequent unlimited. Something, therefore, has been added to the antecedent in order to legitimate the inference, if we are not to hold the consequent itself as absurd ; for, as you will hereafter learn, no conclusion must contain more than was contained in the premises from which it is drawn. What then is this something? If we consider the inductive process, this will be at once apparent.

The affirmation, this, that, and the other body gravitate, is connected with the affirmation, all bodies gravitate, only by inserting between the two a third affirmation, by which the two other affirmations are connected into reason and consequent,—that is, into a logical cause and effect. What that is I shall explain. All scientific induction is founded on the presumption that nature is uniform in her operations. Of the ground and origin of this presumption, I am not now

---

a It may be considered as the one or the other, according as the whole and its parts are viewed in the relations of comprehension or of extension. The latter, however, is the simpler and more convenient point of view; and in this respect Induction is properly synthetic. See the Author's *Discussions*, p. 173.—ED.

to speak. I shall only say, that, as it is a principle which we suppose in all our inductions, it cannot be itself a product of induction. It is, therefore, interpolated in the inductive reasoning by the mind itself. In our example the reasoning will, accordingly, run as follows :

This, that, and the other body, (some bodies), are observed to gravitate ;

But, (as nature is uniform in her operations), this, that, and the other body, (some bodies), represent all bodies ;

Therefore all bodies gravitate.

Now, in this and other examples of induction, it is the mind which binds up the separate substances observed and collected into a whole, and converts what is only the observation of many particulars into a universal law. This procedure is manifestly synthetic.

Now, you will remark that analysis and synthesis are here absolutely dependent on each other. The previous observation and comparison,—the analytic foundation,—are only instituted for the sake of the subsequent induction,—the synthetic consummation. What boots it to observe and to compare, if the uniformities we discover among objects are never generalised into laws ? We have obtained an historical, but not a philosophical knowledge. Here, therefore, analysis without synthesis is incomplete. On the other hand, an induction which does not proceed upon a competent enumeration of particulars, is either doubtful, improbable, or null ; for all synthesis is dependent on a foregone analysis for whatever degree of certainty it may pretend to. Thus, considering philosophy in relation to its second end, unity or system, it is manifest, that the method by which it accomplishes that end, is a method involving both an analytic and a synthetic process.

The history of philosophy manifests the more or less accurate fulfilment of the conditions of the one method. Now, as philosophy has only one possible method, so the History of philosophy only manifests the conditions of this one method, more or less accurately fulfilled. There are aberrations in the method,—no aberrations from it.

Earliest problem of philosophy. "Philosophy commenced with the first act of reflection on the objects of sense or self-consciousness, for the purpose of explaining them. And with that first act of reflection, the method of philosophy began, in its application of an analysis, and in its application of a synthesis, to its object. The first philosophers naturally endeavoured to explain the enigma of external nature. The magnificent spectacle of the material universe, and the marvellous demonstrations of power and wisdom which it everywhere exhibited, were the objects which called forth the earliest efforts of speculation. Philosophy was thus, at its commencement, physical, not psychological ; it was not the problem of the soul, but the problem of the world, which it first attempted to solve.

"And what was the procedure of philosophy in its solution of this problem ? Did it first decompose the whole into its parts, in order again to reconstruct them into a system ? This it could not accomplish ; but still it attempted this, and nothing else. A complete analysis was not to be expected from the first efforts of intelligence ; its decompositions were necessarily partial and imperfect ; a partial and imperfect analysis afforded only hypothetical elements ; and the synthesis of these elements issued, consequently, only in a one-sided or erroneous theory.

"Thales, the founder of the Ionian philosophy, de-

voted an especial study to the phænomena of the
material universe ; and, struck with the appearances
of power which water manifested in the formation of
bodies, he analysed all existences into this element,
which he viewed as the universal principle,—the universal agent of creation.   He proceeded by an incomplete analysis, and generalised by hypothesis the law
which he drew by induction from the observation of a
small series of phænomena.

"The Ionic school continued in the same path. They
limited themselves to the study of external nature, and
sought in matter the principle of existence.   Anaxi-
mander of Miletus, the countryman and disciple of
Thales, deemed that he had traced the primary cause
of creation to an ethereal principle, which occupied
space, and whose different combinations constituted
the universe of matter.   Anaximenes found the original element in air, from which, by rarefaction and
condensation, he educed existences.   Anaxagoras carried his analysis farther, and made a more discreet
use of hypothesis ; he rose to the conception of an
intelligent first cause, distinct from the phænomena
of nature ; and his notion of the Deity was so far
above the gross conceptions of his contemporaries,
that he was accused of atheism.

"Pythagoras, the founder of the Italic school, analysed the properties of number; and the relations which
this analysis revealed, he elevated into principles of
the mental and material universe.   Mathematics were
his only objects ; his analysis was partial, and his
synthesis was consequently hypothetical.   The Italic
school developed the notions of Pythagoras, and, exclusively preoccupied with the relations and harmonies
of existence, its disciples did not extend their specu-

lation to the consideration either of substance or of cause.

"Thus, these earlier schools, taking external nature for their point of departure, proceeded by an imperfect analysis, and a presumptuous synthesis, to the construction of exclusive systems,—in which Idealism, or Materialism, preponderated, according to the kind of data on which they founded.

Eleatic
School.

"The Eleatic school, which is distinguished into two branches, the one of Physical, the other of Metaphysical, speculation, exhibits the same character, the same point of departure, the same tendency, and the same errors.

The Soph-
ists.
Socrates.

"These errors led to the scepticism of the Sophists, which was assailed by Socrates,—the sage who determined a new epoch in philosophy by directing observation on man himself; and henceforward the study of mind becomes the prime and central science of philosophy.

"The point of departure was changed, but not the method. The observation or analysis of the human mind, though often profound, remained always incomplete. Fortunately, the first disciples of Socrates, imitating the prudence of their master, and warned by the downfall of the systems of the Ionic, Italic, and Eleatic schools, made a sparing use of synthesis, and hardly a pretension to system.

Plato and
Aristotle.

"Plato and Aristotle directed their observation on the phænomena of intelligence, and we cannot too highly admire the profundity of their analysis, and even the sobriety of their synthesis. Plato devoted himself more particularly to the higher faculties of intelligence; and his disciples were led, by the love of generalisation, to regard as the intellectual whole those portions of intelligence which their master had analysed; and this exclusive spirit gave birth to systems false, not in themselves, but as resting upon a too narrow basis. Aristotle, on the other hand, whose genius was of a more positive character, analysed with admirable acuteness those operations of mind which stand in more immediate relation to the senses; and this tendency, which among his followers became often exclusive and exaggerated, naturally engendered systems which more or less tended to materialism."[a]

The school of Alexandria, in which the systems resulting from these opposite tendencies were combined, endeavoured to reconcile and to fuse them into a still more comprehensive system. Eclecticism,—conciliation,—union, were, in all things, the grand aim of the Alexandrian school. Geographically situated between Greece and Asia, it endeavoured to ally Greek with Asiatic genius, religion with philosophy. Hence the Neoplatonic system, of which the last great representative is Proclus. This system is the result of the long labour of the Socratic schools. It is an edifice reared by synthesis out of the materials which analysis had collected, proved, and accumulated, from Socrates down to Plotinus.

But a synthesis is of no greater value than its relative analysis; and as the analysis of the earlier Greek philosophy was not complete, the synthesis of the Alexandrian school was necessarily imperfect.

In the scholastic philosophy, analysis and observation were too often neglected in some departments of philosophy, and too often carried rashly to excess in others.

After the revival of letters, during the fifteenth

School of Alexandria.

Proclus.

The Scholastic Philosophy.

a Géruzez, *Nouveau Cours de Philosophie*, p. 4-8. Paris, 1834, (2d ed.)

and sixteenth centuries, the labours of philosophy were principally occupied in restoring and illustrating the Greek systems; and it was not until the seventeenth century, that a new epoch was determined by the genius of Bacon and Descartes.

In Bacon and Descartes our modern philosophy may be said to originate, inasmuch as they were the first who made the doctrine of method a principal object of consideration. They both proclaimed, that, for the attainment of scientific knowledge, it is necessary to observe with care,—that is, to analyse; to reject every element as hypothetical, which this analysis does not spontaneously afford; to call in experiment in aid of observation; and to attempt no synthesis or generalisation, until the relative analysis has been completely accomplished. They showed that previous philosophers had erred, not by rejecting either analysis or synthesis, but by hurrying on to synthetic induction from a limited or specious analytic observation. They propounded no new method of philosophy, they only expounded the conditions of the old. They showed that these conditions had rarely been fulfilled by philosophers in time past; and exhorted them to their fulfilment in time to come. They thus explained the petty progress of the past philosophy; and justly anticipated a gigantic advancement for the future. Such was their precept, but such unfortunately was not their example. There are no philosophers who merit so much in the one respect; none, perhaps, who deserve less in the other.

Of philosophy since Bacon and Descartes we at present say nothing. Of that we shall hereafter have frequent occasion to speak. But to sum up what this historical sketch was intended to illustrate. There is but one possible method of philosophy,—a combina- tion of analysis and synthesis; and the purity and equilibrium of these two elements constitute its perfection. The aberrations of philosophy have been all so many violations of the laws of this one method. Philosophy has erred, because it built its systems upon incomplete or erroneous analysis; and it can only proceed in safety, if, from accurate and unexclusive observation, it rise, by successive generalisation, to a comprehensive system.

LECTURE VII.

THE DIVISIONS OF PHILOSOPHY.

I HAVE already endeavoured to afford you a general notion of what Philosophy comprehends : I now proceed to say something in regard to the Parts into which it has been divided. Here, however, I must limit myself to the most famous distributions, and to those which, as founded on fundamental principles, it more immediately concerns you to know. For, were I to attempt an enumeration of the various Divisions of Philosophy which have been proposed, I should only confuse you with a multitude of contradictory opinions, with the reasons of which you could not, at present, possibly be made acquainted.

Expediency of a division of Philosophy.

Seneca, in a letter to his young friend Lucilius, expresses the wish that the whole of philosophy might, like the spectacle of the universe, be at once submitted to our view. "Utinam, quemadmodum universi mundi facies in conspectum venit, ita philosophia tota nobis posset occurrere, simillimum mundo spectaculum." [a] But as we cannot survey the universe at a glance, neither can we contemplate the whole of philosophy in one act of consciousness. We can only master it gradually and piecemeal; and this is in fact the reason why philosophers have always distributed their

a *Epist.* lxxxix.

science, (constituting, though it does, one organic LECT.
VII. whole), into a plurality of sciences. The expediency, and even necessity, of a division of philosophy, in order that the mind may be enabled to embrace in one general view its various parts, in their relation to each other, and to the whole which they constitute, is admitted by every philosopher. "Res utilis," continues Seneca, "et ad sapientiam properanti utique necessaria, dividi philosophiam, et ingens corpus ejus in membra disponi. Facilius enim per partes in cognitionem totius adducimur." [a]

But although philosophers agree in regard to the utility of such a distribution, they are almost as little at one in regard to the parts, as they are in respect to the definition, of their science ; and, indeed, their differences in reference to the former, mainly arise from their discrepancies in reference to the latter. For they who vary in their comprehension of the whole, cannot agree in their division of the parts.

The most ancient and universally recognised distinction of Philosophy, is into Theoretical and Practical. These are discriminated by the different nature of their ends. Theoretical, called likewise speculative, and contemplative, philosophy has for its highest end mere truth or knowledge. Practical philosophy, on the other hand, has truth or knowledge only as its proximate end,—this end being subordinate to the ulterior end of some practical action. In theoretical philosophy, we know for the sake of knowing, *scimus ut sciamus:* in practical philosophy, we know for the sake of acting, *scimus ut operemur.* [b] I may here

The most ancient division into Theoretical and Practical.

a *Epist.* lxxxix.

β Θεωρητικῆς μὲν ἐπιστήμης τέλος ἀλήθεια, πρακτικῆς δ' ἔργον. Arist. *Metaph.*, A minor, c. 1; "or as Averroes has it, *Per speculativam, scimus ut sciamus, per practicam scimus ut operemur."—Discussions,* p. 134. Cf. *In Metaph.,* lib. ii. com. 3.—ED.

notice the poverty of the English language, in the
want of a word to express that practical activity
which is contradistinguished from mere intellectual
or speculative energy,—what the Greeks express by
πράσσειν, the Germans by *handeln*. The want of
such a word occasions frequent ambiguity; for, to
express the species which has no appropriate word,
we are compelled to employ the generic term *active*.
Thus our philosophers divide the powers of the mind
into Intellectual and Active. They do not, however,
thereby mean to insinuate that the powers called
intellectual are a whit less energetic than those specially denominated active. But, from the want of a
better word, they are compelled to employ a term
which denotes at once much more and much less
than they are desirous of expressing. I ought to
observe that the term *practical* has also obtained
with us certain collateral significations, which render
it in some respects unfit to supply the want.[a] But
to return.

History of
the distinc-
tion of
Theoretical
and Prac-
tical.

This distinction of Theoretical and Practical philosophy was first explicitly enounced by Aristotle;[β]
and the attempts of the later Platonists to carry it up
to Plato, and even to Pythagoras, are not worthy of
statement, far less of refutation. Once promulgated,
the division was, however, soon generally recognised.
The Stoics borrowed it, as may be seen from Seneca:[γ]
—"Philosophia et contemplativa est et activa; spectat,
simulque agit." It was also adopted by the Epicu-
reans; and, in general, by those Greek and Roman

a Cf. *Reid's Works*, p. 511, n. †.—Ed.
β *Metaph.*, v. 1: Πᾶσα διάνοια ἢ
πρακτικὴ ἢ ποιητικὴ ἢ θεωρητική. Cf.
Metaph., x. 7; *Top.*, vi. 6; viii. 3. But
the division had been at least intimat-
ed by Plato; *Politicus*, p. 258: Ταύτη
τοίνυν συμπάσας ἐπιστήμας διαίρει, τὴν
μὲν πρακτικὴν προσειπὼν, τὴν δὲ μόνον
γνωστικήν.—Ed.
γ *Ep.* xcv. 10.

philosophers who viewed their science as versant either
in the contemplation of nature (φυσική), or in the
regulation of human action (ἠθική) ;[a] for by *nature*
they did not denote the material universe alone, but
their Physics included Metaphysics, and their Ethics
embraced Politics and Economics. There was thus
only a difference of nomenclature ; for Physical and
Theoretical,—Ethical and Practical Philosophy, were
with them terms absolutely equivalent.

I regard the division of Philosophy into Theoretical
and Practical as unsound, and this for two reasons.

The first is, that philosophy, as philosophy, is only
cognitive,—only theoretical : whatever lies beyond
the sphere of speculation or knowledge, transcends
the sphere of philosophy ; consequently, to divide
philosophy by any quality ulterior to speculation, is
to divide it by a difference which does not belong to
it.   Now, the distinction of practical philosophy from
theoretical commits this error.   For, while it is admitted that all philosophy, as cognitive, is theoretical,
some philosophy is again taken out of this category
on the ground, that, beyond the mere theory,—the
mere cognition,—it has an ulterior end in its application to practice.

But, in the second place, this difference, even were
it admissible, would not divide philosophy ; for, in
point of fact, all philosophy must be regarded as practical, inasmuch as mere knowledge,—that is, the mere
possession of truth,—is not the highest end of any

The divi-
sion of Phi-
losophy
into Theo-
retical and
Practical
unsound.

a Sextus Empiricus, *Adv. Math.*,
vii. 14 : Τῶν δὲ διμερῆ τὴν φιλοσοφίαν
ὑποστησαμένων Ξενοφάνης μὲν ὁ Κολο-
φώνιος, τὸ φυσικὸν ἅμα καὶ λογικόν,
ὡς φασί τινες, μετήρχετο, Ἀρχέλαος
δὲ ὁ Ἀθηναῖος τὸ φυσικὸν καὶ ἠθικόν
μεθ᾽ οὗ τινὲς καὶ τὸν Ἐπίκουρον τάττου
σιν ὡς καὶ τὴν λογικὴν θεωρίαν ἐκβάλ-
λοντα.   Seneca, *Ep.* lxxxix. : "Epi-
curei duas partes philosophiæ puta-
verunt esse, Naturalem, atque Mora-
lem : Rationalem removerunt."—Ed.

philosophy, but, on the contrary, all truth or knowledge is valuable only inasmuch as it determines the
mind to its contemplation,—that is, to practical energy. Speculation, therefore, inasmuch as it is not a
negation of thought, but, on the contrary, the highest
energy of intellect, is, in point of fact, pre-eminently
practical. The practice of one branch of philosophy
is, indeed, different from that of another; but all are
still practical; for in none is mere knowledge the
ultimate,—the highest end.

Contro-
versy
among
ancients
regarding
the relation
of Logic to
Philoso-
phy.

Among the ancients, the principal difference of
opinion regarded the relation of Logic to Philosophy
and its branches. But as this controversy is of very
subordinate importance, and hinges upon distinctions,
to explain which would require considerable detail, I
shall content myself with saying,—that, by the Platonists, Logic was regarded both as a part, and as
the instrument, of philosophy;—by the Aristotelians,
(Aristotle himself is silent), as an instrument, but not
as a part, of philosophy;—by the Stoics, as forming
one of the three parts of philosophy,—Physics or theoretical, Ethics or practical, philosophy, being the other
two.[a] But as Logic, whether considered as a part of
philosophy proper or not, was by all included under
the philosophical sciences, the division of these sciences
which latterly prevailed among the Academic, the
Peripatetic, and the Stoical sects, was into Logic as
the subsidiary or instrumental doctrine, and into the

---

a Alexander Aphrodisiensis, *In
Anal. Prior.*, p. 2, (ed. 1520); Ammonius, *In Categ.*, c. 4; Philoponus,
In Anal. Prior., f. 4; Cramer's *Anecdota*, vol. iv. p. 417. Compare the
Author's *Discussions*, p. 132. The
division of Philosophy into Logic,
Physics, and Ethics, probably originated with the Stoics. See Laertius,
vii. 39; Pseudo-Plutarch, *De Plac.
Phil.*, Prœm. It is sometimes, but
apparently without much reason, attributed to Plato. See Cicero, *Acad.
Quæst.*, i. 5; Eusebius, *Præp. Evan.*,
xi. 1; Augustin, *De Civ. Dei*, viii. 4.
—ED.

two principal branches of Theoretical and Practical
Philosophy.[a]

It is manifest that in our sense of the term *practical*, Logic, as an instrumental science, would be comprehended under the head of practical philosophy.

I shall take this opportunity of explaining an Application of the terms Art and Science.
anomaly which you will find explained in no work
with which I am acquainted. Certain branches of
philosophical knowledge are called Arts, or Arts and
Sciences indifferently; others are exclusively denominated Sciences. Were this distinction coincident with
the distinction of sciences speculative and sciences
practical,—taking the term practical in its ordinary
acceptation,—there would be no difficulty; for, as
every practical science necessarily involves a theory,
nothing could be more natural than to call the same
branch of knowledge an art, when viewed as relative
to its practical application, and a science, when viewed
in relation to the theory which that application supposes. But this is not the case. The speculative
sciences, indeed, are never denominated arts; we may,
therefore, throw them aside. The difficulty is exclusively confined to the practical. Of these some never
receive the name of arts; others are called arts and
sciences indifferently. Thus the sciences of Ethics,
Economics, Politics, Theology, &c., though all practical, are never denominated arts; whereas this appellation is very usually applied to the practical sciences
of Logic, Rhetoric, Grammar, &c.

That the term art is with us not co-extensive with
practical science, is thus manifest; and yet these are
frequently confounded. Thus, for example, Dr Whately,

in his definition of Logic, thinks that Logic is a science, in so far as it institutes an analysis of the process of the mind in reasoning, and an art, in so far as it affords practical rules to secure the mind from error in its deductions; and he defines an art the application of knowledge to practice.[a] Now, if this view were correct, art and practical science would be convertible terms. But that they are not employed as synonymous expressions is, as we have seen, shown by the incongruity we feel in talking of the art of Ethics, the art of Religion, &c., though these are eminently practical sciences.

The question, therefore, still remains, Is this restriction of the term art to certain of the practical sciences the result of some accidental and forgotten usage, or is it founded on any rational principle which we are able to trace? The former alternative seems to be the common belief; for no one, in so far as I know, has endeavoured to account for the apparently vague and capricious manner in which the terms art and science are applied. The latter alternative, however, is the true; and I shall endeavour to explain to you the reason of the application of the term art to certain practical sciences, and not to others.

Its histori-
cal origin.

You are aware that the Aristotelic philosophy was, for many centuries, not only the prevalent, but, during the middle ages, the one exclusive philosophy in Europe. This philosophy of the middle ages, or, as it is commonly called, the Scholastic Philosophy, has exerted the most extensive influence on the languages of modern Europe; and from this common source has been principally derived that community of expression which these languages exhibit. Now, the peculiar

a See *Discussions*, p. 131.—ED.

application of the term art was introduced into the vulgar tongues from the scholastic philosophy; and was borrowed by that philosophy from Aristotle. This is only one of a thousand instances which might be alleged of the unfelt influence of a single powerful mind, on the associations and habits of thought of generations to the end of time; and of Aristotle is pre-eminently true, what has been so beautifully said of the ancients in general:—

> "The great of old !
> The dead but sceptred sovrans who still rule
> Our spirits from their urns." [a]

Now, then, the application of the term art in the modern languages being mediately governed by certain distinctions which the capacities of the Greek tongue allowed Aristotle to establish, these distinctions must be explained.

In the Aristotelic philosophy, the terms πρᾶξις and πρακτικός,—that is, *practice* and *practical*,— were employed both in a generic or looser, and in a special or stricter, signification. In its generic meaning πρᾶξις, *practice*, was opposed to theory or speculation, and it comprehended under it, practice in its special meaning, and another co-ordinate term to which practice, in this its stricter signification, was opposed. This term was ποίησις, which we may inadequately translate by *production*. The distinction of πρακτικός and ποιητικός consisted in this: the former denoted that action which terminated in action,—the latter, that action which resulted in some permanent product. For example, dancing and music are practical, as leaving no work after their performance; whereas, painting and statuary

a Byron's *Manfred*, Act iii. scene iv.

are productive, as leaving some product over and
above their energy.[a]

Why
Ethics,
Politics,
&c., de-
signated
Sciences;
Logic, Rhe-
toric, &c.,
Arts.

Now Aristotle, in formally defining art, defines it
as a habit productive, and not as a habit practical,
ἕξις ποιητικὴ μετὰ λόγου; and, though he has not
always himself adhered strictly to this limitation, his
definition was adopted by his followers, and the term
in its application to the practical sciences, (the term
practical being here used in its generic meaning),
came to be exclusively confined to those whose end
did not result in mere action or energy. Accordingly
as Ethics, Politics, &c., proposed happiness as their
end, and as happiness was an energy, or at least the
concomitant of energy, these sciences terminated in
action, and were consequently *practical*, not *productive*. On the other hand, Logic, Rhetoric, &c., did
not terminate in a mere,—an evanescent action, but
in a permanent,—an enduring product. For the end
of Logic was the production of a reasoning, the end
of Rhetoric the production of an oration, and so
forth.[β] This distinction is not perhaps beyond the
reach of criticism, and I am not here to vindicate its
correctness. My only aim is to make you aware of
the grounds of the distinction, in order that you may
comprehend the principle which originally determined
the application of the term *art* to some of the practical

a See *Eth. Nic.*, i. 1 : Διαφορὰ δέ
τις φαίνεται τῶν τελῶν τὰ μὲν γάρ
εἰσιν ἐνέργειαι τὰ δὲ παρ' αὐτὰς ἔργα
τινά. *Ibid.*, vi. 4 ; *Magna Moralia*,
i. 35. Cf. Quintilian, *Institut.*, lib.
ii. c. 18.—ED.

β Cf. Burgersdyck, *Institut. Log.*,
lib. i. § 6: "Logica dicitur ποιεῖν, id
est, *facere* sive *efficere* syllogismos,
definitiones, &c. Neque enim verum
est, quod quidam aiunt, ποιεῖν semper
significare ejusmodi actionem, qua

ex palpabili materia opus aliquod
efficitur quod etiam post actionem
permanet. Nam Poetica dicta est
ἀπὸ τοῦ ποιεῖν quæ tamen palpabilem
materiam non tractat, neque opus
facit ipsa Poetæ fictione durabilius.
Quod enim poemata supersint, id non
est ab ea actione qua efficiuntur, sed
a scriptione. Atque hæc de genere."
See also Scheibler, *Opera*, Tract.
Proæm. § iii. p. 6.—ED.

sciences and not to others, and without a knowledge of which principle the various employment of the term must appear to you capricious and unintelligible. It is needless, perhaps, to notice that the rule applies only to the philosophical sciences,—to those which received their form and denominations from the learned. The mechanical dexterities were beneath their notice; and these were accordingly left to receive their appellations from those who knew nothing of the Aristotelic proprieties. Accordingly, the term art is in them applied, without distinction, to productive and unproductive operations. We speak of the art of rope-dancing, equally as of the art of rope-making. But to return.

LECT. VII.

The division of philosophy into Theoretical and Practical is the most important that has been made; and it is that which has entered into nearly all the distributions attempted by modern philosophers. Bacon was the first, after the revival of letters, who essayed a distribution of the sciences and of philosophy. He divided all human knowledge into History, Poetry, and Philosophy. Philosophy he distinguished into branches conversant about the Deity, about Nature, and about Man; and each of these had their subordinate divisions, which, however, it is not necessary to particularise.[a]

Universality of the division of Philosophy into Theoretical and Practical.

Bacon.

Descartes[β] distributed philosophy into theoretical and practical, with various subdivisions; but his followers adopted the division of Logic, Metaphysics, Physics, and Ethics.[γ] Gassendi recognised, like the

Descartes and his followers.

a Advancement of Learning; Works, vol. ii. pp. 100, 124, (ed. Montagu); De Augmentis Scientiarum, lib. ii. c. 1, lib. iii. c. 1; Works, vol. viii. pp. 87, 152.—Ed.

β See the Prefatory Epistle to the Principia.—Ed.

γ See Sylvain Regis, Cours entier de

Philosophie, contenant la Logique, la Metaphysique, la Physique, et la Morale. Cf. Clauberg: "Physica.... Philosophia Naturalis dicitur; distincta a Supernaturali seu Metaphysica, et a Rationali seu Logica, necnon a Morali seu Practica."—Disput. Phys. i., Opera, p. 54.—Ed.

LECT.
VII.

Gassendi.
Locke.
Kant.

Fichte.

ancients, three parts of philosophy, Logic, Physics, and Ethics,[a] and this, along with many other of Gassendi's doctrines, was adopted by Locke.[β] Kant distinguished philosophy into theoretical and practical, with various subdivisions ;[γ] and the distribution into theoretical and practical was also established by Fichte.[δ]

Conclusion
of Intro-
ductory
Lectures.

I have now concluded the Lectures generally introductory to the proper business of the Course. In these Lectures, from the general nature of the subjects, I was compelled to anticipate conclusions, and to depend on your being able to supply a good deal of what it was impossible for me articulately to explain. I now enter upon the consideration of the matters which are hereafter to occupy our attention, with comparatively little apprehension ; for, in these, we shall be able to dwell more upon details, while, at the same time, the subject will open upon us by degrees, so that, every step that we proceed, we shall find the progress easier. But I have to warn you, that you will probably find the very commencement the most arduous, and this not only because you will come less inured to difficulty, but because it will there be necessary to deal with principles, and these of a general and abstract nature ; whereas, having once mastered these, every subsequent step will be comparatively easy.

Order of
the Course.

Without entering upon details, I may now summarily state to you the order which I propose to follow in the ensuing Course. This requires a preliminary exposition of the different departments of

a Syntagma Philosophicum, Lib. Procem. c. 9 (Opera, Lugduni, 1658, vol. i. p. 29.)—Ed.

β Essay, book iv. ch. 21.—Ed.

γ Kritik der reinen Vernunft, Me-

thodenlehre, c. 3.—Ed.

δ Grundlage der gesammten Wissenschaftslehre, § 4 (Werke, vol. i. p. 126.)—Ed.

Philosophy, in order that you may obtain a comprehensive view of the proper objects of our consideration,
and of the relations in which they stand to others.

Science and Philosophy are conversant either about
Mind or about Matter. The former of these is Philosophy properly so called. With the latter we have
nothing to do, except in so far as it may enable us
to throw light upon the former, for Metaphysics, in
whatever latitude the term be taken, is a science,
or complement of sciences, exclusively occupied with
mind. Now the Philosophy of Mind,—Psychology
or Metaphysics, in the widest signification of the
terms,—is *threefold*; for the object it immediately
proposes for consideration may be either, 1°, PHÆNO-
MENA in general ; or, 2°, LAWS ; or, 3°, INFERENCES,—
RESULTS. This I will endeavour to explain.

The whole of philosophy is the answer to these
three questions: 1°, What are the Facts or Phænomena
to be observed ? 2°, What are the Laws which regulate
these facts, or under which these phenomena appear ?
3°, What are the real Results, not immediately manifested, which these facts or phænomena warrant us
in drawing ?

If we consider the mind merely with the view of
observing and generalising the various phænomena it
reveals,—that is, of analysing them into capacities or
faculties,—we have one mental science, or one department of mental science ; and this we may call the
PHÆNOMENOLOGY OF MIND. It is commonly called
PSYCHOLOGY—EMPIRICAL PSYCHOLOGY, or the INDUC-
TIVE PHILOSOPHY OF MIND ; we might call it PHÆ-
NOMENAL PSYCHOLOGY. It is evident that the divisions of this science will be determined by the classes
into which the phænomena of mind are distributed.

If, again, we analyse the mental phænomena with
the view of discovering and considering, not contin-

gent appearances, but the *necessary* and *universal*
facts,—*i.e.*, the Laws by which our faculties are governed, to the end that we may obtain a criterion by
which to judge or to explain their procedures and
manifestations,—we have a science which we may
call the NOMOLOGY OF MIND,—NOMOLOGICAL PSYCHO-

LOGY. Now, there will be as many distinct classes of
Nomological Psychology, as there are distinct classes
of mental phænomena under the Phænomenological
division. I shall, hereafter, show you that there are
Three great classes of these phænomena,—viz., 1°, The
phænomena of our Cognitive faculties, or faculties of
Knowledge ; 2°, The phænomena of our Feelings, or
the phænomena of Pleasure and Pain ; and, 3°, The
phænomena of our Conative powers,—in other words,
the phænomena of Will and Desire. (These you
must, for the present, take upon trust.)[a] Each of
these classes of phænomena has accordingly a science
which is conversant about its laws. For as each proposes a different end, and, in the accomplishment of
that end, is regulated by peculiar laws, each must,
consequently, have a different science conversant about
these laws,—that is, a different Nomology.

There is no one, no Nomological, science of the
Cognitive faculties in general, though we have some
older treatises which, though partial in their subject,
afford a name not unsuitable for a nomology of the
cognitions,—viz., Gnoseologia or Gnostologia. There
is no independent science of the laws of Perception ; if
there were, it might be called Æsthetic, which, however, as we shall see, would be ambiguous. Mnemonic,
or the science of the laws of Memory, has been elabo-

rated at least in numerous treatises ; but the name Anamnestic, the art of Recollection or Reminiscence, might be equally well applied to it.   The laws of the Representative faculty,—that is, the laws of Association,—have not yet been elevated into a separate nomological science.   Neither have the conditions of the Regulative or Legislative faculty, the faculty itself of Laws, been fully analysed, far less reduced to system ; though we have several deservedly forgotten treatises, of an older date, under the inviting name of *Noologies*. The only one of the cognitive faculties, whose laws constitute the object-matter of a separate science, is the Elaborative, — the Understanding Special, the faculty of Relations, the faculty of Thought Proper. This nomology has obtained the name of LOGIC among other appellations, but not from Aristotle.   The best name would have been DIANOETIC.   Logic is the science of the laws of thought, in relation to the end which our cognitive faculties propose,—*i.e.*, the TRUE. To this head might be referred Grammar,—Universal Grammar,—Philosophical Grammar, or the science conversant with the laws of Language as the instrument of thought.

The Nomology of our Feelings, or the science of the laws which govern our capacities of enjoyment, in relation to the end which they propose,—*i.e.*, the PLEASURABLE,—has obtained no precise name in our language.   It has been called the Philosophy of Taste, and, on the Continent especially, it has been denominated Æsthetic.   Neither name is unobjectionable. The first is vague, metaphorical, and even delusive. In regard to the second, you are aware that αἴσθησις in Greek means feeling in general, as well as sense in particular, as our term *feeling* means either the sense of touch in particular, or sentiment and the capacity of the pleasurable and painful in general. Both terms are, therefore, to a certain extent ambiguous; but this objection can rarely be avoided, and Æsthetic, if not the best expression to be found, has already been long and generally employed. It is now nearly a century since Baumgarten, a celebrated philosopher of the Leibnitio-Wolfian school, first applied the term Æsthetic to the doctrine which we vaguely and periphrastically denominate the Philosophy of Taste, the theory of the Fine Arts, the science of the Beautiful and Sublime,[a] &c.; and this term is now in general acceptation, not only in Germany, but throughout the other countries of Europe. The term Apolaustic would have been a more appropriate designation.

3. Nomology of the Conative Powers.

Finally, the Nomology of our Conative powers is Practical Philosophy, properly so called; for practical philosophy is simply the science of the laws regulative of our Will and Desires, in relation to the end which our conative powers propose,—i.e., the GOOD.

Ethics, Politics.

This, as it considers these laws in relation to man as an individual, or in relation to man as a member of society, will be divided into two branches,—Ethics and Politics; and these again admit of various subdivisions.

So much for those parts of the Philosophy of Mind, which are conversant about Phænomena, and about Laws. The Third great branch of this philosophy is that which is engaged in the deduction of Inferences or Results.

III. Ontology, or Metaphysics Proper.

In the First branch,—the Phænomenology of mind, —philosophy is properly limited to the facts afforded in consciousness, considered exclusively in themselves. But these facts may be such as not only to be objects of knowledge in themselves, but likewise to furnish us

---

a Baumgarten's work on this sub-   was published in 1750-58.—ED.
ject, entitled *Æsthetica* (two vols.),

with grounds of inference to something out of themselves.   As effects, and effects of a certain character,
they may enable us to infer the analogous character
of their unknown causes ; as phænomena, and phænomena of peculiar qualities, they may warrant us in
drawing many conclusions regarding the distinctive
character of that unknown principle, of that unknown
substance, of which they are the manifestations.   Although, therefore, existence be only revealed to us in
phænomena, and though we can, therefore, have only
a relative knowledge either of mind or of matter ;
still, by inference and analogy, we may legitimately
attempt to rise above the mere appearances which
experience and observation afford.   Thus, for example,
the existence of God and the immortality of the Soul
are not given us as phænomena, as objects of immediate knowledge ; yet, if the phænomena actually
given do necessarily require, for their rational explanation, the hypotheses of immortality and of God, we
are assuredly entitled, from the existence of the former,
to infer the reality of the latter.   Now, the science
conversant about all such inferences of unknown being
from its known manifestations, is called ONTOLOGY, or
METAPHYSICS PROPER.   We might call it INFERENTIAL
PSYCHOLOGY.

The following is a tabular view of the distribution
of Philosophy as here proposed :—

Mind or Consciousness affords	Facts,—Phænomenology, Empirical Psychology.	Cognitions. Feelings. Conative Powers (Will and Desire).
	Laws, — Nomology, Rational Psychology.	Cognitions,—Logic. Feelings,—Æsthetic. Conative Powers. { Moral Philosophy. Political Philosophy.
	Results, — Ontology, Inferential Psychology.	Being of God. Immortality of the Soul, &c.

LECT.
VII.
———
Meaning of
the term.
In this distribution of the philosophical sciences, you will observe that I take little account of the celebrated division of Philosophy into Speculative and Practical, which I have already explained to you,[a] for I call only one minor division of philosophy practical, —viz., the Nomology of the Conative powers,—not because that science is not equally theoretical with any other, but simply because these powers are properly called practical, as tending to practice or overt action.

Such is the distribution of Philosophy, which I venture to propose as the simplest and most exhaustive; and I shall now proceed, in reference to it, to specify the particular branches which form the objects of our consideration in the present course.

Distribu-
tion of sub-
jects in
Faculty of
Philosophy
in the Uni-
versities of
Europe.
The subjects assigned to the various chairs of the Philosophical Faculty, in the different Universities of Europe, were not calculated upon any comprehensive view of the parts of philosophy, and of their natural connection. Our universities were founded when the Aristotelic philosophy was the dominant, or rather the exclusive, system, and the parts distributed to the different classes, in the faculty of Arts or Philosophy, were regulated by the contents of certain of the Aristotelic books, and by the order in which they were studied. Of these, there were always Four great divisions. There was, first, Logic, in relation to the Organon of Aristotle; secondly, Metaphysics, relative to his books under that title; thirdly, Moral Philosophy, relative to his Ethics, Politics, and Economics; and, fourthly, Physics, relative to his Physics, and the collection of treatises styled in the schools the *Parva Naturalia*. But every university had not a full complement of classes, that is, did not devote a separate year

a See *ante*, p. 113.—Ed.

to each of the four subjects of study; and, accordingly, in those seats of learning where three years formed the curriculum of philosophy, two of these branches were combined. In this university, Logic and Metaphysics were taught in the same year; in others, Metaphysics and Moral Philosophy were conjoined; and, when the old practice was abandoned of the several Regents or Professors carrying on their students through every department, the two branches which had been taught in the same year were assigned to the same chair. What is most curious in the matter is this,—Aristotle's treatise *On the Soul* being, (along with his lesser treatises on *Memory and Reminiscence*, on *Sense and its Objects*, &c.), included in the *Parva Naturalia*, and, he having declared that the consideration of the soul was part of the philosophy of nature,[a] the science of Mind was always treated along with Physics. The Professors of Natural Philosophy have, however, long abandoned the philosophy of mind, and this branch has been, as more appropriate to their departments, taught both by the Professors of Moral Philosophy and by the Professors of Logic and Metaphysics,—for you are not to suppose that metaphysics and psychology are, though vulgarly used as synonymous expressions, by any means the same. So much for the historical accidents which have affected the subjects of the different chairs.

I now return to the distribution of philosophy, which I have given you, and, first, by exclusion, I shall tell you what does not concern us. In this class, we have nothing to do with Practical Philosophy,—that is,

---

a *De Anima*, i. 1: Φυσικοῦ τὸ θεωρῆσαι περὶ ψυχῆς, ἢ πάσης ἢ τῆς τοιαύτης. Cf. *Metaph.*, v. 1: Δῆλον πῶς δεῖ ἐν τοῖς φυσικοῖς τὸ τί ἐστι ζητεῖν καὶ ὁρίζεσθαι, καὶ διότι καὶ περὶ ψυχῆς ἐνίας θεωρῆσαι τοῦ φυσικοῦ, ὅση μὴ ἄνευ τῆς ὕλης ἐστίν.—ED.

LECT.
VII.

Ethics, Politics, Economics. But, with this exception, there is no other branch of philosophy which is not either specially allotted to our consideration, or which does not fall naturally within our sphere. Of the former description, are Logic, and Ontology or Metaphysics Proper. Of the latter, are Psychology, or the Philosophy of Mind in its stricter signification, and Æsthetic.

Comprehension and Order of the Course.

These subjects are, however, collectively too extensive to be overtaken in a single Course, and, at the same time, some of them are too abstract to afford the proper materials for the instruction of those only commencing the study of philosophy. In fact, the department allotted to this chair comprehends the two extremes of philosophy,—Logic, forming its appropriate introduction,—Metaphysics, its necessary consummation. I propose, therefore, in order fairly to exhaust the business of the chair, to divide its subjects between two Courses,—the one on Phænomenology, Psychology, or Mental Philosophy in general; the other on Nomology, Logic, or the laws of the Cognitive Faculties in particular.[a]

a From the following sentences, which appear in the manuscript lecture as superseded by the paragraph given in the text, it is obvious that the Author had originally designed to discuss specifically, and with greater detail, the three grand departments of Philosophy indicated in the distribution proposed by him:—

"The plan which I propose to adopt in the Distribution of the Course, or rather Courses, is the following:

"I shall commence with Mental Philosophy, strictly so called, with the science which is conversant with the Manifestations of Mind,—Phæ-nomenology, or Psychology. I shall then proceed to Logic, the science which considers the Laws of Thought; and finally, to Ontology, or Metaphysics Proper, the philosophy of Results. Æsthetic, or the theory of the Pleasurable, I should consider subsequently to Logic, and previously to Ontology."—On the propriety of according to Psychology the first place in the order of the philosophical sciences, see Cousin, Cours de l'Histoire de la Philosophie, Deuxième Série, tom. ii. p. 71-73 (ed. 1847); Géruzez, Nouveau Cours de Philosophie, pp. 10, 14, 15.—ED.

LECTURE VIII.

PSYCHOLOGY, ITS DEFINITION.   EXPLICATION OF TERMS.

I now pass to the First Division of my subject, which will occupy the present Course, and commence with a definition of PSYCHOLOGY,—THE PHÆNOMENOLOGY OF MIND.

Psychology, or the Philosophy of the Human Mind, strictly so denominated, is the science conversant about the *phænomena*, or *modifications*, or *states* of the *Mind*, or *Conscious-Subject*, or *Soul*, or *Spirit*, or *Self*, or *Ego*.

In this definition, you will observe that I have purposely accumulated a variety of expressions, in order that I might have the earliest opportunity of making you accurately acquainted with their meaning; for they are terms of vital importance and frequent use in philosophy.—Before, therefore, proceeding further, I shall pause a moment in explanation of the terms in which this definition is expressed.   Without restricting myself to the following order, I shall consider the word *Psychology*; the correlative terms *subject* and *substance*, *phænomenon*, *modification*, *state*, &c., and, at the same time, take occasion to explain another correlative, the expression *object*; and, finally, the words *mind*, *soul*, *spirit*, *self*, and *ego*.

Indeed, after considering these terms, it may not be improper to take up, in one series, the philosophical

Definition of Psychology.

Explication of terms.

LECT.
VIII

expressions of principal importance and most ordinary occurrence, in order to render less frequent the necessity of interrupting the course of our procedure, to afford the requisite verbal explanations.

The term Psychology; its use vindicated.

The term *Psychology* is of Greek compound, its elements—ψυχή, signifying *soul* or *mind*, and λόγος, signifying *discourse* or *doctrine*. Psychology, therefore, is the *discourse* or *doctrine treating of the human mind*. But, though composed of Greek elements, it is, like the greater number of the compounds of λόγος, of modern combination. It may be asked,—why use an exotic, a technical name? Why not be contented with the more popular terms, Philosophy of Mind or Mental Philosophy,—Science of Mind or Mental Science?—expressions by which this department of knowledge has been usually designated by those who, in this country, have cultivated it with the most distinguished success. To this there are several answers. In the first place, philosophy itself, and all, or almost all, its branches, have, in our language, received Greek technical denominations;—why not also the most important of all, the science of mind? In the second place, the term psychology is now, and has long been, the ordinary expression for the doctrine of mind in the philosophical language of every other European nation. Nay, in point of fact, it is now naturalised in English, *psychology* and *psychological* having of late years come into common use; and their employment is warranted by the authority of the best English writers. It was familiarly employed by one of our best writers, and most acute metaphysicians, Principal Campbell of Aberdeen;[a] and Dr Beattie, likewise, has entitled the first part of his *Elements of Moral Science*,—that which treats of the mental

a *Philosophy of Rhetoric*, vol. i. p. 143, (1st ed.); p. 123, (ed. 1816.)—ED.

faculties,—Psychology. To say nothing of Coleridge, the late Sir James Mackintosh was also an advocate for its employment, and justly censured Dr Brown for not using it, in place of his very reprehensible expression,—*Physiology of Mind*, the title of his unfinished text-book.[a] But these are reasons in themselves of comparatively little moment: they tend merely to show that, if otherwise expedient, the nomenclature is permissible; and that it is expedient the following reasons will prove. For, in the third place, it is always of consequence for the sake of precision to be able to use one word instead of a plurality of words, — especially, where the frequent occurrence of a descriptive appellation might occasion tedium, distraction, and disgust; and this must necessarily occur in the treatment of any science, if the science be able to possess no single name vicarious of its definition. In this respect, therefore, *Psychology* is preferable to *Philosophy of Mind*. But, in the fourth place, even if the employment of the description for the name could, in this instance, be tolerated, when used substantively, what are we to do when we require, (which we do unceasingly), to use the denomination of the science adjectively? For example, I have occasion to say a psychological fact, a psychological law, a psychological curiosity, &c. How can we express these by the descriptive appellation? A psychological fact may indeed be styled a fact considered relatively to the philosophy of the human mind,—a psychological law may be called a law by which the mental phænomena are governed, —a psychological curiosity may be rendered—by what, I really do not know. But how miserably weak, awkward, tedious,

[a] *Dissertation on the Progress of Ethical Philosophy*, in the Encyclopædia Britannica, vol. i. p. 399, (7th ed.)—ED.

LECT.
VIII.

and affected, is the commutation when it can be made; not only do the vivacity and precision of the original evaporate, the meaning itself is not even adequately conveyed. But this defect is still more manifestly shown when we wish to place in contrast the matters proper to this science, with the matters proper to others. Thus, for example, to say,—this is a psychological, not a physiological, doctrine—this is a psychological observation, not a logical inference. How is the contradistinction to be expressed by a periphrasis? It is impossible,—for the intensity of the contrast consists, first, in the two opposite terms being single words, and second, in their being both even technical and precise Greek. This necessity has, accordingly, compelled the adoption of the terms psychology and psychological into the philosophical nomenclature of every nation, even where the same necessity did not vindicate the employment of a non-vernacular expression. Thus in Germany, though the native language affords a facility of composition only inferior to the Greek, and though it possesses a word (*Seelenlehre*) exactly correspondent to ψυχολογία, yet because this substantive did not easily allow of an adjective flexion, the Greek terms, substantive and adjective, were both adopted, and have been long in as familiar use in the Empire, as the terms geography and geographical,—physiology and physiological, are with us.

The terms Physiology and Physics, as applied to the philosophy of mind, inappropriate.

What I have now said may suffice to show that, to supply a necessity, we must introduce these words into our philosophical vocabulary. But the propriety of this is still further shown by the inauspicious attempts that have been recently made on the name of the science. As I have mentioned before, Dr Brown, in the very title of the abridgment of his lectures on mental philosophy, has styled this philosophy,

"*The Physiology of the Human Mind;*" and I have
also seen two English publications of modern date,—
one entitled the "*Physics of the Soul,*" the other "*Intellectual Physics.*"[a] Now the term *nature*, (φύσις,
natura), though in common language of a more extensive meaning, has, in general, by philosophers, been
applied appropriately to denote the laws which govern
the appearances of the material universe. And the
words Physiology and Physics have been specially
limited to denote sciences conversant about these laws
as regulating the phænomena of organic and inorganic
bodies. The empire of nature is the empire of a mechanical necessity ; the necessity of nature, in philosophy, stands opposed to the liberty of intelligence.
Those, accordingly, who do not allow that mind is
matter,—who hold that there is in man a principle
of action superior to the determinations of a physical
necessity, a brute or blind fate,—must regard the application of the terms Physiology and Physics to the
doctrine of the mind as either singularly inappropriate,
or as significant of a false hypothesis in regard to the
character of the thinking principle.

Mr Stewart objects[β] to the term *Spirit*, as seeming to imply an hypothesis concerning the nature and
essence of the sentient or thinking principle, altogether unconnected with our conclusions in regard to its
phænomena, and their general laws; and, for the same
reason, he is disposed to object to the words Pneumatology and Psychology ; the former of which was
introduced by the schoolmen. In regard to *Spirit*
and *Pneumatology*, Mr Stewart's criticism is perfectly
just. They are unnecessary; and, besides the etymo-

a *Intellectual Physics, an Essay concerning the Nature of Being and the Progression of Existence.* London, 1795. *Intellectual Physics, an Essay concerning the Nature of Being.* 1803. By Governor Pownall.—ED.
β *Philosophical Essays,* Prelim. Dissert. ch. 1; *Works,* vol. v. p. 20.—ED.

logical metaphor, they are associated with a certain
theological limitation, which spoils them as expressions
of philosophical generality.[a]   But this is not the case
with *Psychology*.   For though, in its etymology, it is,
like almost all metaphysical terms, originally of physical application, still this had been long forgotten
even by the Greeks; and, if we were to reject philosophical expressions on this account, we should be
left without any terms for the mental phænomena at
all.   The term *soul* (and what I say of the term soul
is true of the term *spirit*), though in this country less
employed than the term *mind*, may be regarded as
another synonym for the unknown basis of the mental phænomena.   Like nearly all the words significant of the internal world, there is here a metaphor
borrowed from the external; and this is the case not
merely in one, but, as far as we can trace the analogy,
in all languages.   You are aware that ψυχή, the Greek
term for soul, comes from ψύχω, *I breathe* or *blow*,—

as πνεῦμα in Greek, and *spiritus* in Latin, from verbs
of the same signification.   In like manner, *anima* and
animus are words which, though in Latin they have
lost their primary signification, and are only known
in their secondary or metaphorical, yet, in their original physical meaning, are preserved in the Greek
ἄνεμος, *wind* or *air*.   The English *soul*, and the German *Seele*, come from a Gothic root *saivala*,[β] which
signifies *to storm*.   *Ghost*, the old English word for

a [The terms *Psychology* and *Pneumatology*, or *Pneumatic*, are not equivalents.   The latter word was used
for the doctrine of spirit in general,
which was subdivided into three
branches, as it is treated of the three
orders of spiritual substances,—God,
—Angels, and Devils,—and Man.
Thus—

Pneumatolo-
gia or Pneu-
matica,
{ 1. Theologia(Naturalis).
2. Angelographia, Dæ-
      monologia.
3. Psychologia.
—See Theoph. Gale, *Logica*, p. 455,
(1681).]

β See Grimm, *Deutsche Grammatik*,
vol. ii. p. 99.   In Anglo-Saxon, *Sawel*,
Sawal, *Sawl*, *Saul*.—ED.

spirit in general, and so used in our English version of the Scriptures, is the same as the German *Geist*,[a] and is derived from *Gas* or *Gescht*, which signifies *air*. In like manner, the two words in Hebrew for soul or spirit, *nephesh* and *ruach*, are derivatives of a root which means *to breathe;* and in Sanscrit the word *atmā* (analogous to the Greek ἀτμὸς, *vapour* or *air*) signifies both *mind* and *wind* or *air*.[β] *Sapientia*, in Latin, originally meant only the power of tasting ; as *sagacitas* only the faculty of scenting. In French, *penser* comes from the Latin *pendere*, through *pensare*, to weigh, and the terms, *attentio, intentio, (entendement), comprehensio, apprehensio, penetrativ, understanding,* &c., are just so many bodily actions transferred to the expression of mental energies.[γ]

There is, therefore, on this ground, no reason to re- ject such useful terms as *psychology* and *psychological;* terms, too, now in such general acceptation in the philosophy of Europe. I may, however, add an historical notice of their introduction. Aristotle's principal treatise on the philosophy of mind is entitled Περὶ Ψυχῆς ; but the first author who gave a treatise on the subject under the title *Psychologia,* (which I have observed to you is a modern compound), is Otto Casmann, who, in the year 1594, published at Hanau his very curious work, " *Psychologia Anthropologica sive Animæ Humanæ Doctrina.*" This was followed, in two years, by his "*Anthropologiæ Pars II., hoc est, de fabrica Humani Corporis.*" This author

---

a Scotch, *Ghaist, Gastly.*

β [See H. Schmid, *Versuch einer Metaphysik der inneren Natur,* p. 69, note ; Scheidler's *Psychologie,* pp. 299-301, 320 *et seq.* Cf. Theoph. Gale, *Philosophia Generalis,* pp. 321, 322. Pritchard, *Review of the Doctrine of*

a *Vital Principle,* p. 5-6.]

γ [On this point see Leibnitz, *Nouv. Ess.,* liv. iii. ch. i. § 5 ; Stewart, *Phil. Essays — Works,* vol. v. Essay v.; Brown, *Human Understanding,* p. 388 *et seq.*]

had the merit of first giving the name *Anthropologia* to the science of man in general, which he divided into two parts,—the first *Psychologia*, the doctrine of the Human Mind, the second, *Somatologia*, the doctrine of the Human Body ; and these, thus introduced and applied, still continue to be the usual appellations of these branches of knowledge in Germany. I would not say, however, that Casmann was the true author of the term *psychology*, for his master, the celebrated Rudolphus Goclenius of Marburg, published, also in 1594, a work entitled "Ψυχολογία, *hoc est, de Hominis Perfectione, Anima, &c.*," being a collection of dissertations on the subject; in 1596 another, entitled "*De præcipuis Materiis Psychologicis;*" and in 1597 a third, entitled "*Authores Varii de Psychologia,*"—so that I am inclined to attribute the origin of the name to Goclenius.[a] Subsequently, the term became the usual title of the science, and this chiefly through the authority of Wolf, whose two principal works on the subject are entitled "*Psychologia Empirica,*" and "*Psychologia Rationalis.*" Charles Bonnet, in his "*Essai de Psychologie,*"[β] familiarised the name in France; where, as well as in Italy,—indeed, in all the Continental countries,—it is now the common appellation.

In the second place, I said that Psychology is conversant about the *phænomena* of the thinking *subject,* &c., and I now proceed to expound the import of the correlative terms *phænomenon, subject,* &c.

But the meaning of these terms will be best illustrated by now stating and explaining the great axiom, that all human knowledge, consequently that all human philosophy, is only of the relative or phænomenal.[γ] In

a [The term *psychology* is, however, used by Joannes Thomas Freigius in the *Catalogus Locorum Communium,* prefixed to his *Ciceronianus,* 1575.

See also Gale, *Logica,* p. 455.]

β Published in 1755.—ED.

γ Compare *Reid's Works,* (6th edition), pp. 935, 965.—ED.

this proposition, the term *relative* is opposed to the term *absolute;* and, therefore, in saying that we know only the relative, I virtually assert that we know nothing absolute,—nothing existing absolutely; that is, in and for itself, and without relation to us and our faculties. I shall illustrate this by its application. Our knowledge is either of matter or of mind. Now, what is matter? What do we know of matter? Matter, or body, is to us the name either of something known, or of something unknown. In so far as matter is a name for something known, it means that which appears to us under the forms of extension, solidity, divisibility, figure, motion, roughness, smoothness, colour, heat, cold, &c.; in short, it is a common name for a certain series, or aggregate, or complement, of appearances or phænomena manifested in coexistence.

LECT. VIII.

The correlative terms Phænomenon, Subject, illustrated by reference to the relativity of human knowledge.

But as these phænomena appear only in conjunction, we are compelled by the constitution of our nature to think them conjoined in and by something; and as they are phænomena, we cannot think them the phænomena of nothing, but must regard them as the properties or qualities of something that is extended, solid, figured, &c. But this something, absolutely and in itself,—*i.e.*, considered apart from its phænomena,— is to us as zero. It is only in its qualities, only in its effects, in its relative or phænomenal existence, that it is cognisable or conceivable; and it is only by a law of thought, which compels us to think something, absolute and unknown, as the basis or condition of the relative and known, that this something obtains a kind of incomprehensible reality to us. Now, that which manifests its qualities,—in other words, that in which the appearing causes inhere, that to which they belong,— is called their *subject*, or *substance*, or *substratum*. To this subject of the phænomena of extension, solidity,

&c., the term *matter* or *material substance* is commonly given; and, therefore, as contradistinguished from these qualities, it is the name of something unknown and inconceivable.

The same is true in regard to the term *mind*. In so far as mind is the common name for the states of knowing, willing, feeling, desiring, &c., of which I am conscious, it is only the name for a certain series of connected phænomena or qualities, and, consequently, expresses only what is known. But in so far as it denotes that subject or substance in which the phænomena of knowing, willing, &c., inhere,—something behind or under these phænomena,—it expresses what, in itself or in its absolute existence, is unknown.

Thus, mind and matter, as known or knowable, are only two different series of phænomena or qualities ; mind and matter, as unknown and unknowable, are the two substances in which these two different series of phænomena or qualities are supposed to inhere. The existence of an unknown substance is only an inference we are compelled to make, from the existence of known phænomena ; and the distinction of two substances is only inferred from the seeming incompatibility of the two series of phænomena to coinhere in one.

Our whole knowledge of mind and matter is thus, as we have said, only relative ; of existence, absolutely and in itself, we know nothing ; and we may say of man what Virgil says of Æneas, contemplating in the prophetic sculpture of his shield the future glories of Rome,—

"Rerumque ignarus, imagine gaudet." [a]

This is, indeed, a truth, in the admission of which philosophers, in general, have been singularly bar-

a *Æneid*, viii. 730.—ED.

monious; and the praise that has been lavished on Dr
Reid for this observation, is wholly unmerited. In
fact, I am hardly aware of the philosopher who has not General
harmony
of philo-
sophers
regarding
the rela-
tivity of
human
knowledge.
proceeded on the supposition, and there are few who
have not explicitly enounced the observation. It is
only since Reid's death that certain speculators have
arisen, who have obtained celebrity by their attempt
to found philosophy on an immediate knowledge of
the absolute or unconditioned. I shall quote to you
a few examples of this general recognition, as they
happen to occur to my recollection; and, in order to
manifest the better its universality, I purposely overlook the testimonies of a more modern philosophy.

Aristotle, among many similar observations, remarks
in regard to matter, that it is incognisable in itself; [a]
while in regard to mind he says, "that the intellect
does not know itself directly, but only indirectly, in
knowing other things;" [β] and he defines the soul from
its phænomena, "the principle by which we live, and
move, and perceive, and understand." [γ] St Augustin,
the most philosophical of the Christian fathers, admirably says of body,—"Materiam cognoscendo ignorari,
et ignorando cognosci;" [δ] and of mind,—"Mens se
cognoscit cognoscendo se vivere, se meminisse, se intelligere, se velle, cogitare, scire, judicare." [ε] "Non incurrunt," says Melanchthon, "ipsæ substantiæ in oculos,
sed vestitæ et ornatæ accidentibus; hoc est, non pos-

a *Metaph.*, lib. vii. (vi.) c. 10: [ἡ ὕλη
ἄγνωστος καθ᾽ αὑτήν.—ED.]

β *Metaph.*, xii. (xi.) 7: Αὐτὸν δὲ
νοεῖ ὁ νοῦς κατὰ μετάληψιν τοῦ νοητοῦ·
νοητὸς γὰρ γίγνεται θιγγάνων καὶ νοῶν.
Cf. *De Anima*, iii. 4: Καὶ αὐτὸς δὲ
νοητός ἐστω ὥσπερ τὰ νοητά.—ED.

γ *De Anima*, lib. ii. c. 2: Ἡ ψυχὴ
τούτοις ὥρισται, θρεπτικῷ, αἰσθητικῷ,
διανοητικῷ, κινήσει.—ED.

δ *Confess.*, xii. 5: "Dum sibi hæc
dicit humana cogitatio, conetur eam
(materiam) vel nosse ignorando vel
ignorare noscendo."—ED.

ε From the spurious treatise attributed to St Austin, entitled *De
Spiritu et Anima*, c. 32; but see *De
Trinitate*, lib. x. § 16, tom. viii. p.
897, (ed. Benedict.)

sumus, in hac vita, acie oculorum perspicere ipsas sub-
stantias : sed utcunque, ex accidentibus quæ in sensus
exteriores incurrunt, ratiocinamur, quomodo inter se
differant substantiæ." [a]

The elder
Scaliger.

It is needless to multiply authorities, but I cannot
refrain from adducing one other evidence of the general consent of philosophers to the relative character of
our knowledge, as affording a graphic specimen of the
manner of its ingenious author.   " Substantiæ non a
nobis cognoscuntur," says the elder Scaliger, " sed
earum accidentia. Quis enim me doceat quid sit
substantia, nisi miseris illis verbis, *res subsistens?*
Scientiam ergo nostram constat esse umbram in sole.
Et sicut vulpes, elusa a ciconia, lambendo vitreum vas
pultem haud attingit : ita nos externa tantum accidentia percipiendo, formas internas non cognoscimus." [β]
So far there is no difference of opinion among philosophers in general.   We know mind and matter not in
themselves, but in their accidents or phænomena. [γ]

All relative
existence
not com-
prised in
what is re-
lative to
us.

Thus our knowledge is of relative existence only,
seeing that existence in itself, or absolute existence, is
no object of knowledge. [δ]   But it does not follow that
all relative existence is relative *to us;* that all that can
be known, even by a limited intelligence, is actually
cognisable by us.   We must, therefore, more precisely
limit our sphere of knowledge, by adding, that all we
know is known only under the special conditions of
our faculties.   This is a truth likewise generally ac-

a *Erotemata Dialectices*, lib. i., Pr.
Substantia. [This is the text in the
edition of Strigelius. It varies considerably in different editions.—Ed.]

β *De Subtilitate*, Ex. cccvii. § 21.

γ For additional testimonies on this
point, see the Author's *Discussions*,
p. 644.—Ed.

δ [Absolute in two senses : 1°, As
opposed to partial ; 2°, As opposed
to relative. Better if I had said that
our knowledge not of absolute, and,
therefore, only of the partial and relative.]—*Pencil Jotting on Blank Leaf
of Lecture.*

knowledged. " Man," says Protagoras, "is the measure of the universe," (πάντων χρημάτων μέτρον ἄνθρωπος), —a truth which Bacon has well expressed : " Omnes perceptiones tam sensus quam mentis, sunt ex analogia hominis, non ex analogia universi : estque intellectus humanus instar speculi inæqualis ad radios rerum, qui suam naturam naturæ rerum immiscet, eamque distorquet et inficit." [a] " Omne quod cognoscitur," says Boethius, " non secundum sui vim, sed secundum cognoscentium potius comprehenditur facultatem ;" [β] and this is expressed almost in the same terms by the two very opposite philosophers, Kant and Condillac,—" In perception " (to quote only the former) "everything is known according to the constitution of our faculty of sense." [γ]

Now this principle, in which philosophers of the most opposite opinions equally concur, divides itself into two branches. In the first place, it would be unphilosophical to conclude that the properties of existence necessarily are, in number, only as the number of our faculties of apprehending them ; or, in the second, that the properties known, are known in their native purity, and without addition or modification from our organs of sense, or our capacities of intelligence. I shall illustrate these in their order.

In regard to the first assertion, it is evident that nothing exists for us, except in so far as it is known to us, and that nothing is known to us, except certain properties or modes of existence, which are relative or analogous to our faculties. Beyond these modes we know, and can assert, the reality of no existence. But

α *Novum Organum*, lib. i., aph. xli.—ED.

β *De Consol. Phil.*, lib. v. Pr. 4. Quoted in *Discussions*, p. 645.—ED.

γ *Kritik der reinen Vernunft*, Vorrede zur zweiten Auflage. Quoted in *Discussions*, p. 646. Cf. Kant, *ibid.* Transc. Æsth. § 8.—ED.

if, on the one hand, we are not entitled to assert as
actually existent except what we know; neither, on
the other, are we warranted in denying, as possibly
existent, what we do not know.   The universe may be
conceived as a polygon of a thousand, or a hundred
thousand, sides or facets,—and each of these sides or
facets may be conceived as representing one special
mode of existence.   Now of these thousand sides or
modes all may be equally essential, but three or four
only may be turned towards us or be analogous to our
organs.   One side or facet of the universe, as holding
a relation to the organ of sight, is the mode of luminous or visible existence; another, as proportional to
the organ of hearing, is the mode of sonorous or aud-
ible existence ; and so on.   But if every eye to see, if
every ear to hear, were annihilated, the modes of existence to which these organs now stand in relation,—
that which could be seen, that which could be heard,
—would still remain ; and if the intelligences reduced
to the three senses of touch, smell, and taste, were
then to assert the impossibility of any modes of being
except those to which these three senses were analogous, the procedure would not be more unwarranted,
than if we now ventured to deny the possible reality
of other modes of material existence than those to the
perception of which our five senses are accommodated.
I will illustrate this by an hypothetical parallel.   Let
us suppose a block of marble,[a] on which there are four
different inscriptions,—in Greek, in Latin, in Persic,
and in Hebrew, and that four travellers approach, each
able to read only the inscription in his native tongue.
The Greek is delighted with the information the

a This illustration is taken from   *Philosophic—Œuvres Philosophiques*,
F. Hemsterhuis, *Sophyle ou de la*   vol. i. p. 281, (ed. 1792.)—ED.

marble affords him of the siege of Troy.   The Roman
finds interesting matter regarding the expulsion of the
kings.   The Persian deciphers an oracle of Zoroaster.
And the Jew is surprised by a commemoration of the
Exodus.   Here, as each inscription exists or is significant only to him who possesses the corresponding
language ; so the several modes of existence are manifested only to those intelligences who possess the corresponding organs. And as each of the four readers would
be rash if he maintained that the marble could be significant only as significant to him, so should we be
rash, were we to hold that the universe had no other
phases of being, than the few that are turned towards
our faculties, and which our five senses enable us to
perceive.

Voltaire (*aliud agendo*) has ingeniously expressed Illustrated
this truth in one of his philosophical romances. " 'Tell from Vol- taire.
me,' says Micromegas, an inhabitant of one of the planets
of the Dog-Star, to the secretary of the Academy of
Sciences in the planet Saturn, at which he had recently
arrived in a journey through the heavens,—' Tell me,
how many senses have the men on your globe ?' ' We
have seventy-two senses,' answered the academician,
' and we are, every day, complaining of the smallness
of the number.   Our imagination goes far beyond our
wants. What are seventy-two senses! and how pitiful
a boundary, even for beings with such limited perceptions, to be cooped up within our ring and our five
moons.   In spite of our curiosity, and in spite of as
many passions as can result from six dozen of senses,
we find our hours hang very heavily on our hands, and
can always find time enough for yawning.'—' I can
very well believe it,' says Micromegas, ' for, in our
globe, we have very near one thousand senses ; and

yet, with all these, we feel continually a sort of listless inquietude and vague desire, which are for ever telling us that we are nothing, and that there are beings infinitely nearer perfection. I have travelled a good deal in the universe. I have seen many classes of mortals far beneath us, and many as much superior; but I have never had the good fortune to meet with any, who had not always more desires than real necessities to occupy their life. And, pray, how long may you Saturnians live, with your few senses?' continued the Sirian. 'Ah! but a very short time indeed!' said the little man of Saturn, with a sigh. 'It is the same with us,' said the traveller; 'we are for ever complaining of the shortness of life. It must be an universal law of nature.' 'Alas!' said the Saturnian, 'we live only five hundred great revolutions of the sun, (which is pretty much about fifteen thousand years of our counting). You see well, that this is to die almost the moment one is born. Our existence is a point,—our duration an instant,—our globe an atom. Scarcely have we begun to pick up a little knowledge, when death rushes in upon us, before we can have acquired anything like experience. As for me, I cannot venture even to think of any project. I feel myself but like a drop of water in the ocean; and, especially now, when I look to you and to myself, I really feel quite ashamed of the ridiculous appearance which I cut in the universe.'

"'If I did not know you to be a philosopher,' replied Micromegas, 'I should be afraid of distressing you, when I tell you, that our life is seven hundred times longer than yours. But what is even that? and, when we come to the last moment, to have lived a single day, and to have lived a whole eternity, amount to the same thing. I have been in countries where they live a thousand times longer than with us; and I have always found them murmuring, just as we do ourselves. But you have seventy-two senses, and they must have told you something about your globe. How many properties has matter with you?'—'If you mean essential properties,' said the Saturnian, 'without which our globe could not subsist, we count three hundred,—extension, impenetrability, mobility, gravity, divisibility, and so forth.'—'That small number,' replied the gigantic traveller, 'may be sufficient for the views which the Creator must have had with respect to your narrow habitation. Your globe is little; its inhabitants are so too. You have few senses; your matter has few qualities. In all this, Providence has suited you most happily to each other.'

"The academician was more and more astonished with everything which the traveller told him. At length, after communicating to each other a little of what they knew, and a great deal of what they knew not, and reasoning as well and as ill as philosophers usually do, they resolved to set out together on a little tour of the universe." [a]

Before leaving this subject, it is perhaps proper to observe, that had we faculties equal in number to all the possible modes of existence, whether of mind or matter, still would our knowledge of mind or matter be only relative. If material existence could exhibit ten thousand phænomena, and if we possessed ten thousand senses to apprehend these ten thousand phænomena of material existence,—of existence absolutely and in itself, we should be then as ignorant as we are at present.

a *Micromégas,* chap. ii.—ED.

LECT.
VIII.
2. The pro-
perties of
existence
not known
in their
native
purity.

But the consideration that our actual faculties of knowledge are probably wholly inadequate in number to the possible modes of being, is of comparatively less importance than the other consideration to which we now proceed,—that whatever we know is not known as it is, but only as it seems to us to be ; for it is of less importance that our knowledge should be limited than that our knowledge should be pure. It is, therefore, of the highest moment that we should be aware that what we know is not a simple relation apprehended between the object known and the subject knowing,—but that every knowledge is a sum made up of several elements, and that the great business of philosophy is to analyse and discriminate these elements, and to determine from whence these contributions have been derived. I shall explain what I mean

by an example. In the perception of an external object, the mind does not know it in immediate relation to itself, but mediately in relation to the material organs of sense. If, therefore, we were to throw these organs out of consideration, and did not take into account what they contribute to, and how they modify, our knowledge of that object, it is evident, that our conclusion in regard to the nature of external perception would be erroneous. Again, an object of perception may not even stand in immediate relation to the organ of sense, but may make its impression on that organ through an intervening medium. Now, if this medium be thrown out of account, and if it be not considered that the real external object is the sum of all that externally contributes to affect the sense, we shall in like manner run into error. For example, I see a book,—I see that book through an external medium, (what that medium is, we do not now in-

quire),—and I see it through my organ of sight, the
eye. Now, as the full object presented to the mind,
(observe that I say the mind), in perception, is an
object compounded of the external object emitting
or reflecting light, *i.e.*, modifying the external medium,—of this external medium,—and of the living
organ of sense, in their mutual relation,—let us suppose, in the example I have taken, that the full or
adequate object perceived is equal to twelve, and that
this amount is made up of three several parts,—of
four, contributed by the book, of four, contributed
by all that intervenes between the book and the
organ, and of four, contributed by the living organ
itself.[a]

I use this illustration to show that the phænomenon
of the external object is not presented immediately to
the mind, but is known by it only as modified through
certain intermediate agencies ; and to show, that sense
itself may be a source of error, if we do not analyse
and distinguish what elements, in an act of perception,
belong to the outward reality, what to the outward
medium, and what to the action of sense itself. But
this source of error is not limited to our perceptions ;
and we are liable to be deceived, not merely by not
distinguishing in an act of knowledge what is contributed by sense, but by not distinguishing what is contributed by the mind itself. This is the most difficult
and important function of philosophy ; and the greater
number of its higher problems arise in the attempt to
determine the shares to which the knowing subject,
and the object known, may pretend in the total act
of cognition. For according as we attribute a larger

a This illustration is borrowed in an　See his *Sophyle ou de la Philosophie*—
improved form from F. Hemsterhuis.　*Œuvres Philosophiques*, i. 279.—ED.

or a smaller proportion to each, we either run into the
extremes of Idealism and Materialism, or maintain an
equilibrium between the two. But on this subject,
it would be out of place to say anything further at
present.

In what
senses hu-
man know-
ledge is
relative. From what has been said, you will be able, I hope,
to understand what is meant by the proposition, that
all our knowledge is only relative. It is relative, 1°,
Because existence is not cognisable, absolutely and in
itself, but only in special modes; 2°, Because these
modes can be known only if they stand in a certain
relation to our faculties; and, 3°, Because the modes,
thus relative to our faculties, are presented to, and
known by, the mind only under modifications determined by these faculties themselves. This general
doctrine being premised, it will be proper now to take
some special notice of the several terms significant of
the relative nature of our knowledge. And here, there
Two oppo-
site series
of terms as
applied to
human
knowledge. are two opposite series of expressions,—1°, Those
which denote the relative and the known; 2°, Those
which denote the absolute and the unknown. Of the
former class are the words *phænomenon, mode, modification, state,*—words which are employed in the definition of Psychology; and to these may be added the
analogous terms, *quality, property, attribute, accident.* Of the latter class,—that is, the absolute and
the unknown,—is the word *subject,* which we have to
explain as an element of the definition, and its analogous expressions, *substance* and *substratum.* These
opposite classes cannot be explained apart; for, as
each is correlative of the other, each can be comprehended only in and through its correlative.

The term
Subject. The term *subject* (ὑπόστασις, ὑποκείμενον, *subjec-
tum*) is used to denote the unknown basis which lies

under the various phænomena or properties of which
we become aware, whether in our internal or external
experience. In the more recent philosophy, especially
in that of Germany, it has, however, been principally
employed to denote the basis of the various mental
phænomena; but of this special signification we are
hereafter more particularly to speak.[a] The word *sub-*
stance (substantia) may be employed in two—but two
kindred—meanings. It may be used either to denote
that which exists absolutely and of itself; in this sense
it may be viewed as derived from *subsistendo*, and as
meaning *ens per se subsistens;* or it may be viewed as
the basis of attributes, in which sense it may be regarded as derived from *substando*, and as meaning *id*
quod substat accidentibus, like the Greek ὑπόστασις,
ὑποκείμενον. In either case, it will, however, signify
the same thing, viewed in a different aspect. In the
former meaning, it is considered in contrast to, and
independent of, its attributes; in the latter, as conjoined with these, and as affording them the condition
of existence. In different relations a thing may be
at once considered as a *substance*, and as an *attribute*,
quality, or *mode.* This paper is a substance in relation to the attribute of white; but it is itself a mode
in relation to the substance, matter. Substance is thus
a term for the substratum we are obliged to think to
all that we variously denominate a *mode*, a *state*, a
quality, an *attribute*, a *property*, an *accident*, a *phæ-*
nomenon, an *apppearance*, &c. These, though expressions generically the same, are, however, used with
specific distinctions. The terms *mode, state, quality,*

a For the history and various mean- p. 806. See also Trendelenburg,
ings of the terms *Subject* and *Object*, *Elementa Logices Aristotelicæ*, § 1.—
see the Author's note, *Reid's Works*, ED.

attribute, property, accident, are employed in reference
to a substance, as existing; the terms *phænomenon,
appearance,* &c., in reference to it as known. But
each of these expressions has also its peculiar signification.

Mode.
A *mode* is the manner of the existence of a thing.
Take for example, a piece of wax. The wax may
be round, or square, or of any other definite figure; it
may also be solid or fluid. Its existence in any of
these modes is not essential; it may change from one
to the other without any substantial alteration. As the
mode cannot exist without a substance, we can accord
to it only a secondary or precarious existence in relation to the substance, to which we accord the privilege
of existing by itself, *per se existere;* but though the substance be not astricted to any particular mode of existence, we must not suppose that it can exist—or, at
least, be conceived by us to exist—in none. All modes
are, therefore, variable states; and though some mode
is necessary for the existence of a thing, any individual
mode is accidental.

Modifica-
tion.
The word *modification* is properly the bringing a
thing into a certain mode of existence, but it is very
commonly employed for the mode of existence itself.

State.
State is a term nearly synonymous with mode, but
of a meaning more extensive, as not exclusively limited to the mutable and contingent.

Quality,
Essential
and Acci-
dental.
Quality is, likewise, a word of a wider signification,
for there are essential and accidental qualities.[a] The
essential qualities of a thing are those aptitudes, those
manners of existence and action, which it cannot lose
without ceasing to be. For example, in man, the facul-

<hr />

[a] The term *quality* should, in strict-
ness, be confined to accidental attributes. See the Author's note, *Reid's
Works,* p. 836. — ED.

of length, breadth, and thickness; in God, the attri- VIII.
butes of eternity, omniscience, omnipotence, &c. By
accidental qualities, are meant those aptitudes and
manners of existence and action, which substances
have at one time and not at another; or which they
have always, but may lose without ceasing to be. For
example, of the transitory class are the whiteness of a
wall, the health which we enjoy, the fineness of the
weather, &c. Of the permanent class are the gravity
of bodies, the periodical movement of the planets, &c.

The term *attribute* is a word properly convertible Attribute.
with *quality*, for every quality is an attribute, and
every attribute is a quality; but, in our language
custom has introduced a certain distinction in their
application. Attribute is considered as a word of
loftier significance, and is, therefore, conventionally
limited to qualities of a higher application. Thus,
for example, it would be felt as indecorous to speak
of the qualities of God, and as ridiculous to talk of
the attributes of matter.

Property is correctly a synonym for peculiar qua- Property.
lity;[a] but it is frequently used as coextensive with
quality in general. *Accident*, on the contrary, is an Accident.
abbreviated expression for accidental or contingent
quality.

Phænomenon is the Greek word for *that which* Phænomenon.
appears, and may therefore be translated by *appear-* Appearance.
ance. There is, however, a distinction to be noticed.
In the first place, the employment of the Greek term

a In the older and Aristotelian Logicians, the term *property* was less
sense of the term. See *Topics*, i. 5: correctly used to denote a necessary
'Ἴδιον δ' ἐστὶν ὃ μὴ δηλοῖ μὲν τὸ τί ἦν quality, whether peculiar or not.—
εἶναι, μόνῳ δ' ὑπάρχει καὶ ἀντικατηγο- ED.
ρεῖται τοῦ πράγματος. By the later

shows that it is used in a strict and philosophical application. In the second place, the English name is associated with a certain secondary or implied meaning, which, in some degree, renders it inappropriate as a precise and definite expression. For the term *appearance* is used to denote not only that which reveals itself to our observation, as existent, but also to signify that which only seems to be, in contrast to that which truly is. There is thus not merely a certain vagueness in the word, but it even involves a kind of contradiction to the sense in which it is used when employed for *phænomenon*. In consequence of this, the term phænomenon has been naturalised, in our language, as a philosophical substitute for the term appearance.

LECTURE IX.

EXPLICATION OF TERMS—RELATIVITY OF HUMAN KNOWLEDGE.

AFTER giving a definition of Psychology, or the Phi- LECT.
losophy of Mind, in which I endeavoured to comprise IX.
a variety of expressions, the explanation of which Recapitula-
might smooth the way in our subsequent progress, I tion.
was engaged, during my last Lecture, in illustrating
the principle, that all our knowledge of mind and
matter is merely relative. We know, and can know,
nothing absolutely and in itself : all that we know
is existence in certain special forms or modes, and
these, likewise, only in so far as they may be analogous to our faculties. We may suppose existence
to have a thousand modes ;—but these thousand
modes are all to us as zero, unless we possess faculties accommodated to their apprehension. But were
the number of our faculties coextensive with the
modes of being,—had we, for each of these thousand modes, a separate organ competent to make it
known to us,—still would our whole knowledge be,
as it is at present, only of the relative. Of existence,
absolutely and in itself, we should then be as ignorant as we are now. We should still apprehend existence only in certain special modes,—only in certain
relations to our faculties of knowledge.

These relative modes, whether belonging to the

world without or to the world within, are, under different points of view and different limitations, known
under various names, as *qualities, properties, essences,
accidents, phænomena, manifestations, appearances,*
and so forth ; whereas the unknown something of
which they are the modes,—the unknown ground,
which affords them support,—is usually termed their
substance or *subject.* Of the signification and differences of these expressions, I stated only what was
necessary in order to afford a general notion of their
philosophical application. *Substance, (substantia),* I
noticed, is considered either in contrast to its accidents, as *res per se subsistens,* or in connection with
them, as *id quod substat accidentibus.* It, therefore, comprehends both the Greek terms οὐσία and
ὑποκείμενον ;—οὐσία being equivalent to *substantia*
in the meaning of *ens per se subsistens,*—ὑποκείμενον
to it, as *id quod substat accidentibus.*[a] The term
subject is used only for substance in its second meaning, and thus corresponds to ὑποκείμενον ; its literal
signification is, as its etymology expresses, that which
lies, or is placed, *under* the phænomena. So much for
the terms *substance* and *subject,* significant of unknown
or absolute existence.

I then said a few words on the differences of the
various terms expressive of known or relative existence, *mode, modification, state, quality, attribute, pro-*

a Ὑπόστασις, here noted, by way of *interpolation,* as of theological application. [On this point see Melanch-
thon, *Erot. Dial.* (Strigelii) p. 145
et seq.: "In philosophia, generaliter
nomine *Essentiæ* utimur *pro re per
sese considerata,* sive sit in prædica-
mento substantiæ, sive sit accidens.
At *ὑπόστασις* significat *rem subsis-
tentem,* quæ opponitur accidentibus.
Ecclesia vero cum quodam discrimine
his vocabulis utitur. Nam vocabulum
Essentiæ significat *id quod revera est,*
etiamsi est communicatum. Ὑπόστασις autem seu *Persona* est subsistens, vivum, individuum, intelligens,
incommunicabile, non sustentatum in
alio." Compare the relative annota-
tion by Strigelius, and Höcker, *Claris
Phil. Arist.,* p. 301.—ED.]

perty, phænomenon, appearance; but what I stated I
do not think it necessary to recapitulate.

I at present avoid entering into the metaphysics of Philosophers have
fallen into
three dif-
ferent
errors re-
garding
Substance.
substance and phænomenon.   I shall only observe in
general, that philosophers have frequently fallen into
one or other of three different errors.   Some have
denied the reality of any unknown ground of the
known phænomena; and have maintained that mind
and matter have no substantial existence, but are
merely the two complements of two series of associated qualities.   This doctrine, is, however, altogether
futile.   It belies the veracity of our primary beliefs ;
it leaves unsatisfied the strongest necessities of our
intellectual nature ; it admits as a fact that the phænomena are connected, but allows no cause explanatory of the fact of their connection.   Others, again,
have fallen into an opposite error.   They have attempted to speculate concerning the nature of the
unknown grounds of the phænomena of mind and
matter apart from the phænomena, and have, accordingly, transcended the legitimate sphere of philosophy.   A third party have taken some one, or more, of
the phænomena themselves as the basis or substratum
of the others.   Thus Descartes, at least as understood
and followed by Malebranche and others of his disciples, made thought or consciousness convertible
with the substance of mind ;[a] and Bishops Brown
and Law, with Dr Watts, constituted solidity and
extension into the substance of body.   This theory
is, however, liable to all the objections which may
be alleged against the first.[β]

[a] *Principia*, pars i. §§ 8, 51-53.   On
this point see Stewart, *Works*, vol. ii.
p. 473, Note A; also the completed
edition of *Reid's Works*, p. 961.—ED.

[β] *Encyclopædia Britannica*, art.
Metaphysics, pp. 615, 616, (7th ed.)
[Cf. Descartes, *Principia*, pars i. §
53; pars ii. § 4.—ED.]

LECT.
IX.
———
Explana-
tion of
terms—
(contin-
ued.)

I defined Psychology, the science conversant about the *phænomena* of the *mind*, or *conscious-subject*, or *self*, or *ego*. The former parts of the definition have been explained; the terms *mind*, *conscious-subject*, *self*, and *ego*, come now to be considered. These are all only expressions for the unknown basis of the mental phænomena, viewed, however, in different relations.

Of these the word *mind* is the first. In regard to the etymology of this term,[a] it is obscure and doubtful; perhaps, indeed, none of the attempts to trace it to its origin are successful. It seems to hold an analogy with the Latin *mens*, and both are probably derived from the same common root. This root, which is lost in the European languages of Scytho-Indian origin, is probably preserved in the Sanscrit *mena*, *to know* or *understand*. The Greek νοῦς, *intelligence*, is, in like manner, derived from a verb of precisely the same meaning (νοέω). The word mind is of a more limited signification than the term *soul*. In the Greek philosophy, the term ψυχή, *soul*, comprehends, besides the sensitive and rational principle in man, the principle of organic life, both in the animal and vegetable kingdoms; and, in Christian theology, it is likewise used, in contrast to πνεῦμα or *spirit*, in a vaguer and more extensive signification.

Since Descartes limited psychology to the domain of consciousness, the term mind has been rigidly employed for the self-knowing principle alone. Mind, therefore, is to be understood as the subject of the various internal phænomena of which we are conscious, or that subject of which consciousness is the general phænomenon. Consciousness is, in fact, to the mind what extension is to matter or body. Though

———
[a] On etymology of *mind*, &c.—see Scheidler's *Psychologie*, p. 325.

both are phænomena, yet both are essential qualities; LECT.
IX.
for we can neither conceive mind without conscious-
ness, nor body without extension.   Mind can be de- Mind can
be defined
only a pos-
teriori.
fined only *a posteriori*,—that is, only from its mani-
festations.   What it is in itself, that is, apart from its
manifestations,—we, philosophically, know nothing,
and, accordingly, what we mean by mind is simply
that which perceives, thinks, feels, wills, desires, &c.
Mind, with us, is thus nearly coextensive with the
Rational and Animal souls of Aristotle; for the faculty
of voluntary motion, which is a function of the animal
soul in the Peripatetic doctrine, ought not, as is generally done, to be excluded from the phænomena of
consciousness and mind.

The definition of mind from its qualities is given
by Aristotle; it forms the second definition in his
Treatise on the Soul,[a] and after him, it is the one
generally adopted by philosophers, and, among others,
by Dr Reid.[β]   That Reid, therefore, should have been
praised for having thus defined the mind, shows only
the ignorance of his encomiasts.   He has no peculiar
merit in this respect at all.

The next term to be considered is *conscious sub-* Conscious
Subject.
ject.   And first, what is it to be conscious?   Without anticipating the discussion relative to consciousness, as the fundamental function of intelligence, I
may, at present, simply indicate to you what an act
of consciousness denotes.   This act is of the most

---

a *De Anima*, ii. 2: Ἡ ψυχὴ δὲ
τοῦτο ᾧ ζῶμεν καὶ αἰσθανόμεθα καὶ δια-
νοούμεθα πρώτως. Cf. Themistius:
Εἰ δὲ χρὴ λέγειν τί ἕκαστον τούτων,
οἷον τί τὸ νοητικὸν, ἢ τί τὸ αἰσθητικὸν,
πρότερον ἐπισκεπτέον, τί τὸ νοεῖν, καὶ
τί τὸ αἰσθάνεσθαι· πρότεραι γὰρ καὶ σα-
φέστεραι πρὸς ἡμᾶς τῶν δυνάμεών εἰσιν
αἱ ἐνέργειαι· προεντυγχάνομεν γὰρ αὐ-

ταῖς, καὶ τὰς δυνάμεις ἀπὸ τούτων ἐπι-
νοοῦμεν. In lib. ii. *De Anima*, p. 76,
(Ald. Fol.)—Ed.

β *Intellectual Powers*, Essay i. c. 2;
Works, p. 229 : " By the mind of a
man, we understand that in him
which thinks, remembers, reasons,
wills."—Ed.

elementary character; it is the condition of all knowledge; I cannot, therefore, define it to you; but, as
you are all familiar with the thing, it is easy to enable
you to connect the thing with the word.   I know,—
I desire,—I feel.   What is it that is common to all
these?   *Knowing* and *desiring* and *feeling* are not the
same, and may be distinguished.   But they all agree
in one fundamental condition.   Can I know, without
knowing that I know?   Can I desire, without *knowing*
that I desire?   Can I feel, without *knowing* that I
feel?   This is impossible.   Now this knowing that I
know or desire or feel,—this common condition or
self-knowledge, is precisely what is denominated Consciousness.[a]

So much at present for the adjective *conscious:*
now for the substantive, *subject,—conscious-subject.*
Though consciousness be the condition of all internal
phænomena, still it is itself only a phænomenon; and,
therefore, supposes a subject in which it inheres;—
that is, supposes something that is conscious,—something that manifests itself as conscious.   And, since
consciousness comprises within its sphere the whole
phænomena of mind, the expression *conscious-subject*
is a brief, but comprehensive, definition of mind itself.

I have already informed you of the general meaning of the word *subject* in its philosophical application,— viz., the unknown basis of phænomenal or
manifested existence.   It is thus, in its application,
common equally to the external and to the internal
worlds.   But the philosophers of mind have, in a
manner, usurped and appropriated this expression to
themselves.   Accordingly, in their hands, the phrases

---

a Compare *Discussions*, p. 47, and   Note H, p. 929 *et seq.*—ED.
the completed edition of *Reid's Works*,

conscious or *thinking subject,* and *subject* simply, mean precisely the same thing; and custom has prevailed so far, that, in psychological discussions, *the subject* is a term now currently employed, throughout Europe, for the *mind* or *thinking principle.*[a]

The question here occurs, what is the reason of this employment? If mind and subject are only convertible terms, why multiply synonyms? Why exchange a precise and proximate expression for a vague and abstract generality? The question is pertinent, and merits a reply; for unless it can be shown that the word is necessary, its introduction cannot possibly be vindicated. Now, the utility of this expression is founded on two circumstances. The first, that it affords an adjective; the second, that the terms *subject* and *subjective* have opposing relatives in the terms *object* and *objective,* so that the two pairs of words together, enable us to designate the primary and most important analysis and antithesis of philosophy, in a more precise and emphatic manner than can be done by any other technical expressions. This will require some illustration.

Subject, we have seen, is a term for that in which the phænomena revealed to our observation, inhere, —what the schoolmen have designated the *materia in qua.* Limited to the mental phænomena, *subject,* therefore, denotes the mind itself; and *subjective,* that which belongs to, or proceeds from, the thinking subject. *Object,* on the other hand, is a term for that about which the knowing subject is conversant,—what the schoolmen have styled the *materia circa quam;* while *objective* means that which belongs to, or proceeds from, the object known, and not from the subject

a See the Author's note, *Reid's Works,* p. 806.—ED.

knowing; and thus denotes what is real in opposition to what is ideal,—what exists in nature, in contrast to what exists merely in the thought of the individual.

Now the great problem of philosophy is to analyse the contents of our acts of knowledge, or cognitions, —to distinguish what elements are contributed by the knowing subject, what elements by the object known. There must, therefore, be terms adequate to designate these correlative opposites, and to discriminate the share which each has in the total act of cognition. But, if we reject the terms *subject* and *subjective*, —*object* and *objective*, there are no others competent to the purpose.

Errors arising from want of the terms Subject and Object.

At this stage of your progress, Gentlemen, it is not easy to make you aware of the paramount necessity of such a distinction, and of such terms,—or to show you how, from the want of words expressive of this primary antithesis, the mental philosophy of this country has been checked in its development, and involved in the utmost perplexity and misconception. It is sufficient to remark at present, that to this defect in the language of his psychological analysis, is, in a great measure, to be attributed the confusion, not to say the errors, of Reid, in the very cardinal point of his philosophy,—a confusion so great that the whole tendency of his doctrine was misconceived by Brown, who, in adopting a modification of the hypothesis of a representative perception, seems not even to have suspected, that he, and Reid, and modern philosophers in general, were not in this at one.[a] The terms *subjective* and *objective* denote the primary distinction in consciousness of *self* and *not-self*, and this distinction

---

a See on this question the Author's *Supplementary Dissertations* to *Reid's Discussions*, p. 45 *et seq.*, and his *Works*, Notes B and C.—ED.

involves the whole science of mind ; for this science
is nothing more than a determination of the subjective
and objective, in themselves and in their mutual relations. The distinction is of paramount importance, and
of infinite application, not only in Philosophy proper,
but in Grammar, Rhetoric, Criticism, Ethics, Politics,
Jurisprudence, Theology. I will give you an example,
—a philological example. Suppose a lexicographer had
to distinguish the two meanings of the word *certainty.*
Certainty expresses either, the firm conviction which
we have of the truth of a thing ; or the character of
the proof on which its reality rests. The former is the
subjective meaning ; the latter the *objective.* By what
other terms can they be distinguished and described ?

The distinction of subject and object, as marking
out the fundamental and most thorough-going antithesis in philosophy, we owe, among many other
important benefits, to the schoolmen, and from the
schoolmen the terms passed, both in their substantive and adjective forms, into the scientific language
of modern philosophers. Deprived of these terms,
the Critical Philosophy, indeed the whole philosophy
of Germany and France, would be a blank. In this
country, though familiarly employed in scientific language, even subsequently to the time of Locke, the
adjective forms seem at length to have dropt out
of the English tongue. That these words waxed ob-
solete, was, perhaps, caused by the ambiguity which
had gradually crept into the signification of the substantives. Object, besides its proper signification,
came to be abusively applied to denote *motive, end,
final cause,* (a meaning, by the way, not recognised
by Johnson). This innovation was probably borrowed
from the French, in whose language the word had

LECT.
IX.been similarly corrupted, after the commencement
of the last century. Subject in English, as *sujet* in
French, had not been rightly distinguished from
object, taken in its proper meaning, and had thus
returned to the original ambiguity of the corresponding term (ὑποκείμενον) in Greek. It is probable that
the logical application of the word, (subject of pre-
dication), facilitated, or occasioned this confusion.
In using the terms, therefore, we think that an explanation, but no apology, is required. The distinction is expressed by no other terms ; and if these did
not already enjoy a prescriptive right as denizens of
the language, it cannot be denied that, as strictly
analogical, they are well entitled to sue out their
naturalisation. We shall have frequent occasion to
recur to this distinction,—and it is eminently worthy
of your attention.

Self, Ego—
illustrated
from Plato.
The last parallel expressions are the terms *self* and
ego. These we shall take together, as they are absolutely convertible. As the best preparative for a
proper understanding of these terms, I shall translate to you a passage from the *First Alcibiades* of
Plato.[a] The interlocutors are Socrates and Alcibiades.

"*Socr.* Hold, now, with whom do you at present
converse ? Is it not with me ?—*Alcib.* Yes.

Socr. And I also with you ?—*Alcib.* Yes.

Socr. It is Socrates then who speaks ? *Alcib.* Assuredly.

Socr. And Alcibiades who listens ?—*Alcib.* Yes.

a P. 129. The genuineness, however, of this Dialogue is question-
able. See Ritter, *Hist. of Ancient
Philosophy*, vol. ii. p. 164 (English
translation); Schleiermacher's *Introduction*, translated by Dobson, p. 328 ;
Brandis, *Gesch. der Gr.-Röm. Philosophie*, vol. ii. p. 180.—Ed.

Socr. Is it not with language that Socrates speaks?
—*Alcib.* What now? of course.

Socr. To converse, and to use language, are not
these then the same?—*Alcib.* The very same.

Socr. But he who uses a thing, and the thing used,
—are these not different?—*Alcib.* What do you
mean?

Socr. A currier,—does he not use a cutting knife
and other instruments?—*Alcib.* Yes.

Socr. And the man who uses the cutting knife, is
he different from the instrument he uses?—*Alcib.*
Most certainly.

Socr. In like manner, the lyrist, is he not different
from the lyre he plays on?—*Alcib.* Undoubtedly.

Socr. This, then, was what I asked you just now,
—does not he who uses a thing seem to you always different from the thing used?—*Alcib.* Very
different.

Socr. But the currier, does he cut with his instruments alone, or also with his hands?—*Alcib.* Also
with his hands.

Socr. He then uses his hands?—*Alcib.* Yes.

Socr. And in his work he uses also his eyes?—
Alcib. Yes.

Socr. We are agreed, then, that he who uses a
thing, and the thing used, are different?—*Alcib.* We
are.

Socr. The currier and lyrist are, therefore, different
from the hands and the eyes, with which they work?
—*Alcib.* So it seems.

Socr. Now, then, does not a man use his whole
body?—*Alcib.* Unquestionably.

Socr. But we are agreed that he who uses, and that
which is used, are different?—*Alcib.* Yes.

Socr. A man is, therefore, different from his body ?
—*Alcib.* So I think.

Socr. What then is the man ?—*Alcib.* I cannot say.

Socr. You can at least say that the man is that
which uses the body ?—*Alcib.* True.

Socr. Now, does anything use the body but the
mind ?—*Alcib.* Nothing.

Socr. The mind is, therefore, the man ?—*Alcib.* The
mind alone."

To the same effect, Aristotle asserts that the mind
contains the man, not the man the mind.[a] "Thou
art the soul," says Hierocles, " but the body is thine."[β]
So Cicero—" Mens cujusque is est quisque, non ea
figura quæ digito demonstrari potest;"[γ] and Macrobius—" Ergo qui videtur, non ipse verus homo est, sed
verus ille est, a quo regitur quod videtur."[δ]

No one has, however, more beautifully expressed
this truth than Arbuthnot :[ε]

Arbuthnot.

> "What am I, whence produced, and for what end ?
> Whence drew I being, to what period tend ?
> Am I th' abandon'd orphan of blind chance,
> Dropp'd by wild atoms in disorder'd dance ?
> Or, from an endless chain of causes wrought,
> And of unthinking substance, born with thought?
> Am I but what I seem, mere flesh and blood,
> A branching channel with a mazy flood ?
> The purple stream that through my vessels glides,
> Dull and unconscious flows, like common tides,
> The pipes, through which the circling juices stray,
> Are not that thinking I, no more than they :

---

a That the mind is *the man,* is
maintained by Aristotle in several
places. Cf. *Eth. Nic.,* ix. 8; x. 7; but
these do not contain the exact words
of the text.—ED.

β *In Aurea Pythagoreorum Car-
mina,* 26 : Σὺ γὰρ εἶ ἡ ψυχή· τὸ δὴ

σῶμα σόν.—ED.

γ *Somnium Scipionis,* c. 8.—ED.

δ Macrobius, *In Somnium Scipionis,*
lib. ii. c. 12.—ED.

ε *Know thyself.* See Dodsley's *Collection,* vol. i. p. 180.—ED.

This frame, compacted with transcendent skill,
Of moving joints, obedient to my will ;
Nurs'd from the fruitful glebe, like yonder tree,
Waxes and wastes,—I call it mine, not me.
New matter still the mould'ring mass sustains ;
The mansion chang'd, the tenant still remains ;
And, from the fleeting stream repair'd by food,
Distinct, as is the swimmer from the flood."

But let us come to a closer determination of the point ; let us appeal to our experience. " I turn my attention on my being, and find that I have organs, and that I have thoughts. My body is the complement of my organs ; am I then my body, or any part of my body ? This I cannot be. The matter of my body, in all its points, is in a perpetual flux, in a perpetual process of renewal. I,—*I* do not pass away, I am not renewed. None probably of the molecules which constituted my organs some years ago, form any part of the material system which I now call mine. It has been made up anew ; but I am still what I was of old. These organs may be mutilated,—one, two, or any number of them may be removed ; but not the less do I continue to be what I was, one and entire. It is even not impossible to conceive me existing, deprived of every organ,—I, therefore, who have these organs, or this body, *I* am neither an organ nor a body.

" Neither am I identical with my thoughts, for they are manifold and various. I, on the contrary, am one and the same. Each moment they change and succeed each other ; this change and succession takes place in me, but I neither change nor succeed myself in myself. Each moment, I am aware or am conscious of the existence and change of my thoughts : this change is sometimes determined by me, sometimes by

The Self
or Ego in
relation to
bodily or-
gans, and
thoughts.

LECT.
IX.

something different from me; but I always can distinguish myself from them,—I am a permanent being,
an enduring subject, of whose existence these thoughts
are only so many modes, appearances, or phænomena.
I who possess organs and thoughts am, therefore,
neither these organs nor these thoughts.

"I can conceive myself to exist apart from every
organ. But if I try to conceive myself existent without a thought,—without some form of consciousness,
—I am unable. This or that thought may not be
perhaps necessary; but of some thought it is necessary that I should be conscious, otherwise I can no
longer conceive myself to be. A suspension of thought
is thus a suspension of my intellectual existence; I
am, therefore, essentially a thinking,—a conscious
being; and my true character is that of an intelligence,—an intelligence served by organs."[a]

But this thought, this consciousness, is possible
only in, and through, the consciousness of Self. The
Self, the I, is recognised in every act of intelligence,
as the subject to which that act belongs. It is I that
perceive, I that imagine, I that remember, I that
attend, I that compare, I that feel, I that desire, I
that will, I that am conscious. The I, indeed, is only
manifested in one or other of these special modes;
but it is manifested in them all; they are all only
the phænomena of the I, and, therefore, the science
conversant about the phænomena of mind is, most
simply and unambiguously, said to be conversant
about the phænomena of the *I* or *Ego*.

This expression, as that which, in many relations,
best marks and discriminates the conscious mind, has
now become familiar in every country, with the ex-

a Gatien-Arnoult, [*Doct. Phil.*, p. 34-36.—Ed.]

ception of our own. Why it has not been naturalised with us is not unapparent. The French have two words for the Ego or I—*Je* and *Moi*. The former of these is less appropriate as an abstract term, being in sound ambiguous; but *le moi* admirably expresses what the Germans denote, but less felicitously, by their *Das Ich*. In English *the I* could not be tolerated; because, in sound, it would not be distinguished from the word significant of the organ of sight. We must, therefore, either renounce the term, or resort to the Latin *Ego*; and this is perhaps no disadvantage, for, as the word is only employed in a strictly philosophical relation, it is better that this should be distinctly marked, by its being used in that relation alone. The term *Self* is more allowable; yet still the expressions *Ego* and *Non-Ego* are felt to be less awkward than those of *Self* and *Not-Self*.

So much in explanation of the terms involved in the definition which I gave you of Psychology.

LECTURE X.

LECT.
X.

I NOW proceed, as I proposed, to the consideration of a few other words of frequent occurrence in philosophy, and which it is expedient to explain at once, before entering upon discussions in which they will continually recur. I take them up without order, except in so far as they may be grouped together by their meaning ; and the first I shall consider are, the terms *hypothesis* and *theory*.

Hypothe-
sis.

When a phænomenon is presented to us which can be explained by no cause within the sphere of our experience, we feel dissatisfied and uneasy. A desire arises to escape from this unpleasing state ; and the consequence of this desire is an effort of the mind to recall the outstanding phænomenon to unity, by assigning it, *ad interim*, to some cause or class, to which we imagine that it may possibly belong, until we shall be able to refer it, permanently, to that cause, or class, to which we shall have proved it actually to appertain. The judgment by which the phænomenon is thus provisorily referred, is called an *hypothesis,—* a *supposition*.

Hypotheses have thus no other end than to satisfy the desire of the mind to reduce the objects of its knowledge to unity and system ; and they do this in recalling them, *ad interim*, to some principle, through

which the mind is enabled to comprehend them. From
this view of their nature it is manifest, how far they
are permissible, and how far they are even useful and
expedient,—throwing altogether out of account the
possibility that what is at first assumed as hypothetical, may subsequently be proved true.

An hypothesis is allowable only under certain conditions. Of these the first is,—that the phænomenon
to be explained, should be ascertained actually to
exist. It would, for example, be absurd to propose
an hypothesis to account for the possibility of appa-
ritions, until it be proved that ghosts do actually
appear. This precept, to establish your fact before
you attempt to conjecture its cause, may, perhaps,
seem to you too elementary to be worth the statement. But a little longer experience will convince
you of the contrary. That the enunciation of the
rule is not only not superfluous, but even highly
requisite as an admonition, is shown by great and
numerous examples of its violation in the history of
science; and, as Cullen has truly observed, there are
more false facts current in the world than false hypotheses to explain them. There is, in truth, nothing
which men seem to admit so lightly as an asserted
fact. Of this I might adduce to you a host of memorable examples. I shall content myself with one small
but significant illustration.

Charles II., soon after the incorporation of the Royal
Society, which was established under his patronage,
sent to request of that learned body an explanation
of the following phænomenon. When a live fish is
thrown into a basin of water, the basin, water, and
fish do not weigh more than the basin and water
before the fish is thrown in; whereas, when a dead

LECT.
X.

fish is employed, the weight of the whole is exactly equal to the added weights of the basin, the water, and the fish. .Much learned discussion ensued regarding this curious fact, and several elaborate papers, propounding various hypotheses in explanation, were read on the occasion. At length a member, who was better versed in Aristotle than his associates, recollected that the philosopher had laid it down, as a general rule of philosophising, to consider the *an sit* of a fact, before proceeding to investigate the *cur sit;* and he ventured to insinuate to his colleagues, that though the authority of the Stagirite was with them,—the disciples of Bacon,—of small account, it might possibly not be altogether inexpedient to follow his advice on the present occasion; seeing that it did not, in fact, seem at variance with common sense, and that none of the hypotheses proposed were admitted to be altogether satisfactory. After much angry discussion, some members asserting the fact to be in itself notorious, and others declaring that to doubt of its reality was an insult to his majesty, and tantamount to a constructive act of treason, the experiment was made,—when lo! to the confusion of the wise men of Gotham,—the name by which the Society was then popularly known, —it was found that the weight was identical, whether a dead or a living fish were used.

This is only a past and petty illustration. It would be easy to adduce extensive hypotheses, very generally accredited, even at the present hour, which are, however, nothing better than assumptions founded on, or explanatory of, phænomena which do not really exist in nature.

The second.   The second condition of a permissible hypothesis is, —that the phænomenon cannot be explained otherwise, than by an hypothesis. It would, for example, have been absurd, even before the discoveries of Franklin, to account for the phænomenon of lightning by the hypothesis of supernatural agency. These two conditions, of the reality of the phænomenon, and the necessity of an hypothesis for its explanation, being fulfilled, an hypothesis is allowable.[a]

But the necessity of some hypothesis being conceded, how are we to discriminate between a good and a bad, a probable and an improbable, hypothesis ? The comparative excellence of an hypothesis requires, in the first place, that it involve nothing contradictory, either internally or externally,—that is, either between the parts of which it is composed, or between these and any established truths. Thus, the Ptolemaic hypothesis of the heavenly revolutions became worthless, from the moment that it was contradicted by the ascertained phænomena of the planets Venus and Mercury. Thus, the Wernerian hypothesis in geology is improbable, inasmuch as it is obliged to maintain that water was originally able to hold in solution substances which it is now incapable of dissolving. The Huttonian hypothesis, on the contrary, is so far preferable, that it assumes no effect to have been produced by any agent, which that agent is not known to be capable of producing. In the second place, an hypothesis is probable in proportion as the phænomenon in question can be by it more completely explained. Thus, the Copernican hypothesis is more probable than the Tychonic and semi-Tychonic, inasmuch as it enables us to explain a greater number of phænomena. In the third place, an hypothesis is probable, in proportion as it is independent of all subsidiary hypotheses. In this respect, again, the

Criteria of the excellence of an hypothesis.

a [On the conditions of legitimate hypothesis compare John Christopher Sturm, *Physica Electiva*, Diss. Prælim. art. 3, tom. i. p. 28.]

LECT.
X.

Copernican hypothesis is more probable than the Tychonic. For, though both save all the phænomena, the Copernican does this by one principal assumption; whereas the Tychonic is obliged to call in the aid of several subordinate suppositions, to render the principal assumption available. So much for *hypothesis*.

I have dwelt longer on hypothesis than perhaps was necessary; for you must recollect that these terms are, at present, considered only in order to enable you to understand their signification when casually employed. We shall probably, in a subsequent part of the Course, have occasion to treat of them expressly, and with the requisite details. I shall, therefore, be more concise in treating of the cognate expression,—*theory*. This word is employed by English writers, in a very loose and improper sense. It is with them usually convertible with hypothesis, and hypothesis is commonly used as another term for conjecture. Dr Reid, indeed, expressly does this; he identifies the two words, and explains them as philosophical conjectures, as you may see in his First Essay on the *Intellectual Powers*, (Chap. III.)[a] This is, however, wrong; wrong, in relation to the original employment of the terms by the ancient philosophers; and wrong, in relation to their employment by the philosophers of the modern nations.

Theory,
Practice.

The terms *theory* and *theoretical* are properly used in opposition to the terms *practice* and *practical;* in this sense they were exclusively employed by the ancients; and in this sense they are almost exclusively employed by the Continental philosophers. Practice is the exercise of an art, or the application of a science, in life, which application is itself an art, for it

a *Works*, p. 235; see also p. 97.—ED.

is not every one who is able to apply all he knows; there being required, over and above knowledge, a certain dexterity and skill. Theory, on the contrary, is mere knowledge or science. There is a distinction, but no opposition, between theory and practice; each to a certain extent supposes the other. On the one hand, theory is dependent on practice,—practice must have preceded theory; for theory being only a gener-alisation of the principles on which practice proceeds, these must originally have been taken out of, or ab-stracted from, practice. On the other hand, this is true only to a certain extent; for there is no practice without a theory. The man of practice must have always known something, however little, of what he did, of what he intended to do, and of the means by which his intention was to be carried into effect. He was, therefore, not wholly ignorant of the principles of his procedure; he was a limited, he was, in some degree, an unconscious, theorist. As he proceeded, however, in his practice, and reflected on his perfor-mance, his theory acquired greater clearness and ex-tension, so that he became at last distinctly conscious of what he did, and could give, to himself and others, an account of his procedure.

> " Per varios usus artem experientia fecit,
> Exemplo monstrante viam." a

In this view, theory is, therefore, simply a know-ledge of the principles by which practice accomplishes its ends.

The opposition of Theoretical and Practical philo-sophy is somewhat different; for these do not stand simply related to each other as theory and practice. Practical philosophy involves likewise a theory,—a

(marginal notes:) LECT. X. — Theoretical and Practi-cal Philo-sophy.

a [Manilius, i. 62.]

theory, however, subordinated to the practical application of its principles; while theoretical philosophy
has nothing to do with practice, but terminates in
mere speculative or contemplative knowledge.[a]

The next group of associated words to which I
would call your attention, is composed of the terms,—
*power, faculty, capacity, disposition, habit, act, opera-
tion, energy, function,* &c.

Power.
Reid's cri-
ticism of
Locke.

Of these the first is *power,* and the explanation of
this, in a manner, involves that of all the others.

I have, in the first place, to correct an error of
Dr Reid in relation to this term, in his criticism of
Locke's statement of its import.—You will observe
that I do not, at present, enter on the question, How
do we acquire the notion of power? and I defend the
following passage of Locke, only in regard to the
meaning and comprehension of the term. "The
mind," says Locke, "being every day informed, by the
senses, of the alteration of those simple ideas it observes
in things without, and taking notice how one comes to
an end, and ceases to be, and another begins to exist
which was not before; reflecting also on what passes
within itself, and observing a constant change of its
ideas, sometimes by the impression of outward objects
on the senses, and sometimes by the determination of
its own choice; and concluding from what it has so
constantly observed to have been, that the like changes
will, for the future, be made on the same things, by
like agents, and by the like ways; considers, in one
thing, the possibility of having any of its simple ideas
changed, and, in another, the possibility of making
that change; and so comes by that idea which we call
power. Thus we say, fire has a power to melt gold,

a See *ante*, p. 113.—ED.

—that is, to destroy the consistency of its insensible parts, and consequently its hardness, and make it fluid, and gold has a power to be melted: that the sun has a power to blanch wax, and wax a power to be blanched by the sun, whereby the yellowness is destroyed, and whiteness made to exist in its room. In which, and the like cases, the power, we consider, is in reference to the change of perceivable ideas; for we cannot observe any alteration to be made in, or operation upon, anything, but by the observable change of its sensible ideas; nor conceive any alteration to be made, but by conceiving a change of some of its ideas. Power, thus considered, is twofold—viz., as able to make, or able to receive, any change: the one may be called *active*, and the other *passive* power." [a]

I have here only to call your attention to the distinction of power into two kinds, *active* and *passive*— the former meaning *id quod potest facere*, that which *can effect* or *can do*,—the latter *id quod potest fieri*, that which *can be effected* or *can be done*. In both cases the general notion of power is expressed by the verb *potest* or *can*. Now, on this, Dr Reid makes the following strictures.[β] "On this account by Locke," he says, "of the origin of our idea of power, I would beg leave to make two remarks, with the respect that is most justly due to so great a philosopher and so good a man." We are at present concerned only with the first of these remarks by Dr Reid, which is as follows, —"Whereas Locke distinguishes power into *active* and *passive*, I conceive passive power is no power at all. He means by it, the possibility of being changed. To call this *power*, seems to be a misapplication of the

Active and Passive Power.

---

a *Essay*, Book ii. ch. 21, § 1.—ED. *Works*, p. 519.—ED.
β *Active Powers*, Essay i. ch. 3;

word. I do not remember to have met with the phrase *passive power* in any other good author. Mr Locke seems to have been unlucky in inventing it ; and it deserves not to be retained in our language. Perhaps he was unwarily led into it, as an opposite to *active power*. But I conceive we call certain powers *active*, to distinguish them from other powers that are called *speculative*. As all mankind distinguish action from speculation, it is very proper to distinguish the powers by which those different operations are performed, into active and speculative. Mr Locke, indeed, acknowledges that active power is more properly called power : but I see no propriety at all in passive power ; it is a powerless power, and a contradiction in terms."

These observations of Dr Reid are, I am sorry to say, erroneous from first to last. The latter part, in which he attempts to find a reason for Locke being unwarily betrayed into making this distinction, is —supposing the distinction untenable, and Locke its author,—wholly inadequate to account for his hallucination ; for, surely, the powers by which we speculate are, in their operations, not more passive than those that have sometimes been styled *active*, but which are properly denominated *practical*. But in the censure itself on Locke, Reid is altogether mistaken. In the first place, so far was Locke from being unlucky in inventing the distinction, it was invented some two thousand years before. In the second place, to call the *possibility of being changed* a *power*, is no misapplication of the word. In the third place, so far is the phrase *passive power* from not being employed by any good author,—there is hardly a metaphysician previous to Locke, by whom it was

not familiarly used. In fact, this was one of the most
celebrated distinctions in philosophy.   It was first
formally enounced by Aristotle,[a] and from him was
universally adopted.   Active and passive power are
in Greek styled δύναμις ποιητική, and δύναμις παθητική;
in Latin, *potentia activa*, and *potentia passiva*.[β]

Power, therefore, is a word which we may use both
in an active, and in a passive, signification ; and, in
psychology, we may apply it both to the active faculties, and to the passive capacities, of mind.

This leads to the meaning of the terms *faculty* and
capacity.   *Faculty* (*facultas*) is derived from the ob-
solete Latin *facul*,—the more ancient form of *facilis*,
from which again *facilitas* is formed.   It is properly
limited to active power, and, therefore, is abusively
applied to the mere passive affections of mind.

Capacity (*capacitas*), on the other hand, is more
properly limited to these.   Its primary signification,
which is literally *room for*, as well as its employment,
favours this ; although it cannot be denied, that there
are examples of its usage in an active sense.   Leibnitz,
as far as I know, was the first who limited its psychological application to the passivities of mind.   In his
famous *Nouveaux Essais sur l'Entendement Humain*,
a work written in refutation of Locke's *Essay* on the
same subject, he observes :—" We may say that power
(*puissance*), in general, is the possibility of change.

a See *Metaph.*, iv. (v.) 12; viii.
(ix.) 1.—Ed.

β This distinction is, indeed, established in the Greek language itself.
That tongue has, among its other
marvellous perfections, two sets of
potential adjectives, the one for *active*,
the other for *passive* power. Those
for active power are denoted by terminations in τικός, those for passive

power by terminations in τός. Thus
ποιητικός, that which can make; ποιη-
τός, that which can be made; κινητι-
κός, that which can move ; κινητός,
that which can be moved ; and so
πρακτικός and πρακτός, αἰσθητικός and
αἰσθητός, νοητικός and νοητός, οἰκο-
δομητικός and οἰκοδομητός, &c. [Cf.
Lord Monboddo's *Ancient Metaphysics*, vol. i. p. 8.—Ed.]

LECT.
X.

Now the change, or the act of this possibility, being action in one subject and passion in another, there will be two powers (*deux puissances*), the one *passive*, the other *active*. The active may be called *faculty*, and perhaps the passive might be called *capacity*, or receptivity. It is true that the active power is sometimes taken in a higher sense, when, over and above the simple faculty, there is also a tendency, a *nisus*; and it is thus that I have used it in my dynamical considerations. We might give to it in this meaning the special name of *force*." [a] I may notice that Reid seems to have attributed no other meaning to the term power than that of force.

Power, then, is active and passive ; faculty is active power,—capacity is passive power.[b]

Disposition, Habit.

The two terms next in order, are *disposition*, in Greek, διάθεσις; and *habit*, in Greek, ἕξις. I take these together as they are similar, yet not the same. Both are tendencies to action ; but they differ in this, that disposition properly denotes a natural tendency, habit an acquired tendency. Aristotle distinguishes them by another difference. " Habit (ἕξις) is discriminated from disposition (διάθεσις) in this, that the latter is easily movable, the former of longer duration, and more difficult to be moved." [c] I may notice that habit is formed by the frequent repetition of the same action or passion, and that this repetition is called *consuetude*, or *custom*. The latter terms, which properly

a *Nouveaux Essais*, liv. ii. ch. 21. § 2.—ED.

β [Distinction of Faculty and Power,—Faculty being given to self-active forces, Power to both active and passive : see Wolf, *Psych. Emp.*, § 29; *Psych. Rat.*, § 81; Weiss, *Untersuchungen über das Wesen und Wirken der menschlichen Seele*, p. 66; Jouffroy, *Mélanges*, p. 345 et seq.; Daube, *Essai d'Idéologie*, p. 136 ; Fries, *Anthropologie*, i. p. 26, (ed. 1820.)]

γ *Categ.*, c. 8.—ED.

signify the cause, are not unfrequently abusively employed for habit, their effect.

I may likewise observe that the terms *power*,
faculty, *capacity*, are more appropriately applied to
natural, than to acquired, capabilities, and are thus
inapplicable to mere habits. I say *mere* habits, for
where habit is superinduced upon a natural capability,
both terms may be used. Thus we can say both the
faculty of abstraction, and the habit of abstraction,—
the capacity of suffering and the habit of suffering;
but still the meanings are not identical.

The last series of cognate terms are *act, operation,* Act, Ope-
energy. They are all mutually convertible, as all de- Energy.
noting the present exertion or exercise of a power, a
faculty, or a habit. I must here explain to you the
famous distinction of actual and potential existence, Potential
for, by this distinction, act, operation, energy, are con- Existence.
tradiscriminated from power, faculty, capacity, disposition, and habit. This distinction, when divested of
certain subordinate subtleties of no great consequence,
is manifest and simple. Potential existence means
merely that the thing *may be* at some time; actual
existence, that it now *is.*[a] Thus, the mathematician,
when asleep or playing at cards, does not exercise his
skill; his geometrical knowledge is all latent, but he
is still a mathematician,—potentially.

> " Ut quamvis tacet Hermogenes, cantor tamen atque
> Optimus est modulator;—ut Alfenus vafer, omni
> Abjecto instrumento artis, clausaque taberna,
> Sutor erat."[β]

Hermogenes, says Horace, was a singer, even when
silent; how?—a singer, not *in actu* but *in posse.* So

a This distinction is well illustrat- burg on *Arist. de Anima,* ii. 1.—ED.
ed in the learned note of Trendelen- β Horace, *Sat.* i. 3, 129.—ED.

Alfenus was a cobbler, even when not at work ; that
is, he was a cobbler *potential ;* whereas, when busy in
his booth, he was a cobbler *actual.*

In like manner, my sense of sight potentially exists,
though my eyelids are closed ; but when I open them,
it exists actually. Now, *power, faculty, capacity, disposition, habit,* are all different expressions for potential
or possible existence ; *act, operation, energy,* for actual
or present existence. Thus the *power* of imagination
expresses the unexerted capability of imagining ; the
act of imagination denotes that power elicited into
immediate,—into present, existence. The different
synonyms for potential existence, are existence ἐν
δυνάμει, *in potentia, in posse, in power ;* for actual
existence, existence ἐν ἐνεργείᾳ, or ἐν ἐντελεχείᾳ, *in
actu, in esse, in act, in operation, in energy.* The
term *energy* is precisely the Greek term for act or
operation ; but it has vulgarly obtained the meaning
of forcible activity.[a]

Function

The word *functio,* in Latin, simply expresses performance or operation ; *functio muneris* is the exertion of an energy of some determinate kind.[β] But
with us the word *function* has come to be employed
in the sense of *munus* alone, and means not the exercise, but the specific character, of a power. Thus the
function of a clergyman does not mean with us the

a But there is another relation of
potentiality and actuality which I
may notice,—Hermogenes, Alfenus,
before, and after, acquiring the habits
of singer, and cobbler. There is thus
a double kind of potentiality and actuality,—for when Hermogenes has
obtained the habit and power of singing, though not actually exercising,
he is a singer *in actu,* in relation to
himself, before he had acquired the
accomplishment. This affords the
distinction taken by Aristotle of first
and second energy,—the first being
the habit acquired, the second the
immediate exercise of that habit.
[Cf. *De Anima,* lib. ii. c. 1.—Ed.]

β ["Functio est actio qua facultas
vim suam exerit, suumque effectum
producit." Tosca, *Com. Philosoph.,*
vol. vii. p. 156.]

performance of his duties, but the peculiarity of those
duties themselves. The function of nutrition does not mean the operation of that animal power, but its discriminate character.

So much by way of preliminary explanation of the psychological terms in most general and frequent use. Others, likewise, I shall, in the sequel, have occasion to elucidate; but these may, I think, more appropriately be dealt with as they happen to occur.

LECTURE XI.

OUTLINE OF DISTRIBUTION OF MENTAL PHÆNOMENA : CONSCIOUSNESS,—ITS SPECIAL CONDITIONS.

LECT.
XI.

Distribution of the mental phæno-mena.

I NOW proceed to the consideration of the important subject,—the Distribution of the Mental Phænomena into their primary or most general classes. In regard to the distribution of the mental phænomena, I shall not at present attempt to give any history or criticism of the various classifications which have been proposed by different philosophers. These classifications are so numerous, and so contradictory, that, in the present stage of your knowledge, such a history would only fatigue the memory, without informing the understanding ; for you cannot be expected to be as yet able to comprehend, at least many of the reasons which may be alleged for, or against, the different distributions of the human faculties. I shall, therefore, at once proceed to state the classification of these, which I have adopted as the best.

Consciousness,—the one essential element of the mental phæno-mena.

In taking a comprehensive survey of the mental phænomena, these are all seen to comprise one essential element, or to be possible only under one necessary condition. This element or condition is Consciousness, or the knowledge that I,—that the Ego exists, in some determinate state. In this knowledge they appear, or are realised as phænomena, and with this

knowledge they likewise disappear, or have no longer
a phænomenal existence; so that consciousness may
be compared to an internal light, by means of which,
and which alone, what passes in the mind is rendered
visible. Consciousness is simple,—is not composed of
parts, either similar or dissimilar. It always resembles itself, differing only in the degrees of its intensity; thus, there are not various kinds of consciousness, although there are various kinds of mental
modes, or states, of which we are conscious. Whatever division, therefore, of the mental phænomena
may be adopted, all its members must be within consciousness; that is, we must not attempt to divide
consciousness itself, which must be viewed as comprehensive of the whole phænomena to be divided; far
less should we reduce it, as a special phænomenon,
to a particular class. Let consciousness, therefore,
remain one and indivisible, comprehending all the
modifications,—all the phænomena, of the thinking
subject.

But taking, again, a survey of the mental modifications, or phænomena, of which we are conscious,—
these are seen to divide themselves into THREE great
classes. In the first place, there are the phænomena
of Knowledge; in the second place, there are the
phænomena of Feeling, or the phænomena of Pleasure
and Pain; and, in the third place, there are the phænomena of Will and Desire.[a]

Let me illustrate this by an example. I see a picture. Now, first of all, I am conscious of perceiving
a certain complement of colours and figures,—I recognise what the object is. This is the phænomenon of Cognition or Knowledge. But this is not the

Three grand classes of mental phænomena.

a Compare *Stewart's Works*, vol. ii., Advertisement by Editor.—ED.

only phænomenon of which I may be here conscious. I may experience certain affections in the contemplation of this object. If the picture be a masterpiece, the gratification will be unalloyed; but if it be an unequal production, I shall be conscious, perhaps, of enjoyment, but of enjoyment alloyed with dissatisfaction. This is the phænomenon of Feeling,—or of Pleasure and Pain. But these two phænomena do not yet exhaust all of which I may be conscious on the occasion. I may desire to see the picture long,—to see it often,—to make it my own; and, perhaps, I may will, resolve, or determine so to do. This is the complex phænomenon of Will and Desire.

Their nomenclature.

The English language, unfortunately, does not afford us terms competent to express and discriminate, with even tolerable clearness and precision, these classes of phænomena. In regard to the first, indeed, we have comparatively little reason to complain,—the synonymous terms, *knowledge* and *cognition*, suffice to distinguish the phænomena of this class from those of the other two. In the second class, the defect of the language becomes more apparent. The word *feeling* is the only term under which we can possibly collect the phænomena of pleasure and pain, and yet this word is ambiguous. For it is not only employed to denote what we are conscious of as agreeable or disagreeable in our mental states, but it is likewise used as a synonym for the sense of touch.[a] It is, however, principally in relation to the third class that the deficiency is manifested. In English, unfortunately, we have no term capable of adequately expressing what is

a [Brown uses feeling for consciousness.—*Oral Interp.*]; *e.g.*, *Philosophy of the Human Mind*, Lecture xi., p. 66, (ed. 1830): "The mind is susceptible of a variety of feelings, every new feeling being a change of its state."—ED.

common both to will and desire ; that is, the *nisus* or *conatus*,—the tendency towards the realisation of their end. By will is meant a free and deliberate, by desire a blind and fatal, tendency to act.[a] Now, to express, I say, the tendency to overt action,—the quality in which desire and will are equally contained,—we possess no English term to which an exception of more or less cogency may not be taken. Were we to say the phænomena of *tendency*, the phrase would be vague ; and the same is true of the phænomena of *doing*. Again, the term phænomena of *appetency* is objectionable, because (to say nothing of the unfamiliarity of the expression) *appetency*, though perhaps etymologically unexceptionable, has both in Latin and English a meaning almost synonymous with desire. Like the Latin *appetentia*, the Greek ὄρεξις is equally ill-balanced, for, though used by philosophers to comprehend both will and desire, it more familiarly suggests the latter, and we need not, therefore, be solicitous, with Mr Harris and Lord Monboddo, to naturalise in English the term *orectic*.[β] Again, the phrase phænomena of *activity* would be even worse ; every possible objection can be made to the term *active powers*, by which the philosophers of this country have designated the *orectic faculties* of the Aristotelians. For you will observe, that all faculties are equally active ; and it is not the overt performance, but the tendency towards it, for which we are in quest of an expression. The German is the only language I am acquainted with, which is able to supply the term of which philosophy is in want. The expression *Bestrebungs Vermögen*,

---

a Cf. Aristotle, *Rhet.*, i. 10: Βούλη-
σις, μετὰ λόγου ὄρεξις ἀγαθοῦ ἄλογοι
δ' ὀρέξεις, ὀργὴ καὶ ἐπιθυμία.—ED.

β See Lord Monboddo's *Ancient
Metaphysics*, book ii. chaps. vii. ix.
—ED.

which is most nearly, though awkwardly and inadequately, translated by *striving faculties*,—faculties of effort or endeavour,—is now generally employed, in the philosophy of Germany, as the genus comprehending desire and will.　Perhaps the phrase phænomena of *exertion* is, upon the whole, the best expression to denote the manifestations,—and *exertive* faculties, the best expression to denote the faculties, —of will and desire.　*Exero*, in Latin, means literally *to put forth*,—and, with us, *exertion* and *exertive* are the only endurable words that I can find which approximate, though distantly, to the strength and precision of the German expression.　I shall, however, occasionally employ likewise the term *appetency* in the rigorous signification I have mentioned,—as a genus comprehending under it both desires and volitions.[a]

By whom this threefold distribution first made.　　This division of the phænomena of mind into the three great classes of the Cognitive faculties,—the Feelings, or capacities of Pleasure and Pain,—and the Exertive or Conative Powers,—I do not propose as original.　It was first promulgated by Kant;[β] and the felicity of the distribution was so apparent that it has now been long all but universally adopted in Germany by the philosophers of every school ; and, what is curious, the only philosopher of any eminence by whom it has been assailed,—indeed, the only philosopher of

a 1848.　The term *Conative* (from *Conari*) is employed by Cudworth in his *Treatise on Free Will*, published some years ago from his MSS. in the British Museum.　[*A Treatise on Free Will*, by Ralph Cudworth, D.D., edited by John Allen, M.A. (London, 1838), p. 31 : "Notwithstanding which, the hegemonic of the soul may, by conatives and endeavours, acquire more and more power over them."　The terms *Conation* and *Conative* are those finally adopted by the Author, as the most appropriate expressions for the class of phænomena in question.—ED.]

β *Kritik der Urtheilskraft*, Einleitung.　The same division is also adopted as the basis of his *Anthropologie*.—ED.

any reputation by whom it has been, in that country, rejected,—is not an opponent of the Kantian philosophy, but one of its most zealous champions.[a] To the psychologists of this country it is apparently wholly unknown. They still adhere to the old scholastic division into powers of the Understanding and powers of the Will; or, as it is otherwise expressed, into Intellectual and Active powers.[β]

By its author the Kantian classification has received no illustration; and by other German philosophers, it has apparently been viewed as too manifest to require any. Nor do I think it needs much; though a few words in explanation may not be inexpedient. An objection to the arrangement may, perhaps, be taken on the ground that the three classes are not co-ordinate. It is evident that every mental phænomenon is either an act of knowledge, or only possible through an act of knowledge, for consciousness is a knowledge,—a phænomenon of cognition; and, on this principle, many philosophers,—as Descartes, Leibnitz, Spinoza, Wolf, Platner, and others,—have been led to regard the knowing, or representative faculty, as they called it,—the faculty of cognition, as the fundamental power of mind, from which all others are derivative. To this the answer is easy. These philosophers did not observe that, although pleasure and pain—although desire and volition, are only as they are known to be; yet, in these modifications, a quality, a phænomenon of mind, absolutely new, has been superadded, which was never

LECT. XI.

Objection to the classification obviated.

a This philosopher is Krug, who attacked the Kantian Division in his *Grundlage zu einer neuen Theorie der Gefühle und des sogenannten Gefühlsvermögens*, Konigsberg, 1823. See also his *Handworterbuch der Philosophischen Wissenschaften*, art. *Gefühl* and *Seelenkräfte*. A fuller account of this controversy is given by Sir W. Hamilton in a subsequent Lecture. See Lecture XLI., vol. ii. p. 421 et seq.—ED.

β Cf. *Reid's Works*, pp. 242, n. †, 511, nn. * †.—ED.

LECT.
XI. involved in, and could, therefore, never have been evolved out of, the mere faculty of knowledge. The faculty of knowledge is certainly the first in order, inasmuch as it is the *conditio sine qua non* of the others; and we are able to conceive a being possessed of the power of recognising existence, and yet wholly void of all feeling of pain and pleasure, and of all powers of desire and volition. On the other hand, we are wholly unable to conceive a being possessed of feeling and desire, and, at the same time, without a knowledge of any object upon which his affections may be employed, and without a consciousness of these affections themselves.

We can further conceive a being possessed of knowledge and feeling alone — a being endowed with a power of recognising objects, of enjoying the exercise, and of grieving at the restraint, of his activity, and yet devoid of that faculty of voluntary agency—of that conation, which is possessed by man. To such a being would belong feelings of pain and pleasure, but neither desire nor will, properly so called. On the other hand, however, we cannot possibly conceive the existence of a voluntary activity independently of all feeling; for voluntary conation is a faculty which can only be determined to energy through a pain or pleasure,—through an estimate of the relative worth of objects.

In distinguishing the cognitions, feelings, and conations, it is not, therefore, to be supposed that these phænomena are possible independently of each other. In our philosophical systems, they may stand separated from each other in books and chapters;—in nature, they are ever interwoven. In every, the simplest, modification of mind, knowledge, feeling, and desire or will, go to constitute the mental state; and it is only by a scientific abstraction that we are able to analyse the state into elements, which are never really existent but in mutual combination. These elements are found, indeed, in very various proportions in different states, —sometimes one preponderates, sometimes another; but there is no state in which they are not all co-existent.[a]

Let the mental phænomena, therefore, be distributed under the three heads of phænomena of Cognition, or the faculties of Knowledge; phænomena of Feeling, or the capacities of Pleasure and Pain; and phænomena of Desiring or Willing, or the powers of Conation.

The order of these is determined by their relative consecution. Feeling and appetency suppose knowledge. The cognitive faculties, therefore, stand first. But as will, and desire, and aversion, suppose a knowledge of the pleasurable and painful, the feelings will stand second as intermediate between the other two.

Such is the highest or most general classification of the mental phænomena, or of the phænomena of which we are conscious. But as these primary classes are, as we have shown, all included under one universal phænomenon,—the phænomenon of consciousness,—it follows that Consciousness must form the first object of our consideration.

I shall not attempt to give you any preliminary detail of the opinions of philosophers in relation to consciousness. The only effect of this would be to confuse you. It is necessary, in the first place, to obtain correct and definite notions on the subject, and having obtained these, it will be easy for you to understand in what respects the opinions that have been hazarded on the cardinal point of all philosophy, are inadequate or erroneous. I may notice that Dr

Order of the mental phænomena.

Consciousness, the first object of consideration.

No special account of consciousness by Reid or Stewart.

a See below, vol. ii. p. 2 et seq.—ED.

Reid and Mr Stewart have favoured us with no special or articulate account of consciousness. The former, indeed, intended and promised this. In the seventh chapter of the first Essay *On the Intellectual Powers*, which is entitled *Division of the Powers of the Mind*, the concluding paragraph is as follows :—

" I shall not, therefore, attempt a complete enumeration of the powers of the human understanding. I shall only mention those which I propose to explain, and they are the following :

" 1st, The powers we have by means of our External Senses ; 2dly, Memory ; 3dly, Conception ; 4thly, The powers of Resolving and Analysing complex objects, and compounding those that are more simple ; 5thly, Judging; 6thly, Reasoning; 7thly, Taste; 8thly, Moral Perception ; and, last of all, Consciousness."[a]

The work, however, contains no essay upon Consciousness ; but, in reference to this deficiency, the author, in the last paragraph of the book, states,—" As to Consciousness, what I think necessary to be said upon it has been already said; Essay vi., chap. v."[β] —the chapter, to wit, entitled *On the First Principles of Contingent Truths.* To that chapter you may, however, add what is spoken of consciousness in the first chapter of the first Essay, entitled, *Explication of Words*, § 7.[γ] We are, therefore, left to glean the opinion of both Reid and Stewart on the subject of consciousness, from incidental notices in their writings ; but these are fortunately sufficient to supply us with the necessary information in regard to their opinions on this subject.

Conscious-
ness cannot
be defined.
Nothing has contributed more to spread obscurity over a very transparent matter, than the attempts of

a *Works*, p. 244.—ED.   β *Ib.* p. 508.—ED.   γ *Ib.* p. 222.—ED.

philosophers to define consciousness. Consciousness LECT. XI.
cannot be defined,—we may be ourselves fully aware
what consciousness is, but we cannot, without confusion, convey to others a definition of what we
ourselves clearly apprehend. The reason is plain.
Consciousness lies at the root of all knowledge. Consciousness is itself the one highest source of all com-
prehensibility and illustration,—how, then, can we
find aught else by which consciousness may be illustrated or comprehended ? To accomplish this, it would
be necessary to have a second consciousness, through
which we might be conscious of the mode in which
the first consciousness was possible. Many philosophers,—and among others Dr Brown,—have defined
consciousness a *feeling.*[a] But how do they define a
feeling ? They define, and must define it, as something of which we are conscious ; for a feeling of which
we are not conscious, is no feeling at all. Here, therefore, they are guilty of a logical see-saw, or circle.
They define consciousness by feeling, and feeling by
consciousness,—that is, they explain the same by the
same, and thus leave us in the end no wiser than we
were in the beginning. Other philosophers say that
consciousness is a knowledge,—and others, again, that
it is a belief or conviction of a knowledge. Here, again,
we have the same violation of logical law. Is there
any knowledge of which we are not conscious ? Is
there any belief of which we are not conscious? There
is not,—there cannot be ; therefore, consciousness is
not contained under either knowledge or belief, but, on
the contrary, knowledge and belief are both contained
under consciousness. In short, the notion of consciousness is so elementary, that it cannot possibly be re-

a *Philosophy of the Human Mind,* Lecture xi., p. 67 *et seq.*, ed. 1830.—ED.

solved into others more simple.   It cannot, therefore,
be brought under any genus,—any more general conception ; and, consequently, it cannot be defined.

Conscious-
ness admits
of philoso-
phical an-
alysis.

Butthough consciousness cannot be logically defined,
it may, however, be philosophically analysed.   This
analysis is effected by observing and holding fast the
phænomena or facts of consciousness, comparing these,
and, from this comparison, evolving the universal conditions under which alone an act of consciousness is
possible.

It is only in following this method that we can
attain to precise and accurate knowledge of the contents of consciousness ; and it need not afflict us if the
result of our investigation be very different from the
conclusions that have been previously held.

What kind
of act the
word con-
sciousness
is employed
to denote ;
and what
it involves.

But, before proceeding to show you in detail what
the act of consciousness comprises, it may be proper,
in the first place, to recall to you, in general, what
kind of act the word is employed to denote.   I know,
I feel, I desire, &c.   What is it that is necessarily
involved in all these ?   It requires only to be stated
to be admitted, that when I know, I must know that
I know,—when I feel, I must know that I feel,—when
I desire, I must know that I desire.   The knowledge,
the feeling, the desire, are possible only under the
condition of being known, and being known by me.
For if I did not know that I knew, I would not
know,—if I did not know that I felt, I would not
feel,—if I did not know that I desired, I would not
desire.   Now, this knowledge, which I, the subject,
have of these modifications of my being, and through
which knowledge alone these modifications are possible, is what we call *consciousness*.   The expressions
I know that I know,—I know that I feel,—I know

that I desire,—are thus translated by, *I am conscious that I know,*—*I am conscious that I feel,*—*I am conscious that I desire.* Consciousness is thus, on the one hand, the recognition by the mind or ego of its acts and affections;—in other words, the self-affirmation, that certain modifications are known by me, and that these modifications are mine. But, on the other hand, consciousness is not to be viewed as anything different from these modifications themselves, but is, in fact, the general condition of their existence, or of their existence within the sphere of intelligence. Though the simplest act of mind, consciousness thus expresses a relation subsisting between two terms. These terms are, on the one hand, an I or Self, as the subject of a certain modification,—and, on the other, some modification, state, quality, affection, or operation belonging to the subject. Consciousness, thus, in its simplicity, necessarily involves three things,—1°, A recognising or knowing subject; 2°, A recognised or known modification; and, 3°, A recognition or knowledge by the subject of the modification.

From this it is apparent, that consciousness and knowledge each involve the other.[a] An act of knowledge may be expressed by the formula, *I know*; an act of consciousness by the formula, *I know that I know:* but as it is impossible for us to know without at the same time knowing that we know; so it is impossible to know that we know without our actually knowing. The one merely explicitly expresses what the other implicitly contains. Consciousness and knowledge are thus not opposed as really different. Why, then, it may be asked, employ two terms to express notions, which, as they severally infer each other, are

Consciousness and knowledge involve each other.

[a] See *Reid's Works* (completed edition), p. 933.—ED.

really identical? To this the answer is easy. Realities may be in themselves inseparable, while, as objects of our knowledge, it may be necessary to consider them apart. Notions, likewise, may severally imply each other, and be inseparable even in thought; yet, for the purposes of science, it may be requisite to distinguish them by different terms, and to consider them in their relations or correlations to each other. Take

a geometrical example,—a triangle. This is a whole composed of certain parts. Here the whole cannot be conceived as separate from its parts, and the parts cannot be conceived as separate from their whole. Yet it is scientifically necessary to have different names for each, and it is necessary now to consider the whole in relation to the parts, and now the parts in correlation to the whole. Again, the constituent parts of a triangle are sides and angles. Here the sides suppose the angles, the angles suppose the sides, and, in fact, the sides and angles are in themselves, —in reality, one and indivisible. But they are not the same to us,—to our knowledge. For though we cannot abstract, in thought, the sides from the angle, the angle from the sides, we may make one or other the principal object of attention. We may either consider the angles in relation to each other, and to the sides; or the sides in relation to each other, and to the angles. And to express all this, it is necessary to distinguish, in thought and in expression, what, in nature, is one and indivisible.

As it is in geometry, so it is in the philosophy of mind. We require different words, not only to express objects and relations different in themselves, but to express the same objects and relations under the different points of view in which they are placed by

the mind, when scientifically considering them. Thus,
in the present instance, consciousness and knowledge
are not distinguished by different words as different
things, but only as the same thing considered in
different aspects. The verbal distinction is taken for
the sake of brevity and precision, and its convenience
warrants its establishment. Knowledge is a relation,
and every relation supposes two terms. Thus, in the
relation in question, there is, on the one hand, a subject of knowledge,—that is, the knowing mind,—and
on the other, there is an object of knowledge,—that
is, the thing known; and the knowledge itself is
the relation between these two terms. Now, though
each term of a relation necessarily supposes the other,
nevertheless one of these terms may be to us the
more interesting, and we may consider that term as
the principal, and view the other only as subordinate
and correlative. Now, this is the case in the present
instance. In an act of knowledge, my attention may
be principally attracted either to the object known,
or to myself as the subject knowing; and, in the latter
case, although no new element be added to the act,
the condition involved in it,—*I know that I know*,—
becomes the primary and prominent matter of consideration. And when, as in the philosophy of mind,
the act of knowledge comes to be specially considered
in relation to the knowing subject, it is, at last, in
the progress of the science, found convenient, if not
absolutely necessary, to possess a scientific word in
which this point of view should be permanently and
distinctively embodied. But, as the want of a technical and appropriate expression could be experienced
only after psychological abstraction had acquired a
certain stability and importance, it is evident that

the appropriation of such an expression could not, in any language, be of very early date. And this is shown by the history of the synonymous terms for *consciousness* in the different languages,[a]—a history which, though curious, you will find noticed in no

History of the term consciousness.

publication whatever. The employment of the word *conscientia*, of which our term consciousness is a translation, is, in its psychological signification, not older than the philosophy of Descartes. Previously to him this word was used almost exclusively in the ethical sense expressed by our term *conscience*, and in the striking and apparently appropriate dictum of

Its use by St Augustin.

St Augustin,—" certissima scientia et clamante conscientia,"[β]—which you may find so frequently paraded by the Continental philosophers, when illustrating the certainty of consciousness ; in that quotation, the term is, by its author, applied only in its moral or religious signification. Besides the moral application, the words *conscire* and *conscientia* were frequently employed to denote participation in a common knowledge. Thus the members of a conspiracy were said *conscire*,—and *conscius* is even used for conspirator ; and, metaphorically, this community of knowledge is attributed to inanimate objects,—as, wailing to the rocks, a lover says of himself,—

" Et conscia saxa fatigo." [γ]

I would not, however, be supposed to deny that these words were sometimes used, in ancient Latinity, in the modern sense of consciousness, or being conscious. An unexceptionable example is afforded by

a See the completed edition of *Reid's Works*, Note I, p. 942-945.—ED.

β *De Trinitate*, xiii. 1.—ED.

γ Buchanan, *Silvae*, iii. 17. Compare Virgil, *Æneid*, ix. 429: "Cœlum hoc et conscia sidera testor."—ED.

Quintilian in his *Institutiones*, lib. xii. cap. xi. ; [a] and LECT.
XI.
more than one similar instance may be drawn from
Tertullian,[β] and other of the Latin fathers.

Until Descartes, therefore, the Latin terms *conscire* First used
by Des-
cartes in
present
psycholo-
gical mean-
ing.
and *conscientia* were very rarely usurped in their
present psychological meaning,—a meaning which, it
is needless to add, was not expressed by any term in
the vulgar languages : for, besides Tertullian, I am
aware of only one or two obscure instances in which,
as translations of the Greek terms συναισθάνομαι and
συναίσθησις, of which we are about to speak, the
terms *conscio* and *conscientia* were, as the nearest
equivalents, contorted from their established signification to the sense in which they were afterwards employed by Descartes.   Thus, in the philosophy of the
West, we may safely affirm that, prior to Descartes,
there was no psychological term in recognised use for
what, since his time, is expressed in philosophical
Latinity by *conscientia*, in French by *conscience*, in
English by *consciousness*, in Italian by *conscienza*,
and in German by *Bewusstseyn*.   It will be observed
that in Latin, French, and Italian (and I might add
the Spanish and other Romanic languages), the terms
are analogous ; the moral and psychological meaning
being denoted by the same word.

In Greek there was no term for consciousness until

a "Conscius sum mihi, quantum
mediocritate valui, quæque antea
scierim, quæque operis hujusce gra-
tia potuerim inquirere, candide me
atque simpliciter in notitiam eorum,
si qui forte cognoscere voluissent,
protulisse." This sense, however,
is not unusual. Cf. Cicero, *Tusc.
Quæst.*, ii. 4 : "Mihi sum conscius,
nunquam me nimis cupidum fuisse
vitæ."—ED.

β [*De Testimonio Animæ*, c. 5 :
"Sed qui ejusmodi eruptiones animæ
non putavit doctrinam esse naturæ
et congenitæ et ingenitæ conscientiæ
tacita commissa." *De Carne Christi*,
c. 3 : "Sed satis erat illi, inquis,
conscientia sua." Cf. Augustin, *De
Trinitate*, x. c. 7 : "Et quia sibi
bene conscia est principatus sui quo
corpus regit."]

LECT.
XI.

No term
for con-
sciousness
in Greek
until the
decline of
philoso-
phy.

the decline of philosophy, and in the later ages of the language. Plato and Aristotle, to say nothing of other philosophers, had no special term to express the knowledge which the mind affords of the operations of its faculties, though this, of course, was necessarily a frequent matter of their consideration. Intellect was supposed by them to be cognisant of its own operations; it was only doubted whether by a direct or by a reflex act. In regard to sense, the matter was more perplexed; and, on this point, both philosophers seem to vacillate in their opinions. In his *Theætetus*,[a] Plato accords to sense the power of perceiving that it perceives; whereas, in his *Charmides*,[β] this power he denies to sense, and attributes to intelligence, (νοῦς.) In like manner, an apparently different doctrine may be found in different works of Aristotle. In his *Treatise on the Soul* he thus cogently argues :—" When we perceive that we see, hear, &c., it is necessary that by sight itself we perceive that we see, or by another sense. If by another sense, then this also must be a sense of sight, conversant equally about the object of sight, colour. Consequently there must either be two senses of the same object, or every sense must be percipient of itself. Moreover, if the sense percipient of sight be different from sight itself, it follows either that there is a regress to infinity, or we must admit at last some sense percipient of itself; but if so, it is more reasonable to admit this in the original sense at once."[γ] Here a conscious-

a " Accedit testimonium Platonis in Theæteto, ubi ait sensum sentire quod sentit et quod non sentit."— Conimbricenses, *In Arist. De Anim.*, iii. 2. The passage referred to is probably *Theæt.*, p. 192 : 'Αδύνατον . . . ὃ αἰσθάνεταί γε, ἕτερόν τι ἂν αἰσθάνεται οἰηθῆναι εἶναι, καὶ ὃ αἰσθάνεται, ἄν τι

μὴ αἰσθάνεται. This passage, however, is not exactly in point.—Ed.

β P. 167 *et seq.* Cf. Conimbricenses, *l. c.* Plato, however, merely denies that there can be a sense which perceives the act of sensation without perceiving its object.—Ed.

γ *De Anima*, iii. 2.—Ed.

ness is apparently attributed to each several sense. This, however, is expressly denied in his work *On Sleep and Waking,*[a] to say nothing of his *Problems,* which, I am inclined, however, to think, are not genuine. It is there stated that sight does not see that it sees, neither can sight or taste judge that sweet is a quality different from white; but that this is the function of some common faculty, in which they both converge. The apparent repugnance may, however, easily be reconciled. But—what concerns us at present, in all these discussions by the two philosophers—there is no single term employed to denote that special aspect of the phænomenon of knowledge, which is thus by them made matter of consideration. It is only under the later Platonists and Aristotelians that peculiar terms, tantamount to our consciousness, were adopted into the language of philosophy. In the text of Diogenes Laertius, indeed, (vii. 85), I find συνείδησις manifestly employed in the sense of *consciousness.* This, however, is a corrupt reading; and the authority of the best manuscripts and of the best critics shows that σύνδεσις is the true lection.[β] The Greek Platonists and Aristotelians, in general, did not allow that the recognition that we know, that we feel, that we desire, &c., was the act of any special faculty, but the general attribute of intellect; and the power of reflecting, of turning back upon itself, was justly viewed as the distinctive quality of intelligence. It

---

a *De Somno,* c. 2, § 4. The passage in the *Problems,* which may perhaps have the same meaning, though it admits of a different interpretation, is sect. xi. § 33 : Χωρισθεῖσα δὲ αἴσθησις διανοίας καθάπερ ἀναίσθητον πόνον ἔχει, ὥσπερ εἴρηται τὸ, Νοῦς ὁρᾷ, καὶ νοῦς ἀκούει. See further, *Discussions,* p. 51.—ED.

β The correction σύνδεσις is made by Menage on the authority of Suidas, v. ὁρμή. Kuster, on the other hand, proposes, on the authority of Laertius, to read συνείδησις for σύνδεσις in Suidas.—ED.

was, however, necessary to possess some single term expressive of this intellectual retortion,—of this ἐπιστροφὴ πρὸς ἑαυτόν, and the term συναίσθησις was adopted. This I find employed particularly by Proclus, Plotinus, and Simplicius.[a] The term συνείδησις, the one equivalent to the *conscientia* of the Latins, remained like *conscientia* itself, long exclusively applied to denote conscience or the moral faculty ; and it is only in Greek writers who, as Eugenius of Bulgaria, have flourished since the time of Descartes and Leibnitz, that συνείδησις has, like the *conscientia* of the Latins, been employed in the psychological meaning of consciousness.[β] I may notice that the word συνεπίγνωσις, in the sense of consciousness, is also to be occasionally met with in the later authors on philosophy in the Greek tongue. The expression συναίσθησις, which properly denotes the self-recognition of sense and feeling, was, however, extended to mark consciousness in general. Some of the Aristotelians, however, like certain philosophers in this country, attributed this recognition to a special faculty. Of these I have been able to discover only three : Philoponus, in his commentary on Aristotle's treatise *Of the Soul* ;[γ]

[a Plotinus, *Enn.*, v. lib. iii. c. 2. Proclus, *Inst. Theol.*, c. 39. Simplicius, *In Epict. Enchir.*, p. 28, Heins. —(p. 49, Schweigh.)] In the two first of these passages, συναίσθησις appears to be used merely in its etymological sense of perception of an object in conjunction with other objects. In the last, however, it seems to be fully equivalent to the modern *consciousness;* as also in Hierocles, *In Aurea Pyth. Carm.*, 41, p. 213, ed. 1654. Sextus Empiricus, *Adv. Math.*, ix. 68 (p. 407, Bekker). Michael Ephesius, *In Arist. de Memoria*, p. 134. Plutarch, *De Profectibus in Virtute*, c. 1, 3. Plotinus, *Enn.*, iii. lib. 4, c. 4. Simplicius, *In Arist. Categ.*, p. 83, *b.* ed. 1551.—Ed.

β See the *Logic* of Eugenius, p. 113. He also uses συνεπίγνωσις in the same sense. The title of his work is, Ἡ λογικὴ ἐκ παλαιῶν τε καὶ νεωτέρων συνερανισθεῖσα· ὑπὸ Εὐγενίου διακόνου τοῦ Βουλγαρέως· ἐν Λειψίᾳ τῆς Σαξονίας. Ἔτει ͵αψξϛ. (1766.)— Ed.

γ On lib. iii. c. 2. He mentions this as the opinion of the more recent interpreters. See *Reid's Works*, p. 942 (completed edition), where the passage in question is translated by Sir W. Hamilton.—Ed.

Michael Ephesius, in his commentary on Aristotle's LECT.
treatise of *Memory and Reminiscence;* [a] and Michael ———
Psellus, in his work on *Various Knowledge.* [β] It is
doubted, however, whether the two last be not the
same person ; and their remarkable coincidence in the
point under consideration, is even a strong argument
for their identity.   They assign this recognition to a
faculty which they call τὸ προσεκτικόν,—that is τὸ
προσεκτικὸν μέρος, the attentive part or function of
mind.   This is the first indication in the history of
philosophy of that false analysis which has raised attention into a separate faculty.   I beg you, however,
to observe, that Philoponus and his follower, Michael
Ephesius, do not distinguish attention from consciousness.   This is a point we are hereafter especially to
consider, when perhaps it may be found that, though
wrong in making consciousness or attention a peculiar
faculty, they were right, at least, in not dividing consciousness and attention into different faculties.

But to return from our historical digression.   We The most
general
may lay it down as the most general characteristic of characteristic of
consciousness, that it is the recognition by the thinking consciousness.
subject of its own acts or affections.

So far there is no difficulty and no dispute.   In this The special
conditions
all philosophers are agreed.   The more arduous task of consciousness.
remains of determining the special conditions of consciousness. [γ]   Of these, likewise, some are almost too
palpable to admit of controversy.   Before proceeding
to those in regard to which there is any doubt or diffi-

---

a Rather in the Commentary on
the *Nicomachean Ethics,* usually attributed to Eustratius, p. 160, *b.* It
is not mentioned in the Commentary
on the *De Memoria.*—ED.

β [Psellus, *De Omnifaria Doctrina,*
§ 46 :] Προσοχὴ δέ ἐστι καθ' ἣν προσ-

ἔχομεν τοῖς ἔργοις οἷς πράττομεν καὶ
τοῖς λόγοις οἷς λέγομεν.—ED.

γ On the conditions and limitations
of consciousness, see *Reid's Works,*
(completed edition), p. 932 *et seq.*—
ED.

LECT.
XI.
___

1. Those
generally
admitted.

Conscious-
ness im-
plies, 1,
actual
knowledge.

culty, it will be proper, in the first place, to state and dispose of such determinations as are too palpable to be called in question. Of these admitted limitations, the first is, that consciousness is an actual and not a potential knowledge.[a] Thus a man is said to know,— *i.e.* is able to know, that $7 + 9$ are $= 16$, though that equation be not, at the moment, the object of his thought; but we cannot say that he is conscious of this truth unless while actually present to his mind.

The second limitation is, that consciousness is an immediate, not a mediate knowledge. We are said, for example, to know a past occurrence when we represent it to the mind in an act of memory. We know the mental representation, and this we do immediately and in itself, and are also said to know the past occurrence, as mediately knowing it through the mental modification which represents it. Now, we are conscious of the representation as immediately known, but we cannot be said to be conscious of the thing represented, which, if known, is only known through its representation. If, therefore, mediate knowledge be in propriety a knowledge, consciousness is not coextensive with knowledge. This is, however, a problem we are hereafter specially to consider. I may here also observe, that, while all philosophers agree in making consciousness an immediate knowledge, some, as Reid and Stewart, do not admit that all immediate knowledge is consciousness. They hold that we have an immediate knowledge of external objects, but they hold that these objects are beyond the sphere of consciousness.[β] This is an opinion we are, likewise, soon to canvass.

a Compare *Reid's Works,* p. 810. —Ed.

β See Reid, *Intellectual Powers,* Essay vi. ch. 5, §§ 1, 5; *Works,* pp.

The third condition of consciousness, which may be held as universally admitted, is, that it supposes a contrast,—a discrimination ; for we can be couscious only inasmuch as we are conscious of something ; and we are conscious of something only inasmuch as we are conscious of what that something is,—that is, distinguish it from what it is not. This discrimination is of different kinds and degrees.

In the first place, there is the contrast between the two grand opposites, self and not-self,—ego and non-ego,—mind and matter ; (the contrast of subject and object is more general.) We are conscious of self only in and by its contradistinction from not-self ; and are conscious of not-self only in and by its contradistinction from self. In the second place, there is the discrimination of the states or modifications of the internal subject of self from each other. We are conscious of one mental state only as we contradistinguish it from another ; where two, three, or more such states are confounded, we are conscious of them as one ; and were we to note no difference in our mental modifications, we might be said to be absolutely unconscious.[a] Hobbes has truly said, " Idem semper sentire, et non sentire, ad idem reciduut."[β] In the third place, there is the distinction between the parts and qualities of the outer world. We are conscious of an external object only as we are conscious of it as distinct from others ; where several distinguishable objects are confounded, we are conscious of them as one ; where no object is discriminated, we are not conscious of any.

Marginal notes: LECT. XI. 3. Contrast,—discrimination of one object from another. This discrimination of various kinds and degrees.

442, 445. Stewart, *Outlines of Moral Philosophy*, part i. §§ 1, 2 ; *Collected Works*, vol. ii. p. 12.—ED.

a [Cf. Aristotle, *Phys. Auscult.*, lib. iv. c. 16, § 1, (ed. Pacii).]

β *Elementa Philosophiæ*, part iv. c. 25, § 5. *Opera*, ed. Molesworth, vol. i. p. 321. *English Works*, vol. i. p. 394.—ED.

LECT.
XI.

Before leaving this condition, I may parenthetically
state, that, while all philosophers admit that consciousness involves a discrimination, many do not allow it
any cognisance of aught beyond the sphere of self. The
great majority of philosophers do this because they
absolutely deny the possibility of an immediate knowledge of external things, and, consequently, hold that
consciousness, in distinguishing the non-ego from the
ego, only distinguishes self from self; for they maintain, that what we are conscious of as something different from the perceiving mind, is only, in reality, a
modification of that mind, which we are condemned to
mistake for the material reality. Some philosophers,
however, (as Reid and Stewart), who hold, with mankind at large, that we do possess an immediate knowledge of something different from the knowing self,
still limit consciousness to a cognisance of self; and,
consequently, not only deprive it of the power of distinguishing external objects from each other, but even
of the power of discriminating the ego and non-ego.
These opinions we are afterwards to consider. With
this qualification, all philosophers may be viewed as
admitting that discrimination is an essential condition
of consciousness.

4. Judg-
ment.

The fourth condition of consciousness, which may
be assumed as very generally acknowledged, is, that
it involves judgment. A judgment is the mental
act by which one thing is affirmed or denied of another. This fourth condition is in truth only a
necessary consequence of the third,—for it is impossible to discriminate without judging,—discrimination,
or contradistinction, being in fact only the denying
one thing of another. It may to some seem strange
that consciousness, the simple and primary act of in-

telligence, should be a judgment, which philosophers, in general, have viewed as a compound and derivative operation. This is, however, altogether a mistake. A judgment is, as I shall hereafter show you, a simple act of mind, for every act of mind implies a judgment. Do we perceive or imagine without affirming, in the act, the external or internal existence of the object ?[a] Now these fundamental affirmations are the affirmations,—in other words, the judgments,—of consciousness.

The fifth undeniable condition of consciousness is memory. This condition also is a corollary of the third. For without memory our mental states could not be held fast, compared, distinguished from each other, and referred to self. Without memory, each indivisible, each infinitesimal, moment in the mental succession, would stand isolated from every other,— would constitute, in fact, a separate existence. The notion of the ego or self, arises from the recognised permanence and identity of the thinking subject in contrast to the recognised succession and variety of its modifications. But this recognition is possible only through memory. The notion of self is, therefore, the result of memory. But the notion of self is involved in consciousness, so consequently is memory.

5. Memory.

a See *Reid's Works* (completed with the Editor's Notes.—ED. edition), pp. 243, 414, 878, 933-4,

LECTURE XII.

CONSCIOUSNESS,—ITS SPECIAL CONDITIONS : RELATION
TO COGNITIVE FACULTIES IN GENERAL.

LECT.
XII.
—————
Recapitu-
lation.

So far as we have proceeded, our determination of the contents of consciousness may be viewed as that universally admitted; for though I could quote to you certain counter-doctrines, these are not of such importance as to warrant me in perplexing the discussion by their refutation, which would indeed be nothing more than the exposition of very palpable mistakes. Let us, therefore, sum up the points we have established. We have shown, in general, that consciousness is the self-recognition that we know, or feel, or desire, &c. We have shown, in particular, 1°, That consciousness is an actual or living, and not a potential or dormant, knowledge ;—2°, That it is an immediate and not a mediate knowledge ;—3°, That it supposes a discrimination ;—4°, That it involves a judgment ;—and, 5°, That it is possible only through memory.

II. Special
conditions
of con-
sciousness
not gener-
ally ad-
mitted.

We are now about to enter on a more disputed territory; and the first thesis I shall attempt to establish, involves several subordinate questions.

1. Our con-
sciousness
coextensive
with our
knowledge.

I state, then, as the first contested position which I am to maintain, that our consciousness is coextensive with our knowledge. But this assertion, that we have no knowledge of which we are not conscious, is tan-

tamount to the other, that consciousness is coextensive with our cognitive faculties,—and this again is
convertible with the assertion, that consciousness is
not a special faculty, but that our special faculties of
knowledge are only modifications of consciousness.[a]
The question, therefore, may be thus stated,—Is consciousness the genus under which our several faculties of knowledge are contained as species,—or, is
consciousness itself a special faculty co-ordinate with,
and not comprehending, these?

Before proceeding to canvass the reasonings of those
who have reduced consciousness from the general
condition, to a particular variety, of knowledge, I may
notice the error of Dr Brown, in asserting that, "in
the systems of philosophy which have been most generally prevalent, especially in this part of the island,
consciousness has always been classed as one of the
intellectual powers of the mind, differing from its
other powers, as these mutually differ from each
other."[β] This statement, in so far as it regards the
opinion of philosophers in general, is not only not true,
but the very reverse of truth. For, in place of consciousness being, "in the systems most generally prevalent," classed as a special faculty, it has, in all the
greater schools of philosophy, been viewed as the universal attribute of the intellectual arts. Was consciousness degraded to a special faculty in the Platonic,
in the Aristotelian, in the Cartesian, in the Lockian,
in the Leibnitian, in the Kantian philosophies? These
are the systems which have obtained a more general
authority than any others, and yet in none of these is
the supremacy of consciousness denied ; in all of them

a Compare *Reid's Works* (completed
edition), p. 929-30.—ED.

β *Philosophy of the Human Mind*,
Lecture xi., p. 67, ed. 1830.—ED.

it is either expressly or implicitly recognised. Dr
Brown's assertion is so far true in relation to this
country, that by Hutcheson,[a] Reid, and Stewart,—to
say nothing of inferior names,—consciousness has been
considered as nothing higher than a special faculty.
As I regard this opinion to be erroneous, and as the
error is one affecting the very cardinal point of philosophy,—as it stands opposed to the peculiar and
most important principles of the philosophy of Reid
and Stewart themselves, and has even contributed to
throw around their doctrine of perception an obscurity that has caused Dr Brown actually to mistake it
for its converse, and as I have never met with any
competent refutation of the grounds on which it rests,
—I shall endeavour to show you that, notwithstanding
the high authority of its supporters, this opinion is
altogether untenable.

As I previously stated to you, neither Dr Reid
nor Mr Stewart has given us any regular account
of consciousness; their doctrine on this subject is
to be found scattered in different parts of their
works. The two following brief passages of Reid
contain the principal positions of that doctrine.
The first is from the first chapter of the first
Essay *On the Intellectual Powers :*[β]—" Consciousness
is a word used by philosophers to signify that immediate knowledge which we have of our present
thoughts and purposes, and, in general, of all the present operations of our minds. Whence we may observe that consciousness is only of things present.
To apply consciousness to things past, which sometimes is done in popular discourse, is to confound

a See *Reid's Works* (completed
edition), p. 930.—Ed.   β *Works*, p. 222.

consciousness with memory; and all such confusion of words ought to be avoided in philosophical discourse. It is likewise to be observed, that consciousness is only of things in the mind, and not of external things. It is improper to say, I am conscious of the table which is before me. I perceive it, I see it; but do not say I am conscious of it. As that consciousness by which we have a knowledge of the operations of our own minds, is a different power from that by which we perceive external objects, and as these different powers have different names in our language, and, I believe, in all languages, a philosopher ought carefully to preserve this distinction, and never to confound things so different in their nature." The second is from the fifth chapter of the sixth Essay *On the Intellectual Powers:*[a]—"Consciousness is an operation of the understanding of its own kind, and cannot be logically defined. The objects of it are our present pains, our pleasures, our hopes, our fears, our desires, our doubts, our thoughts of every kind; in a word, all the passions and all the actions and operations of our own minds, while they are present. We may remember them when they are past; but we are conscious of them only while they are present." Besides what is thus said in general of consciousness, in his treatment of the different special faculties Reid contrasts consciousness with each. Thus in his essays on Perception, on Conception or Imagination, and on Memory, he specially contradistinguishes consciousness from each of these operations;[β] and it is also incidentally by Reid,[γ] but more articulately by

a *Works,* p. 442.
β See *Intellectual Powers,* Essay i., *Works,* p. 222, and Essay ii., *Works,* p. 297; Essay iii., *Works,* pp. 340, 351; Essay iv., *Works,* p. 368.—Ed.
γ See *Works,* p. 239. Compare pp. 240, 258, 347, 419-20, 443.—Ed.

Stewart,[a] discriminated from Attention and Reflection.

Conscious-
ness a spe-
cial faculty,
according
to Reid and
Stewart.

According to the doctrine of these philosophers, consciousness is thus a special faculty,[β] co-ordinate with the other intellectual powers, having like them a particular operation and a peculiar object. And what is the peculiar object which is proposed to consciousness?[γ] The peculiar objects of consciousness, says Dr Reid, are all the present passions and operations of our minds. Consciousness thus has for its objects, among the other modifications of the mind, the acts of our cognitive faculties. Now here a doubt arises. If consciousness has for its object the cognitive operations, it must know these operations, and, as it knows these operations, it must know their objects : consequently, consciousness is either not a special faculty, but a faculty comprehending every cognitive act ; or it must be held that there is a double knowledge of every object,—first, the knowledge of that object by its particular faculty, and second, a knowledge of it by consciousness as taking cognisance of every mental operation. But the former of these alternatives is a surrender of consciousness as a co-ordinate and special faculty, and the latter is a supposition not only unphilosophical but absurd. Now, you will attend to the mode in which Reid escapes, or endeavours to escape, from this dilemma. This he does by assigning to consciousness, as its object, the various intellectual operations to the exclusion of their several objects. " I am conscious," he says, " of perception, but not of the object I perceive ; I am conscious of memory, but

a *Coll. Works*, vol. ii. p. 134, and pp. 122, 123.—Ed.

β On Reid's reduction of consciousness to a special faculty, compare the Author's edition of his *Works*,

Note H, p. 929 *et seq.*, completed edition.—Ed.

γ See the same argument in the Author's *Discussions*, p. 47.—Ed.

not of the object I remember." By this limitation, if
tenable, he certainly escapes the dilemma, for he would
thus disprove the truth of the principle on which it
proceeds—viz., that to be conscious of the operation
of a faculty is, in fact, to be conscious of the object
of that operation. The whole question, therefore, turns
upon the proof or disproof of this principle,—for if it
can be shown that the knowledge of an operation necessarily involves the knowledge of its object, it follows
that it is impossible to make consciousness conversant
about the intellectual operations to the exclusion of
their objects. And that this principle must be admitted, is what, I hope, it will require but little argument
to demonstrate.

Some things can be conceived by the mind each
separate and alone; others only in connection with
something else. The former are said to be things
absolute; the latter, to be things relative. Socrates,
and Xanthippe, may be given as examples of the former; husband and wife, of the latter. Socrates, and
Xanthippe, can each be represented to the mind without the other; and if they are associated in thought,
it is only by an accidental connection. Husband and
wife, on the contrary, cannot be conceived apart. As
relative and correlative, the conception of husband
involves the conception of wife, and the conception
of wife involves the conception of husband. Each is
thought only in and through the other, and it is impossible to think of Socrates as the husband of Xanthippe, without thinking of Xanthippe as the wife of
Socrates. We cannot, therefore, know what a husband
is without also knowing what is a wife, as, on the other
hand, we cannot know what a wife is without also
knowing what is a husband. You will, therefore, understand from this example the meaning of the logical

axiom, that the knowledge of relatives is one,—or that the knowledge of relatives is the same.

This being premised, it is evident that if our intellectual operations exist only in relation, it must be impossible that consciousness can take cognisance of one term of this relation without also taking cognisance of the other. Knowledge, in general, is a relation between a subject knowing and an object known, and each operation of our cognitive faculties only exists by relation to a particular object,—this object at once calling it into existence, and specifying the quality of its existence. It is, therefore, palpably impossible that we can be conscious of an act without being conscious of the object to which that act is relative. This, however, is what Dr Reid and Mr Stewart maintain. They maintain that I can know *that* I know, without knowing *what* I know,—or that I can know the knowledge without knowing what the knowledge is about; for example, that I am conscious of perceiving a book without being conscious of the book perceived,—that I am conscious of remembering its contents without being conscious of these contents remembered,—and

Shown in detail with respect to the different cognitive faculties.

so forth. The unsoundness of this opinion must, however, be articulately shown by taking the different faculties in detail, which they have contradistinguished from consciousness, and by showing, in regard to each, that it is altogether impossible to propose the operation of that faculty to the consideration of consciousness, and to withhold from consciousness its object.

Imagination.

I shall commence with the faculty of imagination, to which Dr Reid and Mr Stewart have chosen, under various limitations, to give the name of Conception.[a]

a Reid, *Intellectual Powers*, Essay Elements, vol. i. ch. 3 ; *Works*, vol. iv. ch. 1 ; *Works*, p. 360. Stewart, ii. p. 145.—ED.

This faculty is peculiarly suited to evince the error of holding that consciousness is cognisant of acts, but not of the objects of these acts.

"Conceiving, Imagining, and Apprehending," says Dr Reid, "are commonly used as synonymous in our language, and signify the same thing which the logicians call Simple Apprehension. This is an operation of the mind different from all those we have mentioned [Perception, Memory, &c.] Whatever we perceive, whatever we remember, whatever we are conscious of, we have a full persuasion or conviction of its existence. What never had an existence cannot be remembered ; what has no existence at present cannot be the object of perception or of consciousness ; but what never had, nor has any existence, may be conceived. Every man knows that it is as easy to conceive a winged horse or a centaur, as it is to conceive a horse or a man. Let it be observed, therefore, that to conceive, to imagine, to apprehend, when taken in the proper sense, signify an act of the mind which implies no belief or judgment at all. It is an act of the mind by which nothing is affirmed or denied, and which therefore can neither be true nor false."[a] And again : "Consciousness is employed solely about objects that do exist, or have existed. But conception is often employed about objects that neither do, nor did, nor will, exist. This is the very nature of this faculty, that its object, though distinctly conceived, may have no existence. Such an object we call a creature of imagination, but this creature never was created.

"That we may not impose upon ourselves in this matter, we must distinguish between that act or operation of the mind, which we call conceiving an

a *Works*, p. 223.

object, and the object which we conceive. When we conceive anything, there is a real act or operation of the mind; of this we are conscious, and can have no doubt of its existence. But every such act must have an object; for he that conceives must conceive something. Suppose he conceives a centaur, he may have a distinct conception of this object, though no centaur ever existed."[a] And again : " I conceive a centaur. This conception is an operation of the mind of which I am conscious, and to which I can attend. The sole object of it is a centaur, an animal which, I believe, never existed."[β]

Now, here it is admitted by Reid, that imagination has an object, and in the example adduced, that this object has no existence out of the mind. The object of imagination is, therefore, in the mind,—is a modification of the mind. Now, can it be maintained that there can be a modification of mind,—a modification of which we are aware, but of which we are not conscious ? But let us regard the matter in another aspect. We are conscious, says Dr Reid, of the imagination of a centaur, but not of the centaur imagined. Now, nothing can be more evident than that the object and the act of imagination are identical. Thus, in the example alleged, the centaur imagined and the act of imagining it, are one and indivisible. What is the act of imagining a centaur but the centaur imaged, or the image of the centaur ? what is the image of the centaur but the act of imagining it ? The centaur is both the object and the act of imagination : it is the same thing viewed in different relations. It is called the object of imagination, when considered as representing a possible existence ; for everything that can

a Works, p. 368.　　　　　β Works, p. 373.

be construed to the mind,—everything that does not   LECT.
violate the laws of thought,—in other words, every-   XII.
thing that does not involve a contradiction, may be
conceived by the mind as possible.   I say, therefore,
that the centaur is called the object of imagination,
when considered as representing a possible existence ;
whereas the centaur is called the act of imagination,
when considered as the creation, work, or operation, of
the mind itself.   The centaur imagined and the imagination of the centaur, are thus as much the same
indivisible modification of mind as a square is the
same figure, whether we consider it as composed of
four sides, or as composed of four angles,—or as pater-
nity is the same relation whether we look from the
son to the father, or from the father to the son.   We
cannot, therefore, be conscious of imagining an object
without being conscious of the object imagined, and,
as regards imagination, Reid's limitation of consciousness is, therefore, futile.

   I proceed next to Memory :—" It is by Memory,"   Memory.
says Dr Reid, " that we have an immediate knowledge
of things past.   The senses give us information of
things only as they exist in the present moment ; and
this information, if it were not preserved by memory,
would vanish instantly, and leave us as ignorant as if
it had never been.   Memory must have an object.
Every man who remembers must remember something, and that which he remembers is called the
object of his remembrance.   In this, memory agrees
with perception, but differs from sensation, which has
no object but the feeling itself.   Every man can distinguish the thing remembered from the remembrance
of it.   We may remember anything which we have
seen, or heard, or known, or done, or suffered ; but the

remembrance of it is a particular act of the mind which now exists, and of which we are conscious. To confound these two is an absurdity which a thinking man could not be led into, but by some false hypothesis which hinders him from reflecting upon the thing which he would explain by it."[a] "The object of memory, or thing remembered, must be something that is past ; as the object of perception and of consciousness, must be something which is present. What now is, cannot be an object of memory ; neither can that which is past and gone be an object of perception, or of consciousness."[b] To these passages, which are taken from the first chapter of the third Essay *On the Intellectual Powers*, I must add another from the sixth chapter of the same Essay,—the chapter in which he criticises Locke's doctrine in regard to our Personal Identity. " Leaving," he says, " the consequences of this doctrine to those who have leisure to trace them, we may observe, with regard to the doctrine itself, first, that Mr Locke attributes to consciousness the conviction we have of our past actions, as if a man may now be conscious of what he did twenty years ago. It is impossible to understand the meaning of this, unless by consciousness be meant memory, the only faculty by which we have an immediate knowledge of our past actions. Sometimes, in popular discourse, a man says he is conscious that he did such a thing, meaning that he distinctly remembers that he did it. It is unnecessary, in common discourse, to fix accurately the limits between consciousness and memory. This was formerly shown to be the case with regard to sense and memory. And, therefore, distinct remembrance is sometimes called

a *Works*, p. 339.   β *Works*, p. 340.

sense, sometimes consciousness, without any inconvenience. But this ought to be avoided in philosophy, otherwise we confound the different powers of the mind, and ascribe to one what really belongs to another. If a man be conscious of what he did twenty years or twenty minutes ago, there is no use for memory, nor ought we to allow that there is any such faculty. The faculties of consciousness and memory are chiefly distinguished by this, that the first is an immediate knowledge of the present, the second an immediate knowledge of the past." [a]

From these quotations it appears that Reid distinguishes memory from consciousness in this,—that memory is an immediate knowledge of the past, consciousness an immediate knowledge of the present. We may, therefore, be conscious of the act of memory as present, but of the object of memory as past, consciousness is impossible. Now, if memory and consciousness be, as Reid asserts, the one an immediate knowledge of the past, the other an immediate knowledge of the present, it is evident that memory is a faculty whose object lies beyond the sphere of consciousness ; and, consequently, that consciousness cannot be regarded as the general condition of every intellectual act. We have only, therefore, to examine whether this attribution of repugnant qualities to consciousness and memory be correct,—whether there be not assigned to one or other a function which does not really belong to it.

Now, in regard to what Dr Reid says of consciousness, I admit that no exception can be taken. Consiousness is an immediate knowledge of the present. We have, indeed, already shown that consciousness is

a *Works,* p. 351.

LECT.
XII.

an immediate knowledge, and, therefore, only of the actual or now-existent. This being admitted, and professing, as we do, to prove that consciousness is the one generic faculty of knowledge, we, consequently, must maintain that all knowledge is immediate, and only of the actual or present,—in other words, that what is called mediate knowledge, knowledge of the past, knowledge of the absent, knowledge of the non-actual or possible, is either no knowledge at all, or only a knowledge contained in, and evolved out of, an immediate knowledge of what is now existent and actually present to the mind. This, at first sight, may appear like paradox; I trust you will soon admit that the counter doctrine is self-repugnant.

Memory not an immediate knowledge of the past.

I proceed, therefore, to show that Dr Reid's assertion of memory being an immediate knowledge of the past, is not only false, but that it involves a contradiction in terms. [a]

Conditions of immediate knowledge.

Let us first determine what immediate knowledge is, and then see whether the knowledge we have of the past, through memory, can come under the conditions of immediate knowledge. Now nothing can be more evident than the following positions: 1°, An object to be known immediately must be known in itself,—that is, in those modifications, qualities, or phænomena, through which it manifests its existence, and not in those of something different from itself; for, if we suppose it known not in itself, but in some other thing, then this other thing is what is immediately known, and the object known through it is only an object mediately known.

But, 2°, If a thing can be immediately known only if known in itself, it is manifest, that it can only be known in itself, if it be itself actually in existence, and actually in immediate relation to our faculties of knowledge.

Such are the necessary conditions of immediate knowledge; and they disprove at once Dr Reid's assertion, that memory is an immediate knowledge of the past. An immediate knowledge is only conceivable of the now existent, as the now existent alone can be known in itself. But the past is only past, inasmuch as it is not now existent; and as it is not now existent, it cannot be known in itself. The immediate knowledge of the past is, therefore, impossible.

We have, hitherto, been considering the conditions of immediate knowledge in relation to the object; let us now consider them in relation to the cognitive act. Every act, and consequently every act of knowledge, exists only as it now exists; and as it exists only in the *now*, it can be cognisant only of a now-existent object. Memory is an act,—an act of knowledge; it can, therefore, be cognisant only of a now-existent object. But the object known in memory is, *ex hypothesi*, past; consequently, we are reduced to the dilemma, either of refusing a past object to be known in memory at all, or of admitting it to be only mediately known, in and through a present object. That the latter alternative is the true one, it will require a very few explanatory words to convince you. What are the contents of an act of memory? An act of memory is merely a present state of mind, which we are conscious of not as absolute, but as relative to, and representing, another state of mind, and accompanied with the belief that the state of mind, as now represented, has actually been. I remember an event I saw,—tho

Application of these conditions to the knowledge we have in Memory.

landing of George IV. at Leith. This remembrance is only a consciousness of certain imaginations, involving the conviction that these imaginations now represent ideally what I formerly really experienced. All that is immediately known in the act of memory, is the present mental modification ; that is, the representation and concomitant belief. Beyond this mental modification, we know nothing ; and this mental modification is not only known to consciousness, but only exists in and by consciousness. Of any past object, real or ideal, the mind knows and can know nothing, for, *ex hypothesi*, no such object now exists ; or if it be said to know such an object, it can only be said to know it mediately, as represented in the present mental modification. Properly speaking, however, we know only the actual and present, and all real knowledge is an immediate knowledge. What is said to be mediately known, is, in truth, not known to be, but only believed to be ; for its existence is only an inference resting on the belief, that the mental modification truly represents what is in itself beyond the sphere of knowledge. What is immediately known must be ; for what is immediately known is supposed to be known as existing. The denial of the existence,—and of the existence within the sphere of consciousness,—involves, therefore, a denial of the immediate knowledge of an object. We may, accordingly, doubt the reality of any object of mediate knowledge, without denying the reality of the immediate knowledge on which the mediate knowledge rests. In memory, for instance, we cannot deny the existence of the present representation and belief, for their existence is the consciousness of their existence itself. To doubt their existence, therefore, is, for us, to doubt the existence of our consciousness. But as this doubt itself exists only through consciousness, it would, consequently, annihilate itself. But, though in memory we must admit the reality of the representation and belief, as facts of consciousness, we may doubt, we may deny, that the representation and belief are true. We may assert that they represent what never was, and that all beyond their present mental existence is a delusion. This, however, could not be the case if our knowledge of the past were immediate. So far, therefore, is memory from being an immediate knowledge of the past, that it is at best only a mediate knowledge of the past; while, in philosophical propriety, it is not a knowledge of the past at all, but a knowledge of the present and a belief of the past. But in whatever terms we may choose to designate the contents of memory, it is manifest that these contents are all within the sphere of consciousness.[a]

[a] What I have said in regard to Dr Reid's doctrine of memory as an immediate knowledge of the past, applies equally to his doctrine of conception or imagination, as an immediate knowledge of the distant, — a case which I deferred noticing, when I considered his contradistinction of that faculty from consciousness. " I can conceive," he says, " an individual object that really exists, such as St Paul's Church in London. I have an idea of it; that is, I conceive it. The immediate object of this conception is four hundred miles distant; and I have no reason to think that it acts upon me, or that I act upon it; but I can think of it notwithstanding." This requires no comment. I shall, subsequently, have occasion to show how Reid confused himself about the term object, — this being part and parcel of his grand error in confounding representative or mediate, and intuitive or immediate knowledge.

LECTURE XIII.

CONSCIOUSNESS,—ITS SPECIAL CONDITIONS: RELATION
TO COGNITIVE FACULTIES IN GENERAL.

LECT.
XIII.
___

Our con-
sciousness
coextensive
with our
knowledge.

Reid con-
tradistin-
guishes
conscious-
ness from
perception.

WE now proceed to consider the third faculty which
Dr Reid specially contradistinguishes from Consciousness,—I mean Perception, or that faculty through
which we obtain a knowledge of the external world.
Now, you will observe that Reid maintains against
the immense majority of all, and the entire multitude
of modern, philosophers, that we have a direct and
immediate knowledge of the external world. He thus
vindicates to mind not only an immediate knowledge
of its own modifications, but also an immediate knowledge of what is essentially different from mind or
self,—the modifications of matter. He did not, however, allow that these were known by any common
faculty, but held that the qualities of mind were
exclusively made known to us by Consciousness, the
qualities of matter exclusively made known to us by
Perception. Consciousness was, thus, the faculty of
immediate knowledge, purely subjective; perception,
the faculty of immediate knowledge, purely objective.
The Ego was known by one faculty, the Non-Ego by
another. "Consciousness," says Dr Reid, " is only of
things in the mind, and not of external things. It is
improper to say, I am conscious of the table which is
before me. I perceive it, I see it, but do not say I

am conscious of it. As that consciousness by which we have a knowledge of the operations of our own minds, is a different power from that by which we perceive external objects, and as these different powers have different names in our language, and, I believe, in all languages, a philosopher ought carefully to preserve this distinction, and never to confound things so different in their nature." [a] And in another place he observes :—" Consciousness always goes along with perception ; but they are different operations of the mind, and they have their different objects. Consciousness is not perception, nor is the object of consciousness the object of perception." [β]

Dr Reid has many merits as a speculator, but the only merit which he arrogates to himself,—the principal merit accorded to him by others, is, that he was the first philosopher, in more recent times, who dared, in his doctrine of immediate perception, to vindicate, against the unanimous authority of philosophers, the universal conviction of mankind. But this doctrine he has at best imperfectly developed, and, at the same time, has unfortunately obscured it, by errors of so singular a character that some acute philosophers,—for Dr Brown does not stand alone,—have never even suspected what his doctrine of perception actually is. One of these errors is the contradistinction of perception from consciousness.

Principal merit accorded to Reid as a philosopher.

I may here notice, by anticipation, that philosophers, at least modern philosophers, before Reid, allowed to the mind no immediate knowledge of the external reality. They conceded to it only a representative or mediate knowledge of external things. Of these some,

Modern philosophers before Reid held a doctrine of representative perception, in one or other of two forms.

[a] *Intellectual Powers*, Essay i., chap. i. *Works*, p. 223.   [β] *Ibid.*, Essay ii., chap. iii. *Works*, p. 297.

however, held that the representative object,—the
object immediately known,—was different from the
mind knowing, as it was also different from the reality
it represented; while others, on a simple hypothesis,
maintained that there was no intermediate entity, no
tertium quid, between the reality and the mind, but
that the immediate or representative object was itself
a mental modification.[a] The latter thus granting to
mind no immediate knowledge of aught beyond its
own modification, could, consequently, only recognise
a consciousness of self. The former, on the contrary,
could, as they actually did, accord to consciousness

Reid ex-
empts the
object of
perception
from the
sphere of
conscious-
ness.

a cognisance of not-self. Now, Reid, after asserting
against the philosophers the immediacy of our knowledge of external things, would almost appear to have
been startled by his own boldness; and, instead of
carrying his principle fairly to its issue, by according
to consciousness on his doctrine that knowledge of the
external world as existing, which, in the doctrine of
the philosophers, it obtained of the external world as
represented, he inconsistently stopped short, split immediate knowledge into two parts, and bestowed the
knowledge of material qualities on perception alone,
allowing that of mental modifications to remain exclusively with consciousness. Be this, however, as it
may, the exemption of the objects of perception from
the sphere of consciousness, can be easily shown to be
self-contradictory.

What! say the partisans of Dr Reid, are we not to
distinguish, as the product of different faculties, the
knowledge we obtain of objects in themselves the

a For a full discussion of the various theories of knowledge and perception, see the Author's *Supplemen-*
tary Dissertations to Reid's Works,
Notes B and C.—ED.

most opposite? Mind and matter are mutually separated by the whole diameter of being. Mind and matter are, in fact, nothing but words to express two series of phænomena known less in themselves, than in contradistinction from each other. The difference of the phænomena to be known, surely legitimates a difference of faculty to know them. In answer to this we admit at once, that were the question merely whether we should not distinguish, under consciousness, two special faculties,—whether we should not study apart, and bestow distinctive appellations on, consciousness considered as more particularly cognisant of the external world, and consciousness considered as more particularly cognisant of the internal,—this would be highly proper and expedient. But this is not the question. Dr Reid distinguishes consciousness as a special faculty from perception as a special faculty, and he allows to the former the cognisance of the latter in its operation, to the exclusion of its object. He maintains that we are conscious of our perception of a rose, but not of the rose perceived—that we know the ego by one act of knowledge, the non-ego by another. This doctrine I hold to be erroneous, and it is this doctrine I now proceed to refute.

In the first place, it is not only a logical axiom, but a self-evident truth, that the knowledge of opposites is one. Thus, we cannot know what is tall without knowing what is short,—we know what is virtue only as we know what is vice,—the science of health is but another name for the science of disease. Nor do we know the opposites, the I and Thou, the ego and non-ego, the subject and object, mind and matter, by a different law. The act which affirms that this par-

That in this Reid is wrong shown, 1°. From the principle, that the knowledge of opposites is one.

ticular phænomenon is a modification of Me, virtually
affirms that the phænomenon is not a modification of
anything different from Me, and, consequently, implies a
common cognisance of self and not-self; the act which
affirms that this other phænomenon is a modification
of something different from Me, virtually affirms that
the phænomenon is not a modification of Me, and,
consequently, implies a common cognisance of not-self
and self. But unless we are prepared to maintain
that the faculty cognisant of self and not-self is different from the faculty cognisant of not-self and self,
we must allow that the ego and non-ego are known
and discriminated in the same indivisible act of knowledge.   What, then, is the faculty of which this act
of knowledge is the energy ? It cannot be Reid's consciousness, for that is cognisant only of the ego or
mind,—it cannot be Reid's perception, for that is cognisant only of the non-ego or matter. But as the
act cannot be denied, so the faculty must be admitted.
It is not, however, to be found in Reid's catalogue.
But though not recognised by Reid in his system, its
necessity may, even on his hypothesis, be proved.
For if with him we allow only a special faculty immediately cognisant of the ego, and a special faculty immediately cognisant of the non-ego, we are at once met
with the question,—By what faculty are the ego and
non-ego discriminated ? We cannot say by consciousness, for that knows nothing but mind,—we cannot
say by perception, for that knows nothing but matter.
But as mind and matter are never known apart
and by themselves, but always in mutual correlation
and contrast, this knowledge of them in connection
must be the function of some faculty, not like Reid's
consciousness and perception, severally limited to mind

and matter as exclusive objects, but cognisant of them as the ego and non-ego,—as the two terms of a relation. It is thus shown that an act and a faculty, must, perforce, on Reid's own hypothesis, be admitted in which these two terms shall be comprehended together in the unity of knowledge,—in short, a higher consciousness, embracing Reid's consciousness and perception, and in which the two acts, severally cognitive of mind and matter, shall be comprehended, and reduced to unity and correlation. But what is this but to admit at last, in an unphilosophical complexity, the common consciousness of subject and object, of mind and matter, which we set out with denying in its philosophical simplicity?

But, in the second place, the attempt of Reid to make consciousness conversant about the various cognitive faculties to the exclusion of their objects, is equally impossible in regard to Perception, as we have shown it to be in relation to Imagination and Memory; nay, the attempt, in the case of perception, would, if allowed, be even suicidal of his great doctrine of our immediate knowledge of the external world.

Reid's assertion that we are conscious of the act of perception, but not of the object perceived, involves, first of all, a general absurdity. For it virtually asserts that we can know what we are not conscious of knowing. An act of perception is an act of knowledge; what we perceive, that we know. Now, if in perception there be an external reality known, but of which external reality we are, on Reid's hypothesis, not conscious, then is there an object known, of which we are not conscious. But as we know only inasmuch as we know that we know,—in other words, inasmuch as we are conscious that we know,—we cannot know an object without being conscious of that object as known ; consequently, we cannot perceive an object without being conscious of that object as perceived.

And, secondly, it destroys the distinction of consciousness itself.
But, again, how is it possible that we can be conscious of an operation of perception, unless consciousness be coextensive with that act ; and how can it be coextensive with the act, and not also conversant with its object ?   An act of knowledge is only possible in relation to an object, and it is an act of one kind or another only by special relation to a particular object.   Thus the object at once determines the existence, and specifies the character of the existence, of the intellectual energy.   An act of knowledge existing and being what it is only by relation to its object, it is manifest that the act can be known only through the object to which it is correlative ; and Reid's supposition that an operation can be known in consciousness to the exclusion of its object, is impossible.   For example, I see the inkstand.   How can I be conscious that my present modification exists,—that it is a perception, and not another mental state,—that it is a perception of sight to the exclusion of every other sense,—and, finally, that it is a perception of the inkstand, and of the inkstand only ; unless my consciousness comprehend within its sphere the object which at once determines the existence of the act, qualifies its kind, and distinguishes its individuality ?   Annihilate the inkstand, you annihilate the perception ; annihilate the consciousness of the object, you annihilate the consciousness of the operation.

Whence the apparent incongruity of the expression, "Consciousness of the object in perception."
It undoubtedly sounds strange to say, I am conscious of the inkstand, instead of saying, I am conscious of the perception of the inkstand.   This I admit, but the admission can avail nothing to Dr Reid, for the apparent incongruity of the expres-

LECT.
XIII.

sion arises only from the prevalence of that doctrine
of perception in the schools of philosophy, which it
is his principal merit to have so vigorously assailed.
So long as it was universally assumed by the learned,
that the mind is cognisant of nothing beyond, either,
on one theory, its own representative modifications,
or, on another, the species, ideas, or representative
entities, different from itself, which it contains, and
that all it knows of a material world is only an
internal representation which, by the necessity of its
nature, it mistakes for an external reality,—the supposition of an immediate knowledge of material phænomena was regarded only as a vulgar, an unphilosophical illusion, and the term consciousness, which was
exclusively a learned or technical expression for all immediate knowledge, was, consequently, never employed
to express an immediate knowledge of aught beyond
the mind itself; and thus, when at length, by Reid's
own refutation of the prevailing doctrine, it becomes
necessary to extend the term to the immediate knowledge of external objects, this extension, so discordant
with philosophic usage, is, by the force of association
and custom, felt at first as strange and even contradictory.   A slight consideration, however, is sufficient to
reconcile us to the expression, in showing, if we hold
the doctrine of immediate perception, the necessity of
not limiting consciousness to our subjective states. In
fact, if we look beneath the surface, consciousness was
not, in general, restricted, even in philosophical usage,
to the modifications of the conscious self.   That great
majority of philosophers who held that, in perception,
we know nothing of the external reality as existing,
but that we are immediately cognisant only of a representative something, different both from the object
represented, and from the percipient mind,—these

philosophers, one and all, admitted that we are conscious of this *tertium quid* present to, but not a modification of, mind ; for, except Reid and his school, I
am aware of no philosophers who denied that consciousness was coextensive or identical with immediate knowledge.

3°, A sup-
position
on which
some of the
self-contra-
dictions of
Reid's doc-
trine may
be avoided. But, in the third place, we have previously reserved
a supposition on which we may possibly avoid some
of the self-contradictions which emerge from Reid's
proposing as the object of consciousness the act, but
excluding from its cognisance the object, of perception,—that is, the object of its own object. The supposition is, that Dr Reid committed the same error in
regard to perception, which he did in regard to memory and imagination, and that in maintaining our
immediate knowledge in perception, he meant nothing
more than to maintain, that the mind is not, in that
act, cognisant of any representative object different
from its own modification, of any *tertium quid* minis-
tering between itself and the external reality ; but
that, in perception, the mind is determined itself to
represent the unknown external reality, and that, on
this self-representation, he abusively bestowed the
name of immediate knowledge, in contrast to that more
complex theory of perception, which holds that there
intervenes between the percipient mind and the external existence an intermediate something, different
from both, by which the former knows, and by which
the latter is represented. On the supposition of this
mistake, we may believe him guiltless of the others ;
and we can certainly, on this ground, more easily conceive how he could accord to consciousness a knowledge only of the percipient act,—meaning by that act
the representation of the external reality ; and how he

could deny to consciousness a knowledge of the object of perception,—meaning by that object the unknown reality itself. This is the only opinion which Dr Brown and others ever suspect him of maintaining; and a strong case might certainly be made out to prove that this view of his doctrine is correct. But if such were, in truth, Reid's opinion, then has he accomplished nothing,—his whole philosophy is one mighty blunder. For, as I shall hereafter show, idealism finds in this simpler hypothesis of representation even a more secure foundation than on the other; and, in point of fact, on this hypothesis, the most philosophical scheme of idealism that exists, the Egoistic or Fichtean, is established.

Taking, however, the general analogy of Reid's system, and a great number of unambiguous passages into account, I am satisfied that this view of his doctrine is erroneous; and I shall endeavour, when we come to treat of mediate and immediate knowledge, to explain how, from his never having formed to himself an adequate conception of these under all their possible forms, and from his historical ignorance of them as actually held by philosophers, he often appears to speak in contradiction of the vital doctrine which, in equity, he must be held to have steadily maintained.

This supposition untenable.

Besides the operations we have already considered, —Imagination or Conception, Memory, and Perception, —which Dr Reid and Mr Stewart have endeavoured to discriminate from Consciousness, there are further to be considered Attention and Reflection, which, in like manner, they have maintained to be an act or acts, not subordinate to, or contained in, Consciousness. But, before proceeding to show that their doctrine on this point is almost equally untenable as on

Reid and Stewart maintain, that Attention and Reflection are acts not subordinate to, or contained in, consciousness.

LECT.
XIII.

Certain
collateral
errors no-
ticed.
Stewart
misrepre-
sents Reid's
doctrine of
the mean-
ing and
difference
of Atten-
tion and
Reflection.

the preceding, it is necessary to clear up some confusion, and to notice certain collateral errors.

In the first place, on this head, these philosophers
are not at one; for Mr Stewart seems inadvertently
to have misrepresented the opinion of Dr Reid in regard to the meaning and difference of Attention and
Reflection. Reid either employs these terms as synonymous expressions, or he distinguishes them only by
making attention relative to the consciousness and
perception of the present; reflection, to the memory
of the past. In the fifth chapter of the second Essay
on the *Intellectual Powers,*[a] he says :—" In order,
however, to our having a distinct notion of any of the
operations of our own minds, it is not enough that
we be conscious of them; for all men have this consciousness. It is farther necessary that we attend to
them while they are exerted, and reflect upon them
with care while they are recent and fresh in our
memory. It is necessary that, by employing ourselves
frequently in this way, we get the habit of this attention and reflection," &c. And in the first chapter of
the sixth Essay, " Mr Locke," he says, " has restricted
the word *reflection* to that which is employed about
the operations of our minds, without any authority,
as I think, from custom, the arbiter of language. For,
surely, I may reflect upon what I have seen or heard,
as well as upon what I have thought. The word, in
its proper and common meaning, is equally applicable
to objects of sense and to objects of consciousness.
He has likewise confounded reflection with consciousness, and seems not to have been aware that they are
different powers, and appear at very different periods
of life."[b] In the first of these quotations, Reid might

use *attention* in relation to the consciousness of the
present, *reflection*, to the memory of the past; but
in the second, in saying that reflection "is equally
applicable to objects of sense and to objects of consciousness," he distinctly indicates that the two terms
are used by him as convertible.   Reid (I may notice
by the way) is wholly wrong in his strictures on
Locke for his restricted usage of the term *reflection* ;
for it was not until after his time that the term came,
by Wolf, to be philosophically employed in a more
extended signification than that in which Locke correctly applies it.ᵃ   Reid is likewise wrong, if we
literally understand his words, in saying that reflection is employed in common language in relation to
objects of sense.   It is never employed except upon
the mind and its contents.   We cannot be said to
reflect upon any external object, except in so far as
that object has been previously perceived, and its
image become part and parcel of our intellectual
furniture.   We may be said to reflect upon it in
memory, but not in perception.   But to return.

Reid, therefore, you will observe, identifies attention
and reflection.   Now, Mr Stewart, in the chapter
on Attention in the first volume of his *Elements*,ᵝ
says:—"Some important observations on the subject
of attention occur in different parts of Dr Reid's writings; particularly in his *Essays on the Intellectual
Powers of Man*, p. 62, and his *Essays on the Active
Powers of Man*, p. 78 *et seq.*   To this ingenious author we are indebted for the remark, that attention to

Side notes: Reid wrong in his censure of Locke's usage of the term Reflection. And in saying that Reflection is employed in relation to objects of sense.

ᵃ [Wolf, *Psychologia Empirica*, § 257: "Attentionis successiva directio ad ea quæ in re percepta insunt, dicitur *Reflexio*. Unde simul liquet quid sit facultas reflectendi, scilicet quod sit facultas attentionem suam successive ad ea quæ in re percepta insunt, pro arbitrio dirigendi."] Reid is further criticised in the Author's edition of his *Works*, pp. 347, 420.—En.

ᵝ *Works*, vol. ii. pp. 122, 123.

Locke not
the first to
use the
term Re-
flection in
its psycho-
logical ap-
plication.

things external is properly called *observation;* and attention to the subjects of our consciousness, *reflection.*"[a]

I may, however, notice a more important inadvertence of Mr Stewart, and this it is the more requisite to do, as his authority is worthy of high respect, not only on account of philosophical talent, but of historical accuracy. In various passages of his writings, Mr Stewart states that Locke seems to have considered the employment of the term reflection, in its psychological acceptation, as original to himself; and he notices it as a curious circumstance that Sir John Davies, Attorney-General to Queen Elizabeth, should, in his poem on the *Immortality of the Soul,* have employed this term in the same signification. How Mr Stewart could have fallen into this error, is wholly inconceivable. The word, as employed by Locke, was in common use in every school of philosophy for fifteen hundred years previous to the publication of the *Essay on the Human Understanding.*[β] It was a term in the philosophy both of Descartes[γ] and of Gassendi;[δ] and it was borrowed by them from the schoolmen, with whom it was a household word.[ε] From the schoolmen, indeed, Locke seems to have adopted

a This distinction has been attempted by others. [See Keckermann, *Opera,* tom. i. p. 1612, where he distinguishes *reflection,*—"*intellectio reflexa, interna,* per quam homo intelligit suum intellectum,"—from "*intellectio externa,* qua intellectus alias res extra se positas percipit." See also Mazure, *Cours de Philosophie,* tom. i. p. 381.—Ed.]

β For historical notices of the use of the term, see *Reid's Works,* (completed edition), pp. 946, 947.—Ed.

γ [Descartes, *Epist.,* P. ii., Ep. vi. (See Gruyer, *Essais Philosophiques,* tom. iv. p. 118.) De la Forge, *Traité*

de *l'Esprit de l'Homme,* préface, p. xi.]

δ [Gassendi, *Physica,* Sect. III. Memb. Post., lib. ix. c. 3. (*Opera,* Leyden, 1658, vol. ii. p. 451.) "Ad secundam vero operationem praesertim spectat ipsa intellectus ad suam operationem attentio, reflexione illa supra actionem propriam, qua se intelligere intelligit, cogitatve se cogitare."]

ε [We have the Scholastic brocard pointing to the difficulties of the study of self: "Reflexiva cogitatio facile fit deflexiva." See Keckermann, *Opera,* tom. i. p. 406.]

the fundamental principle of his philosophy, the derivation of our knowledge through the double medium
of sense and reflection,—at least, some of them had
in terms articulately enounced this principle five
centuries previous to the English philosopher, and
enounced it also in a manner far more correct than
was done by him ; [a] for they did not, like Locke, regard reflection itself as a source of knowledge,—thus
reducing all our knowledge to experience and its generalisation, but viewed in reflection only the channel
through which, along with the contingent phænomena
of our internal experience, we discover the necessary
judgments which are original or native to the mind.

There is, likewise, another oversight of Mr Stewart
which I may notice. "Although," he says, "the connection between attention and memory has been frequently remarked in general terms, I do not recollect that the power of attention has been mentioned
by any of the writers on pneumatology, in their enumeration of the faculties of the mind ; nor has it been
considered by any one, so far as I know, as of sufficient
importance to deserve a particular examination."[β] So
far is this from being the case that there are many
previous authors who have considered attention as a
separate faculty, and treated of it even at greater

a [See Scotus, *Super Universalibus
Porphyrii*, Qu. iii. : "Ad tertium
dico quod illa propositio Aristotelis,
nihil est in intellectu quin prius fuerit
in sensu, vera est de eo quod est
primum intelligibile, quod est scilicet quod quid est rei materialis, non
autem de omnibus per se intelligibi-
libus ; quia multa per se intelligun-
tur, non quia speciem faciunt in sensu,
sed per reflexionem intellectus." (By
the Scotists the act of intellect was
regarded as threefold: *Rectus,—Col-*
lativus,—Reflexus. See Constantius,
(a Sarnano), *Tract. de Secundis In-
tentionibus*, ad calcem *Scoti Operum*,
p. 452.) See also Philip Mocenicus,
Contemplationes (1581), *passim ;* Goclenius, *Lexikon Philosophicum*, v.
Reflexus ; Keckermann, *Opera*, tom.
i. pp. 1600, 1612 ; Conimbricenses,
In Arist. De Anima, pp. 370, 373.]
[Compare *Reid's Works*, (completed
edition), pp. 777, 778, 946.—ED.]

β *Elements*, i. c. 2. *Collected Works*,
vol. ii. p. 122.—ED.

length than Mr Stewart himself. This is true not only of the celebrated Wolf,[a] but of the whole Wolfian school; and to these I may add Condillac,[β] Contzen,[γ] Tiedemann,[δ] Irwing,[ε] Malebranche,[ζ] and many others.[η] But this by the way.

Taking, however, Attention and Reflection for acts of the same faculty, and supposing, with Mr Stewart, that reflection is properly attention directed to the phænomena of mind, observation, attention directed to the phænomena of matter; the main question comes to be considered,—Is attention a faculty different from consciousness, as Reid and Stewart maintain? As the latter of these philosophers has not argued the point himself, but merely refers to the arguments of the former in confirmation of their common doctrine, it will be sufficient to adduce the

following passage from Reid, in which his doctrine on this head is contained. " I return," he says, " to what I mentioned as the main source of information on this subject—attentive reflection upon the operations of our own minds.

" All the notions we have of mind and its operations, are, by Mr Locke, called *ideas of reflection*. A man may have as distinct notions of remembrance, of judgment, of will, of desire, as he has of any object whatever. Such notions, as Mr Locke justly observes, are got by the power of reflection. But what is this

[a] *Psychologia Empirica*, § 234 et seq.—ED.
[β] *Origine des Connoissances Humaines*, part i. § ii. ch. 2.—ED.
[γ] *Prelectiones Logicæ et Metaphysicæ*, auctore Adamo Contzen, (Mechlin, 1830), vol. iii. p. 31. (Originally published in 1775-1780.)—ED.
[δ] *Handbuch der Psychologie*, p. 121.—ED.
[ε] *Erfahrungen und Untersuchungen über den Menschen*, von Karl Franz von Irwing, Berlin, 1777, b. i. p. 411; b. ii. p. 209.—ED.
[ζ] *De la Recherche de la Vérité*, lib. iii. ch. 4; lib. vi. ch. 2. *Traité de Morale*, ch. 5.—ED.
[η] Compare *Reid's Works*, (completed edition), p. 945-46.—ED.

power of reflection? 'It is,' says the same author, 'that power by which the mind turns its view inward, and observes its own actions and operations.' He observes elsewhere, 'That the understanding, like the eye, whilst it makes us see and perceive all other things, takes no notice of itself; and that it requires art and pains to set it at a distance, and make it its own object.'

" This power of the understanding to make its own operations its object, to attend to them, and examine them on all sides, is the power of reflection, by which alone we can have any distinct notion of the powers of our own or of other minds.

" This reflection ought to be distinguished from consciousness, with which it is too often confounded, even by Mr Locke. All men are conscious of the operations of their own minds, at all times while they are awake; but there are few who reflect upon them, or make them objects of thought." [a]

Dr Reid has rightly said that attention is a volun- *What Attention is.* tary act. This remark might have led him to the observation, that attention is not a separate faculty, or a faculty of intelligence at all, but merely an act of will or desire, subordinate to a certain law of intelligence. This law is, that the greater the number of objects to which our consciousness is simultaneously extended, the smaller is the intensity with which it is able to consider each, and consequently the less vivid and distinct will be the information it obtains of the several objects.[β] This law is expressed in the old adage

" Pluribus intentus, minor est ad singula sensus."

---

a *Intellectual Powers,* Essay i., chap. v. *Works,* p. 239.
β [Cf. Steeb, *Über den Menschen,* ii. 673; Fries, *Anthropologie,* i. 83; and Schulze, *Über die menschliche Erkenntniss,* p. 65.]

Such being the law, it follows that, when our interest in any particular object is excited, and when we wish to obtain all the knowledge concerning it in our power, it behoves us to limit our consideration to that object, to the exclusion of others. This is done by an act of volition or desire, which is called *attention*. But to view attention as a special act of intelligence, and to distinguish it from consciousness, is utterly inept. Consciousness may be compared to a telescope, attention to the pulling out or in of the tubes in accommodating the focus to the object; and we might, with equal justice, distinguish, in the eye, the adjustment of the pupil from the general organ of vision, as, in the mind, distinguish attention from consciousness as separate faculties. Not, however, that they are to be accounted the same. Attention is consciousness and something more. It is consciousness voluntarily applied, under its law of limitations, to some determinate object; it is consciousness concentrated. In this respect, attention is an interesting subject of consideration; and having now finished what I proposed in proof of the position, that consciousness is not a special faculty of knowledge, but coextensive with all our cognitions, I shall proceed to consider it in its various aspects and relations; and

Attention as a general phænomenon of consciousness. having just stated the law of limitation, I shall go on to what I have to say in regard to attention as a general phænomenon of consciousness.

Can we attend to more than a single object at once ? And here, I have first to consider a question in which I am again sorry to find myself opposed to many distinguished philosophers, and, in particular, to one whose opinion on this, as on every other point of psychological observation, is justly entitled to the highest consideration. The philosopher I allude to is Mr Stewart. The question is, Can we attend to more

than a single object at once ?   For if attention be
nothing but the concentration of consciousness on a
smaller number of objects than constitute its widest
compass of simultaneous knowledge, it is evident that,
unless this widest compass of consciousness be limited
to only two objects, we do attend when we converge
consciousness on any smaller number than that total
complement of objects which it can embrace at once.
For example, if we suppose that the number of objects
which consciousness can simultaneously apprehend be
six, the limitation of consciousness to five, or four, or
three, or two, or one, will all be acts of attention, different in degree, but absolutely identical in kind.

Mr Stewart's doctrine is as follows :—" Before," he
says, " we leave the subject of Attention, it is proper to
take notice of a question which has been stated with
respect to it ; whether we have the power of attending
to more than one thing at one and the same instant ;
or, in other words, whether we can attend, at one and
the same instant, to objects which we can attend to
separately ?   This question has, if I am not mistaken,
been already decided by several philosophers in the
negative ; and I acknowledge, for my own part, that
although their opinion has not only been called in
question by others, but even treated with some degree
of contempt as altogether hypothetical, it appears to
me to be the most reasonable and philosophical that
we can form on the subject.

" There is, indeed, a great variety of cases in which
the mind apparently exerts different acts of attention
at once ; but from the instances which have already
been mentioned, of the astonishing rapidity of thought,
it is obvious that all this may be explained without
supposing those acts to be coexistent ; and I may
even venture to add, it may all be explained in the

Stewart
quoted in
reference
to this
question.

most satisfactory manner, without ascribing to our intellectual operations a greater degree of rapidity than that with which we know, from the fact, that they are sometimes carried on. The effect of practice in increasing this capacity of apparently attending to different things at once, renders this explanation of the phenomenon in question more probable than any other.

"The case of the equilibrist and rope-dancer already mentioned, is particularly favourable to this explanation, as it affords direct evidence of the possibility of the mind's exerting different successive acts in an interval of time so short, as to produce the same sensible effect as if they had been exerted at one and the same moment. In this case, indeed, the rapidity of thought is so remarkable, that if the different acts of the mind were not all necessarily accompanied with different movements of the eye, there can be no reason for doubting that the philosophers whose doctrine I am now controverting, would have asserted that they are all mathematically coexistent.

"Upon a question, however, of this sort, which does not admit of a perfectly direct appeal to the fact, I would by no means be understood to decide with confidence; and, therefore, I should wish the conclusions I am now to state, to be received as only conditionally established. They are necessary and obvious consequences of the general principle, 'that the mind can only attend to one thing at once;' but must stand or fall with the truth of that supposition.

"It is commonly understood, I believe, that in a concert of music, a good ear can attend to the different parts of the music separately, or can attend to them all at once, and feel the full effect of the harmony. If the doctrine, however, which I have endeavoured to establish be admitted, it will follow that in the latter case the mind is constantly varying its attention from the one part of the music to the other, and that its operations are so rapid as to give us no perception of an interval of time.

" The same doctrine leads to some curious conclusions with respect to vision. Suppose the eye to be fixed in a particular position, and the picture of an object to be painted on the retina. Does the mind perceive the complete figure of the object at once, or is this perception the result of the various perceptions we have of the different points in the outline ? With respect to this question, the principles already stated lead me to conclude, that the mind does at one and the same time perceive every point in the outline of the object, (provided the whole of it be painted on the retina at the same instant,) for perception, like consciousness, is an involuntary operation. As no two points, however, of the outline are in the same direction, every point by itself constitutes just as distinct an object of attention to the mind, as if it were separated by an interval of empty space from all the rest. If the doctrine, therefore, formerly stated be just, it is impossible for the mind to attend to more than one of these points at once ; and as the perception of the figure of the object implies a knowledge of the relative situation of the different points with respect to each other, we must conclude that the perception of figure by the eye is the result of a number of different acts of attention. These acts of attention, however, are performed with such rapidity, that the effect with respect to us, is the same as if the perception were instantaneous.

" In farther confirmation of this reasoning, it may

be remarked, that if the perception of visible figure were an immediate consequence of the picture on the retina, we should have, at the first glance, as distinct an idea of a figure of a thousand sides, as of a triangle or a square. The truth is, that when the figure is very simple, the process of the mind is so rapid that the perception seems to be instantaneous : but when the sides are multiplied beyond a certain number, the interval of time necessary for these different acts of attention becomes perceptible.

" It may, perhaps, be asked what I mean by a *point* in the outline of a figure, and what it is that constitutes this point *one* object of attention. The answer, I apprehend, is, that this point is the *minimum visibile*. If the point be less, we cannot perceive it ; if it be greater, it is not all seen in one direction.

" If these observations be admitted, it will follow that, without the faculty of memory, we could have had no perception of visible figure."[a]

On this point, Dr Brown not only coincides with Mr Stewart in regard to the special fact of attention, but asserts in general that the mind cannot exist at the same moment in two different states, that is, in two states in either of which it can exist separately. "If the mind of man," he says, " and all the changes which take place in it, from the first feeling with which life commenced to the last with which it closes, could be made visible to any other thinking being, a certain series of feelings alone,—that is to say, a certain number of successive states of mind, would be distinguishable in it, forming indeed a variety of sensations, and thoughts, and passions, as momentary states of the mind, but all of them existing individu-

a *Elements*, vol. i. chap. 2. *Works*, vol. ii. p. 140-143.

ally and successively to each other. To suppose the
mind to exist in two different states, in the same
moment, is a manifest absurdity."[a]

Criticism
of Stewart's
doctrine.
His first
illustration
from the
pheno-
mena of
sound.

I shall consider these statements in detail. Mr
Stewart's first illustration of his doctrine is drawn
from a concert of music, in which, he says, " a good
ear can attend to the different parts of the music
separately, or can attend to them all at once, and feel
the full effect of the harmony." This example, however, appears to me to amount to a reduction of his
opinion to the impossible. What are the facts in this
example? In a musical concert, we have a multitude
of different instruments and voices, emitting at once
an infinity of different sounds. These all reach the
ear at the same indivisible moment in which they
perish, and, consequently, if heard at all, much more
if their mutual relation or harmony be perceived, they
must be all heard simultaneously. This is evident.
For if the mind can attend to each minimum of sound
only successively, it, consequently, requires a minimum
of time in which it is exclusively occupied with each
minimum of sound. Now, in this minimum of time,
there coexist with it, and with it perish, many minima
of sound which, *ex hypothesi*, are not perceived,—are
not heard, as not attended to. In a concert, therefore,
on this doctrine, a small number of sounds only could
be perceived, and above this petty maximum, all sounds
would be to the ear as zero. But what is the fact?
No concert, however numerous its instruments, has
yet been found to have reached, far less to have surpassed, the capacity of mind and its organ.

But it is even more impossible, on this hypothesis,

a *Lectures on the Philosophy of the* 1830).—ED.
Human Mind, Lect. xi. p. 67, (ed.

LECT.
XIII.

Impossible, on Stewart's doctrine, to understand how we can perceive the relation of different sounds.

to understand how we can perceive the relation of different sounds, that is, have any feeling of the harmony of a concert. In this respect, it is, indeed, *felo de se.* It is maintained that as we cannot attend at once to two sounds, we cannot perceive them as coexistent; consequently, the feeling of harmony of which we are conscious, must proceed from the feeling of the relation of these sounds as successively perceived in different points of time. We must, therefore, compare the past sound, as retained in memory, with the present, as actually perceived. But this is impossible on the hypothesis itself. For we must, in this case, attend to the past sound in memory, and to the present sound in sense at once, or they will not be perceived in mutual relation as harmonic. But one sound in memory and another sound in sense, are as much two different objects as two different sounds in sense. Therefore, one of two conclusions is inevitable : either we can attend to two different objects at once, and the hypothesis is disproved ; or we cannot, and all knowledge of relation and harmony is impossible, which is absurd.

The consequences of this doctrine are equally startling, as taken from Mr Stewart's second illustration, from the phænomena of vision. He holds that the perception of figure by the eye is the result of a number of separate acts of attention, and that each act of attention has for its object a point the least that can be seen,—the *minimum visibile.* On this hypothesis, we must suppose that, at every instantaneous opening of the eyelids, the moment sufficient for us to take in the figure of the objects comprehended in the sphere of vision, is subdivided into almost infinitesimal parts, in each of which a separate act of

attention is performed. This is, of itself, sufficiently
inconceivable. But this being admitted, no difficulty
is removed. The separate acts must be laid up in
memory, in imagination. But how are they there to
form a single whole, unless we can, in imagination,
attend to all the *minima visibilia* together, which in
perception we could only attend to severally? On
this subject I shall, however, have a more appropriate
occasion of speaking, when I consider Mr Stewart's
doctrine of the relation of colour to extension.[a]

a See *infra*, vol. ii. p. 144 *et seq.*

LECTURE XIV.

CONSCIOUSNESS,—ATTENTION IN GENERAL.

LECT.
XIV.
———
Recapitu-
lation.

IN the former part of our last Lecture, I concluded the argument against Reid's analysis of Consciousness into a special faculty, and showed you that, even in relation to Perception, (the faculty by which we obtain a knowledge of the material universe), Consciousness is still the common ground in which every cognitive operation has its root. I then proceeded to prove the same in regard to Attention. After some observations touching the confusion among philosophers, more or less extensive, in the meaning of the term *reflection*, as a subordinate modification of attention, I endeavoured to explain to you what attention properly is, and in what relation it stands to consciousness. I stated that attention is consciousness applied by an act of will or desire under a particular law. In so far as attention is an act of the conative faculty, it is not an act of knowledge at all, for the mere will or desire of knowing is not an act of cognition. But the act of the conative faculty is exerted by relation to a certain law of consciousness, or knowledge, or intelligence. This law, which we call the Law of Limitation, is, that the intension of our knowledge is in the inverse ratio of its extension,—in other words, that the fewer objects we consider at once,

the clearer and more distinct will be our knowledge
of them.   Hence the more vividly we will or desire
that a certain object should be clearly and distinctly
known, the more do we concentrate consciousness
through some special faculty upon it.   I omitted, I
find, to state that I think Reid and Stewart incorrect
in asserting that attention is only a voluntary act,
meaning by the expression *voluntary*, an act of free-
will.   I am far from maintaining, as Brown and others
do, that all will is desire ; but still I am persuaded
that we are frequently determined to an act of attention, as to many other acts, independently of our free
and deliberate volition.   Nor is it, I conceive, possible
to hold that, though immediately determined to an
act of attention by desire, it is only by the permission
of our will that this is done ; consequently, that every
act of attention is still under the control of our volition.   This I cannot maintain.   Let us take an example :—When occupied with other matters, a person
may speak to us, or the clock may strike, without our
having any consciousness of the sound ;[a] but it is
wholly impossible for us to remain in this state of unconsciousness intentionally and with will.   We cannot
determinately refuse to hear by voluntarily withhold-
ing our attention; and we can no more open our eyes,
and, by an act of will, avert our mind from all perception of sight, than we can, by an act of will, cease
to live.   We may close our ears or shut our eyes, as we
may commit suicide ; but we cannot, with our organs
unobstructed, wholly refuse our attention at will.   It,
therefore, appears to me the more correct doctrine to
hold that there is no consciousness without attention,
—without concentration, but that attention is of three

a See Reid, *Active Powers*, Essay ii. ch. 3.   *Works*, p. 537.—Ed.

LECT.
XIV.

Attention
of three
degrees or
kinds.

degrees or kinds. The first, a mere vital and irresistible act; the second, an act determined by desire,
which, though involuntary, may be resisted by our
will; the third, an act determined by a deliberate volition. An act of attention,—that is, an act of concentration,—seems thus necessary to every exertion of
consciousness, as a certain contraction of the pupil is
requisite to every exercise of vision. We have formerly
noticed, that discrimination is a condition of consciousness; and a discrimination is only possible by
a concentrative act, or act of attention. This, however, which corresponds to the lowest degree,—to the
mere vital or automatic act of attention, has been
refused the name; and *attention*, in contradistinction
to this mere automatic contraction, given to the two
other degrees, of which, however, Reid only recognises
the third.

Attention, then, is to consciousness, what the contraction of the pupil is to sight; or to the eye of the
mind, what the microscope or telescope is to the
bodily eye. The faculty of attention is not, therefore,
a special faculty, but merely consciousness acting
under the law of limitation to which it is subjected.
But whatever be its relations to the special faculties,
attention doubles all their efficiency, and affords them
a power of which they would otherwise be destitute.
It is, in fact, as we are at present constituted, the
primary condition of their activity. ·

Can we
attend to
more than
a single
object at
once?

Having thus concluded the discussion of the question regarding the relation of consciousness to the
other cognitive faculties, I proceeded to consider
various questions which, as not peculiar to any of the
special faculties, fall to be discussed under the head
of consciousness, and I commenced with the curious

problem, Whether we can attend to more than a single object at once. Mr Stewart maintains, though not without hesitation, the negative. I endeavoured to show you that his arguments are not conclusive, and that they even involve suppositions which are so monstrous as to reduce the thesis he supports *ad impossibile.* I have now only to say a word in answer to Dr Brown's assertion of the same proposition, though in different terms. In the passage I adduced in our last Lecture, he commences by the assertion, that the mind cannot exist, at the same moment, in two different states,—that is, in two states in either of which it can exist separately, and concludes with the averment that the contrary supposition is a manifest absurdity. I find the same doctrine maintained by Locke in that valuable, but neglected, treatise entitled *An Examination of Père Malebranche's Opinion of Seeing all Things in God.* In the thirty-ninth section he says :—" Different sentiments are different modifications of the mind. The mind or soul that perceives, is one immaterial, indivisible substance. Now, I see the white and black on this paper, I hear one singing in the next room, I feel the warmth of the fire I sit by, and I taste an apple I am eating, and all this at the same time. Now, I ask, take modification for what you please, can the same unextended, indivisible substance have different, nay, inconsistent and opposite, (as these of white and black must be), modifications at the same time ? Or must we suppose distinct parts in an indivisible substance, one for black, another for white, and another for red ideas, and so of the rest of those infinite sensations which we have in sorts and degrees ; all which we can distinctly perceive, and so are distinct ideas, some where-

Brown's doctrine, that the mind cannot exist at the same moment in two different states.

This doctrine maintained by Locke.

of are opposite as heat and cold, which yet a man may feel at the same time?" Leibnitz has not only given a refutation of Locke's *Essay*, but likewise of his *Examination of Malebranche*. In reference to the passage I have just quoted, Leibnitz says: " Mr Locke asks, 'Can the same unextended, indivisible substance, have different, nay, inconsistent and opposite modifications, at the same time?' I reply, it can. What is inconsistent in the same object, is not inconsistent in the representation of different objects which we conceive at the same moment. For this there is no necessity that there should be different parts in the soul, as it is not necessary that there should be different parts in the point on which, however, different angles rest."[a] The same thing had, however, been even better said by Aristotle, whose doctrine I prefer translating to you, as more perspicuous, in the following passage from Joannes Grammaticus, (better known by the surname Philoponus),—a Greek philosopher, who flourished towards the middle of the sixth century. It is taken from the Prologue to his valuable commentary on the *De Anima* of Aristotle; and, what is curious, the very supposition which on Locke's doctrine would infer the corporeal nature of mind, is alleged by the Aristotelians and Condillac, in proof of its immateriality. " Nothing bodily, says Aristotle, can, at the same time, in the same part, receive contraries. The finger cannot at once be wholly participant of white and of black, nor can it, at once and in the same place, be both hot and cold. But the sense at the same moment apprehends contraries. Wherefore, it knows that this is first, and that second, and that it discriminates the black from the white. In what

Opposed by Leibnitz.

Aristotle opposed to foregoing doctrine.

His view, as paraphrased by Philoponus.

a *Remarques sur le Sentiment du   phica*, edit. Erdmann, p. 451.—ED.
Père Malebranche; Opera Philoso-

manner, therefore, does sight simultaneously perceive contraries? Does it do so by the same? or does it by one part apprehend black, by another, white? If it does so by the same, it must apprehend these without parts, and it is incorporeal. But if by one part it apprehends this quality, and by another that,—this, he says, is the same as if I perceived this, and you that. But it is necessary that that which judges should be one and the same, and that it should even apprehend by the same the objects which are judged. Body cannot, at the same moment and by the same part, apply itself to contraries or things absolutely different. But sense at once applies itself to black and to white; it, therefore, applies itself indivisibly. It is thus shown to be incorporeal. For if by one part it apprehended white, by another part apprehended black, it could not discern the one colour from the other; for no one can distinguish that which is perceived by himself as different from that which is perceived by another."[a] So far Philoponus.

Dr Brown calls the sensation of sweet one mental state, the sensation of cold another; and as the one of these states may exist without the other, they are consequently different states. But will it be maintained

---

[a] The text of Aristotle here partially paraphrased, (Procem. f. 3b ed. 1535), and more fully in Commentary on texts 144, 149, is as follows:—Ἦ καὶ δῆλον ὅτι ἡ σάρξ οὐκ ἔστι τὸ ἔσχατον αἰσθητήριον· ἀνάγκη γὰρ ἦν ἀπτόμενον αὐτοῦ κρίνειν τὸ πρῖνον. Οὔτε δὴ κεχωρισμένοις δέχεται κρίνειν ὅτι ἕτερον τὸ γλυκὺ τοῦ λευκοῦ, ἀλλὰ δεῖ ἐπί τινι ἀμφω δῆλα εἶναι. Οὔτω μὲν γὰρ κἂν εἰ τοῦ μὲν ἐγὼ τοῦ δὲ σὺ αἴσθοιο, δῆλον ἂν εἴη ὅτι ἕτερα ἀλλήλων. Δεῖ δὲ τὸ ἓν λέγειν ὅτι ἕτερον· ἕτερον γὰρ τὸ γλυκὺ τοῦ λευκοῦ. Λέγει ἄρα τὸ αὐτό· Ὥστε

ὡς λέγει, οὔτω καὶ νοεῖ καὶ αἰσθάνεται. Ὅτι μὲν οὖν οὐχ οἷόν τε κεχωρισμένοις κρίνειν τὰ κεχωρισμένα, δῆλον· ὅτι δ' οὐδ' ἐν κεχωρισμένῳ χρόνῳ, ἐντεῦθεν, Ὥσπερ γὰρ τὸ αὐτὸ λέγει ὅτι ἕτερον, τὸ ἀγαθὸν καὶ τὸ κακόν, οὔτω καὶ ὅτε θάτερον λέγει ὅτι ἕτερον καὶ θάτερον, οὐ κατὰ συμβεβηκὸς τὸ ὅτι· λέγω δ', οἷον νῦν λέγω ὅτι ἕτερον, οὐ μέντοι ὅτι νῦν ἕτερον. Ἀλλ' οὔτω λέγει, καὶ νῦν, καὶ ὅτι νῦν ἅμα ἄρα. Ὥστε ἀχώριστον καὶ ἐν ἀχωρίστῳ χρόνῳ. De Anima, lib. iii. c. 2, § 11. Cf. §§ 9, 10, 12, 13, 14, with the relative commentary by Philoponus.—Ed.

that we cannot, at one and the same moment, feel the
sensations of sweet and cold, or that sensations forming apart different states, do, when coexistent in the
same subject, form only a single state?

*On this
view com-
parison im-
possible.*

The doctrine that the mind can attend to, or be
conscious of, only a single object at a time, would, in
fact, involve the conclusion that all comparison and
discrimination are impossible; but comparison and
discrimination being possible, this possibility disproves
the truth of the counter-proposition. An act of comparison or discrimination supposes that we are able to
comprehend, in one indivisible consciousness, the different objects to be compared or discriminated. Were
I only conscious of one object at one time, I could
never possibly bring them into relation; each could be
apprehended only separately, and for itself. For in
the moment in which I am conscious of the object A,
I am, *ex hypothesi*, unconscious of the object B; and
in the moment I am conscious of the object B, I am
unconscious of the object A. So far, in fact, from consciousness not being competent to the cognisance of
two things at once, it is only possible under that cognisance as its condition. For without discrimination
there could be no consciousness; and discrimination
necessarily supposes two terms to be discriminated.

No judgment could be possible were not the subject
and predicate of a proposition thought together by the
mind, although expressed in language one after the
other. Nay, as Aristotle has observed, a syllogism
forms in thought one simultaneous act;[a] and it is only
the necessity of retailing it piecemeal and by succes-

a This is said by Aristotle of the
act of judgment; but the remark applies to that of reasoning also. See
De Anima, iii. 6: 'Ἐν οἷς τὸ ψεῦδος
καὶ τὸ ἀληθὲς, σύνθεσίς τις ἤδη νοημά-
των ὥσπερ ἓν ὄντων. . . . . Τὸ
δὲ ἓν ποιοῦν, τοῦτο ὁ νοῦς ἕκαστον.—
Ed.

sion, in order to accommodate thought to the imperfection of its vehicle,—language, that affords the appearance of a consecutive existence. Some languages, as the Sanscrit, the Latin, and the Greek, express the syntactical relations by flexion, and not by mere juxtaposition. Their sentences are thus bound up into one organic whole, the preceding parts remaining suspended in the mind, till the meaning, like an electric spark, is flashed from the conclusion to the commencement. This is the reason of the greater rhetorical effect of terminating the Latin period by the verb. And to take a mere elementary example,—" How could the mind comprehend these words of Horace,

> ' Bacchum in remotis carmina rupibus
> Vidi docentem,'

unless it could seize at once those images in which the adjectives are separated from their substantives ?" [a]

The modern philosophers who have agitated this question, are not aware that it was one canvassed likewise in the schools of the middle ages. It was there expressed by the proposition, *Possitne intellectus noster plura simul intelligere.*[β] Maintaining the negative, we find St Thomas, Cajetanus, Ferrariensis, Capreolus, Hervæus, Alexander Alensis, Albertus Magnus, and Durandus; while the affirmative was asserted by Scotus, Occam, Gregorius, Ariminensis, Lichetus, Marsilius, Biel, and others.[7]

This question canvassed in the schools of the middle ages.

Supposing that the mind is not limited to the simultaneous consideration of a single object, a question arises, How many objects can it embrace at once ?

How many objects can the mind embrace at once?

a [Bonstetten, *Etudes de l'Homme,* tom. ii. p. 377, note.]

β [See Aquinas, *Summa,* pars i., qu. 85, art. 4. Cf. Alex. Aphrodisiensis, *De Anima,* lib. i. c. 22, f. 134 a (ed.

Ald.) Nemesius, *De Natura Hominis,* c. vii. p. 184, ed. Matthæi.]

7 For these authorities, see Conimbricenses, *In De Anima,* lib. iii. c. viii. qu. 6, p. 499 *et seq.*—ED.

LECT.
XIV.
You will recollect that I formerly stated that the greater the number of objects among which the attention of the mind is distributed, the feebler and less distinct will be its cognisance of each.

" Pluribus intentus, minor est ad singula sensus."

Consciousness will thus be at its maximum of intensity when attention is concentrated on a single object; and the question comes to be, how many several objects can the mind simultaneously survey, not with vivacity, but without absolute confusion? I find this problem stated and differently answered, by different philosophers, and apparently without a knowledge of each other. By Charles Bonnet[a] the mind is allowed to have a distinct notion of six objects at once; by Abraham Tucker[β] the number is limited to four; while Destutt-Tracy[γ] again amplifies it to six. The opinion of the first and last of these philosophers appears to me correct. You can easily make the experiment for yourselves, but you must beware of grouping the objects into classes. If you throw a handful of marbles on the floor, you will find it difficult to view at once more than six, or seven at most, without confusion; but if you group them into twos, or threes, or fives, you can comprehend as many groups as you can units; because the mind considers these groups only as units, —it views them as wholes, and throws their parts out of consideration. You may perform the experiment also by an act of imagination.

a [*Essai de Psychologie*, c. xxxviii. p. 132. Compare his *Essai Analytique sur l'Ame*, tom. i. c. xiii. p. 163 *et seq.*]

β [*Light of Nature*, c. xiv. § 5.]

γ [*Idéologie*, tom. i. p. 453. Compare Degerando, *Des Signes*, i. 167, who allows us to embrace, at one view, five unities. D'Alembert, *Mélanges*, vol. iv. pp. 40, 151. Ancillon, *Nouveaux Mélanges*, tom. ii. p. 135. Malebranche, *Recherche*, liv. iii. c. 2, tom. i. p. 191.]

Before leaving this subject, I shall make some observations on the value of attention, considered in its
highest degree as an act of will, and on the importance of forming betimes the habit of deliberate concentration.

The greater capacity of continuous thinking that a
man possesses, the longer and more steadily can he
follow out the same train of thought,—the stronger is
his power of attention; and in proportion to his power
of attention will be the success with which his labour
is rewarded. All commencement is difficult; and this
is more especially true of intellectual effort. When we
turn for the first time our view on any given object,
a hundred other things still retain possession of our
thoughts. Even when we are able, by an arduous
exertion, to break loose from the matters which have
previously engrossed us, or which every moment force
themselves on our consideration,—even when a reso-
lute determination, or the attraction of the new object,
has smoothed the way on which we are to travel;
still the mind is continually perplexed by the glimmer
of intrusive and distracting thoughts, which prevent
it from placing that which should exclusively occupy
its view, in the full clearness of an undivided light.
How great soever may be the interest which we take
in the new object, it will, however, only be fully established as a favourite when it has been fused into
an integral part of the system of our previous knowledge, and of our established associations of thoughts,
feelings, and desires. But this can only be accomplished by time and custom. Our imagination and
our memory, to which we must resort for materials
with which to illustrate and enliven our new study,
accord us their aid unwillingly, — indeed, only by

compulsion.   But if we are vigorous enough to pursue
our course in spite of obstacles, every step, as we advance, will be found easier ; the mind becomes more
animated and energetic ; the distractions gradually diminish ; the attention is more exclusively concentrated
upon its object ; the kindred ideas flow with greater
freedom and abundance, and afford an easier selection
of what is suitable for illustration.   At length, our
system of thought harmonises with our pursuit.   The
whole man becomes, as it may be, philosopher, or historian, or poet ; he lives only in the trains of thought
relating to this character.   He now energises freely,
and, consequently, with pleasure ; for pleasure is the
reflex of unforced and unimpeded energy.   All that is
produced in this state of mind, bears the stamp of excellence and perfection.   Helvetius justly observes,
that the very feeblest intellect is capable of comprehending the inference of one mathematical position
from another, and even of making such an inference
itself.[a]   Now, the most difficult and complicate demonstrations in the works of a Newton or a Laplace,
are all made up of such immediate inferences.   They
are like houses composed of single bricks.   No greater
exertion of intellect is required to make a thousand
such inferences than is requisite to make one ; as the
effort of lying a single brick is the maximum of any
individual effort in the construction of such a house.
Thus, the difference between an ordinary mind and
the mind of a Newton, consists principally in this, that
the one is capable of the application of a more continuous attention than the other,—that a Newton is able
without fatigue to connect inference with inference in
one long series towards a determinate end ; while the
man of inferior capacity is soon obliged to break or let

a *De l'Esprit*, Discours iii. c. iv.—Ed.

fall the thread which he had begun to spin. This is, in fact, what Sir Isaac, with equal modesty and shrewdness, himself admitted. To one who complimented him on his genius, he replied that if he had made any discoveries, it was owing more to patient attention than to any other talent.[a] There is but little analogy between mathematics and play-acting; but I heard the great Mrs Siddons, in nearly the same language, attribute the whole superiority of her unrivalled talent to the more intense study which she bestowed upon her parts. If what Alcibiades, in the *Symposium*[β] of Plato, narrates of Socrates were true, the father of Greek philosophy must have possessed this faculty of meditation or continuous attention in the highest degree. The story, indeed, has some appearance of exaggeration; but it shows what Alcibiades, or rather Plato through him, deemed the requisite of a great thinker. According to this report, in a military expedition which Socrates made along with Alcibiades, the philosopher was seen by the Athenian army to stand for a whole day and a night, until the breaking of the second morning, motionless, with a fixed gaze,—thus showing that he was uninterruptedly engrossed with the consideration of a single subject: "And thus," says Alcibiades, "Socrates is ever wont to do when his mind is occupied with inquiries in which there are difficulties to be overcome. He then never interrupts his meditation, and forgets to eat, and drink, and sleep,—everything, in short, until his inquiry has reached its termination, or, at least, until he has seen some light in it." In this history there may be, as I have said, exaggeration; but still the truth of the principle is undeniable. Like Newton, Descartes arro-

Marginal notes: LECT. XIV. — Sir Isaac Newton. — Socrates. — Descartes.

---

[a] See *Reid's Works*, p. 537.   [β] P. 220.—ED.

LECT.
XIV.

Bacon.

Helvetius.

Buffon.

Cuvier.

Chester-field.

Instances
of the
power of
Abstrac-
tion.

gated nothing to the force of his intellect. What he had accomplished more than other men, that he attributed to the superiority of his method;[a] and Bacon, in like manner, eulogises his method, in that it places all men with equal attention upon a level, and leaves little or nothing to the prerogatives of genius.[β] Nay, genius itself has been analysed by the shrewdest observers into a higher capacity of attention. "Genius," says Helvetius,[γ] whom we have already quoted, "is nothing but a continued attention," (*une attention suivie*.) "Genius," says Buffon,[δ] "is only a protracted patience," (*une longue patience*.) "In the exact sciences, at least," says Cuvier,[ε] "it is the patience of a sound intellect, when invincible, which truly constitutes genius." And Chesterfield has also observed, that "the power of applying an attention, steady and undissipated, to a single object, is the sure mark of a superior genius."[ζ]

These examples and authorities concur in establishing the important truth, that he who would, with success, attempt discovery, either by inquiry into the works of nature, or by meditation on the phænomena of mind, must acquire the faculty of abstracting himself, for a season, from the invasion of surrounding objects, must be able even, in a certain degree, to emancipate himself from the dominion of the body, and live, as it were, a pure intelligence, within the circle of his thoughts. This faculty has been manifested, more or less, by all whose names are associated

a *Discours de la Méthode*, p. 1.—ED.
β *Nov. Org.*, lib. i. aph. 61.—ED.
γ *De l'Esprit*, Discours iii. chap. iv.—ED.
δ [Quoted by Ponelle, *Manuel*, p. 371.]

ε [*Eloge Historique de M. Haüy*, quoted by Toussaint, *De la Pensée*, p. 219.]
ζ *Letters to his Son.* Letter lxxxix. [Compare Bonnet, *Essai Analytique*, tom. i., préface, p. 8.]

with the progress of the intellectual sciences. In some, LECT. XIV.
indeed, the power of abstraction almost degenerated
into a habit akin to disease, and the examples which
now occur to me, would almost induce me to retract
what I have said about the exaggeration of Plato's
history of Socrates.

Archimedes,[a] it is well known, was so absorbed in a Archime-des.
geometrical meditation, that he was first aware of the
storming of Syracuse by his own death-wound, and
his exclamation on the entrance of Roman soldiers
was,—*Noli turbare circulos meos.* In like manner,
Joseph Scaliger,[β] the most learned of men, when a Joseph Scaliger.
Protestant student in Paris, was so engrossed in the
study of Homer, that he became aware of the mas-
sacre of St Bartholomew, and of his own escape, only
on the day subsequent to the catastrophe. The philosopher Carneades[γ] was habitually liable to fits of medi- Carneades.
tation so profound, that, to prevent him sinking from
inanition, his maid found it necessary to feed him like
a child. And it is reported of Newton, that, while Newton.
engaged in his mathematical researches, he sometimes
forgot to dine. Cardan,[δ] one of the most illustrious Cardan.
of philosophers and mathematicians, was once, upon a
journey, so lost in thought, that he forgot both his way
and the object of his journey. To the questions of his
driver whither he should proceed, he made no answer;
and when he came to himself at nightfall, he was surprised to find the carriage at a stand-still, and directly
under a gallows. The mathematician Vieta[ε] was some- Vieta.

a See Valerius Maximus, lib. viii. c. 7.—ED.

β See D. Heinsius, *In Josephi Scaligeri Obitum Funebris Oratio,* (1609), p. 15.—ED.

γ Valerius Maximus, *loc. cit.*—ED.

δ [Steeb, *Über den Menschen,* ii. 671.]

ε See Thuanus, *Historiæ sui tem-poris,* lib. cxxix., tom. v. p. 1045, ed. 1630.—ED.

LECT.
XIV.

Budæus.

Malebranche
quoted on
place and
importance
of attention.

times so buried in meditation, that for hours he bore
more resemblance to a dead person than to a living,
and was then wholly unconscious of everything going
on around him. On the day of his marriage, the great
Budæus forgot everything in his philological speculations, and he was only awakened to the affairs of the
external world by a tardy embassy from the marriage-
party, who found him absorbed in the composition of
his *Commentarii.*

It is beautifully observed by Malebranche, "that
the discovery of truth can only be made by the labour
of attention; because it is only the labour of attention which has light for its reward;"[a] and, in another place :[b]—" The attention of the intellect is a natural prayer by which we obtain the enlightenment of
reason. But since the Fall, the intellect frequently experiences appalling droughts; it cannot pray; the labour
of attention fatigues and afflicts it. In fact, this labour
is at first great, and the recompense scanty; while, at
the same time, we are unceasingly solicited, pressed,
agitated by the imagination and the passions, whose
inspiration and impulses it is always agreeable to
obey. Nevertheless, it is a matter of necessity; we
must invoke reason to be enlightened; there is no
other way of obtaining light and intelligence but by
the labour of attention. Faith is a gift of God which
we earn not by our merits; but intelligence is a gift
usually only conceded to desert. Faith is a pure
grace in every sense; but the understanding of a
truth is a grace of such a character that it must be
merited by labour, or by the co-operation of grace.
Those, then, who are capable of this labour, and who

a *Traité de Morale,* partie i. chap.   β *Ibid.,* partie i. chap. v. § 4.—
vi. § 1.   ED.

are always attentive to the truth which ought to guide them, have a disposition which would undoubtedly deserve a name more magnificent than those bestowed on the most splendid virtues. But although this habit or this virtue be inseparable from the love of order, it is so little known among us that I do not know if we have done it the honour of a particular name. May I, therefore, be pardoned in calling it by the equivocal name of *force of intellect.* To acquire this true force by which the intellect supports the labour of attention, it is necessary to begin betimes to labour; for, in the course of nature, we can only acquire habits by acts, and can only strengthen them by exercise. But perhaps the only difficulty is to begin. We recollect that we began, and that we were obliged to leave off. Hence we get discouraged; we think ourselves unfit for meditation; we renounce reason. If this be the case, whatever we may allege to justify our sloth and negligence, we renounce virtue, at least in part. For without the labour of attention, we shall never comprehend the grandeur of religion, the sanctity of morals, the littleness of all that is not God, the absurdity of the passions, and of all our internal miseries. Without this labour, the soul will live in blindness and in disorder; because there is naturally no other way to obtain the light that should conduct us : we shall be eternally under disquietude and in strange embarrassment; for we fear everything when we walk in darkness and surrounded by precipices. It is true that faith guides and supports; but it does so only as it produces some light by the attention which it excites in us; for light alone is what can assure minds, like ours, which have so many enemies to fear."

I have translated a longer extract than I intended
when I began; but the truth and importance of the
observations are so great, and they are so admirably
expressed in Malebranche's own inimitable style, that
it was not easy to leave off. They are only a fragment of a very valuable chapter on the subject, to
which I would earnestly refer you,—indeed, I may
take this opportunity of saying, that there is no philosophical author who can be more profitably studied
than Malebranche. As a thinker, he is perhaps the
most profound that France has ever produced; and
as a writer on philosophical subjects, there is not another European author who can be placed before him.
His style is a model at once of dignity and of natural
ease; and no metaphysician has been able to express
himself so clearly and precisely without resorting to
technical and scholastic terms. That he was the author
of a celebrated, but exploded hypothesis, is, perhaps,
the reason why he is far less studied than he otherwise
deserves. His works are of principal value for the
admirable observations on human nature which they
embody; and were everything to be expunged from
them connected with the *Vision of all Things in the
Deity*, and even with the Cartesian hypothesis in general, they would still remain an inestimable treasury of
the acutest analyses, expressed in the most appropriate, and, therefore, the most admirable, eloquence. In
the last respect, he is only approached, certainly not
surpassed, by Hume and Mendelssohn.

I have dwelt at greater length upon the practical
bearings of Attention, not only because this principle
constitutes the better half of all intellectual power,
but because it is of consequence that you should be
fully aware of the incalculable importance of acquir-

Study of
the writ-
ings of
Male-
branche
recom-
mended.

ing, by early and continued exercise, the habit of
attention.   There are, however, many points of great
moment on which I have not touched, and the dependence of Memory upon Attention might alone form
an interesting matter of discussion. You will find
some excellent observations on this subject in the first
and third volumes of Mr Stewart's *Elements*.[a]

a See *Coll. Works*, ii. p. 122 *et seq.*, and p. 352.—ED.

LECTURE XV.

CONSCIOUSNESS,—ITS EVIDENCE AND AUTHORITY

LECT.
XV.
———
Conscious-
ness the
source of
Philoso-
phy.

HAVING now concluded the discussion in regard to
what Consciousness is, and shown you that it constitutes the fundamental form of every act of knowledge ;—I now proceed to consider it as the source
from whence we must derive every fact in the Philosophy of Mind.   And, in prosecution of this purpose,
I shall, in the first place, endeavour to show you that
it really is the principal, if not the only source, from
which all knowledge of the mental phænomena must
be obtained ;[a] in the second place, I shall consider
the character of its evidence, and what, under different relations, are the different degrees of its authority ; and, in the last place, I shall state what, and of
what nature, are the more general phænomena which

a Under the first head here specified, the Author occasionally delivered from the Chair three lectures,
which contained "a summary view
of the nervous system in the higher
animals, more especially in man ;
and a statement of some of the results obtained [by him] from an extensive and accurate induction on
the size of the Encephalus and its
principal parts both in man and the
lower animals,—serving to prove that
no assistance is afforded to Mental
Philosophy by the examination of
the Nervous System, and that the
doctrine, or doctrines, which found
upon the supposed parallelism of
brain and mind, are, as far as observation extends, wholly groundless." These lectures, as foreign in
their details from the general subject
of the Course, are omitted in the
present publication. A general summary of the principal conclusions to
which the researches of the Author
on this subject conducted him, will
be found in Appendix II.—ED.

it reveals. Having terminated these, I shall then descend to the consideration of the special faculties of knowledge, that is, to the particular modifications of which consciousness is susceptible.

We proceed to consider, in the first place, the authority,—the certainty, of this instrument. Now, it is at once evident, that philosophy, as it affirms its own possibility, must affirm the veracity of consciousness; for, as philosophy is only a scientific development of the facts which consciousness reveals, it follows, that philosophy, in denying or doubting the testimony of consciousness, would deny or doubt its own existence. If, therefore, philosophy be not *felo de se*, it must not invalidate the integrity of that which is ; as it were, the heart, the *punctum saliens*, of its being, and as it would actively maintain its own credit, it must be able positively to vindicate the truth of consciousness : for, as Lucretius[a] well observes,

The possibility of Philosophy implies the veracity of consciousness.

> ". . . Ut in Fabrica, si prava est Regula prima,
> Normaque si fallax rectis regionibus exit,
> Omnia mendose fieri, atque obstipa necessum est ;
> Sic igitur Ratio tibi rerum prava necesse est,
> Falsaque sit, falsis quacunque ab Sensibus orta est."

And Leibnitz[β] truly says—"If our immediate internal experience could possibly deceive us, there could no longer be for us any truth of fact (*vérité de fait*), nay, nor any truth of reason (*vérité de raison*)."

So far there is, and can be, no dispute ; if philosophy is possible, the evidence of consciousness is authentic. No philosopher denies its authority, and even the Sceptic can only attempt to show, on the hypothesis of the Dogmatist, that consciousness, as at

---

a *De Rerum Natura*, lib. iv. 516.
—Ed.

β *Nouveaux Essais*, liv. ii. c. 27,
§ 13.—Ed.

variance with itself, is, therefore, on that hypothesis, mendacious.

But if the testimony of consciousness be in itself confessedly above all suspicion, it follows, that we inquire into the conditions or laws which regulate the legitimacy of its applications. The conscious mind being at once the source from which we must derive our knowledge of its phænomena, and the mean through which that knowledge is obtained, Psychology is only an evolution, by consciousness, of the facts which consciousness itself reveals. As every system of Mental Philosophy is thus only an exposition of these facts, every such system, consequently, is true and complete, as it fairly and fully exhibits what, and what only, consciousness exhibits.

Conscious-
ness, as the
criterion of
philoso-
phy natur-
ally clear
and unerr-
ing.

But, it may be objected,—if consciousness be the only revelation we possess of our intellectual nature, and if consciousness be also the sole criterion by which we can interpret the meaning of what this revelation contains, this revelation must be very obscure,—this criterion must be very uncertain, seeing that the various systems of philosophy all equally appeal to this revelation, and to this criterion, in support of the most contradictory opinions. As to the fact of the variety and contradiction of philosophical systems,—this cannot be denied, and it is also true that all these systems either openly profess allegiance to consciousness, or silently confess its authority. But admitting all this, I am still bold enough to maintain, that consciousness affords not merely the only revelation, and only criterion of philosophy, but that this revelation is naturally clear,—this criterion, in itself, unerring. The history of philosophy, like the history of theology, is only, it is too true, the history

of variations, and we must admit of the book of con- LECT.
sciousness what a great Calvinist divine[a] bitterly XV.
confessed of the book of Scripture,—

" Hic liber est in quo quærit sua dogmata quisque ;
Invenit et pariter dogmata quisque sua."

In regard, however, to either revelation, it can be Cause of
shown that the source of this diversity is not in the in philoso-
book, but in the reader. If men will go to the Bible, phy.
not to ask of it what they shall believe, but to find in
it what they believe already, the standard of unity and
truth becomes in human hands only a Lesbian rule.[β]
And if philosophers, in place of evolving their doctrines out of consciousness, resort to consciousness
only when they are able to quote its authority in
confirmation of their preconceived opinions, philosophical systems, like the sandals of Theramenes,[γ] may
fit any feet, but can never pretend to represent the
immutability of nature. And that philosophers have
been, for the most part, guilty of this, it is not extremely difficult to show. They have seldom or never
taken the facts of consciousness, the whole facts of consciousness, and nothing but the facts of consciousness.
They have either overlooked, or rejected, or interpolated.

Before we are entitled to accuse consciousness of We are
being a false, or vacillating, or ill-informed witness, bound to
we are bound, first of all, to see whether there be any there be
rules by which, in employing the testimony of con- any rules
sciousness, we must be governed ; and whether philo- in employ-
sophers have evolved their systems out of conscious- ing the
testimony
of con-
sciousness,
we must be
governed.

[a] S. Werenfels, *Dissertationes*, Am- κατόν.—ED.
stel, 1716, vol. ii. p. 391.—ED. [γ] Θηραμένης διὰ τὸ μὴ μόνιμον ἀλλὰ
[β] Aristotle, *Eth. Nic.*, v. 10 : Τοῦ καὶ ἐπαμφοτερίζον ἀεὶ τῇ προαιρέσει τῆς
γὰρ ἀορίστου ἀόριστος καὶ ὁ κανών ἐστιν, πολιτείας, ἐπεκλήθη Κόθορνος. Plu-
ὥσπερ καὶ τῆς Λεσβίας οἰκοδομῆς ὁ tarch, *Nicias,—Opera*, vol. i. p. 524
μολίβδινος κανών· πρὸς γὰρ τὸ σχῆμα (ed. 1599).—ED.
τοῦ λίθου μετακινεῖται καὶ οὐ μένει ὁ

ness in obedience to these rules. For if there be
rules under which alone the evidence of consciousness
can be fairly and fully given, and, consequently, under
which alone consciousness can serve as an infallible
standard of certainty and truth, and if philosophers
have despised or neglected these, then must we
remove the reproach from the instrument, and affix it
to those blundering workmen who have not known
how to handle and apply it. In attempting to vindicate the veracity and perspicuity of this, the natural,
revelation of our mental being, I shall, therefore, first,
endeavour to enumerate and explain the general rules
by which we must be governed in applying consciousness as a mean of internal observation, and thereafter show how the variations and contradictions of
philosophy have all arisen from the violation of one
or more of these laws. If I accomplish this at present but imperfectly, I may at least plead in excuse,
that the task I undertake is one that has not been
previously attempted. I, therefore, request that you
will view what I am to state to you on this subject
rather as the outline of a course of reasoning, than as
anything pretending to finished argument.

Three
grand
Laws, un-
der which
conscious-
ness can be
legitimate-
ly applied
to the con-
sideration
of its own
phæno-
mena.
In attempting a scientific deduction of the philosophy of mind from the data of consciousness, there are,
in all, if I generalise correctly, three laws which afford
the exclusive conditions of psychological legitimacy.
These laws, or regulative conditions, are self-evident,
and yet they seem never to have been clearly proposed to
themselves by philosophers,—in philosophical speculation, they have certainly never been adequately obeyed.

1. The Law
of Parci-
mony.
The First of these rules is,—That no fact be assumed
as a fact of consciousness but what is ultimate and
simple. This I would call the law of Parcimony.

The Second,—that which I would style the law of Integrity, is—That the whole facts of consciousness be taken without reserve or hesitation, whether given as constituent, or as regulative, data.

The Third is,—That nothing but the facts of consciousness be taken, or, if inferences of reasoning be admitted, that these at least be recognised as legitimate only as deduced from, and in subordination to, the immediate data of consciousness, and every position rejected as illegitimate, which is contradictory of these. This I would call the law of Harmony.

I shall consider these in their order.

I. The first law, that of Parcimony, is,—That no fact be assumed as a fact of consciousness but what is ultimate and simple. What is a fact of consciousness? This question of all others requires a precise and articulate answer, but I have not found it adequately answered in any psychological author.

In the first place, every mental phænomenon may be called a fact of consciousness. But as we distinguish consciousness from the special faculties, though these are all only modifications of consciousness,—only branches of which consciousness is the trunk; so we distinguish the special and derivative phænomena of mind from those that are primary and universal, and give to the latter the name of *facts of consciousness*, as more eminently worthy of that appellation. In an act of perception, for example, I distinguish the pen I hold in my hand, and my hand itself, from my mind perceiving them. This distinction is a particular fact,— the fact of a particular faculty, perception. But there is a general fact, a general distinction, of which this is only a special case. This general fact is the distinction of the Ego and non-Ego, and it belongs to consciousness as the general faculty. Whenever, there-fore, in our analysis of the intellectual phænomena, we arrive at an element which we cannot reduce to a gene-ralisation from experience, but which lies at the root of all experience, and which we cannot, therefore, re-solve into any higher principle,—this we properly call a fact of consciousness. Looking to such a fact of consciousness as the last result of an analysis, we call it an *ultimate* principle ; looking from it as the first constituent of all intellectual combination, we call it a *primary* principle. A fact of consciousness is, thus, a simple, and, as we regard it, either an ultimate, or a primary, datum of intelligence. It obtains also various denominations ; sometimes it is called an *a priori principle*, sometimes a *fundamental law* of mind, some-times a *transcendental condition* of thought,[a] &c. &c.

2. Neces-sary.

But, in the second place, this, its character of ulti-mate priority, supposes its character of necessity. It must be impossible not to think it. In fact, by its necessity alone can we recognise it as an original datum of intelligence, and distinguish it from any mere result of generalisation and custom.

3. Given with a mere belief of its re-ality.

In the third place, this fact, as ultimate, is also given to us with a mere belief of its reality; in other words, consciousness reveals that it is, but not why or how it is. This is evident. Were this fact given us, not only with a belief, but with a knowledge of how or why it is, in that case it would be a derivative, and not a primary, datum. For that whereby we were thus enabled to comprehend its how and why,—in other words, the reason of its existence,—this would be relatively prior, and to it or to its antecedent must we ascend, until we arrive at that primary fact, in which

a See *Reid's Works*, p. 755 *et seq.*—Ed.

we must at last believe,—which we must take upon trust, but which we could not comprehend, that is, think under a higher notion.

A fact of consciousness is thus,—that whose existence is given and guaranteed by an original and necessary belief. But there is an important distinction to be here made, which has not only been overlooked by all philosophers, but has led some of the most distinguished into no inconsiderable errors.

The facts of consciousness are to be considered in two points of view; either as evidencing their own ideal or phænomenal existence, or as evidencing the objective existence of something else beyond them.[a] A belief in the former is not identical with a belief in the latter. The one cannot, the other may possibly, be refused. In the case of a common witness, we cannot doubt the fact of his personal reality, nor the fact of his testimony as emitted, but we can always doubt the truth of that which his testimony avers. So it is with consciousness. We cannot possibly refuse the fact of its evidence as given, but we may hesitate to admit that beyond itself of which it assures us. I shall explain by taking an example. In the act of External Perception, consciousness gives as a conjunct fact, the existence of Me or Self as perceiving, and the existence of something different from Me or Self as perceived. Now the reality of this, as a subjective datum,—as an ideal phænomenon, it is absolutely impossible to doubt without doubting the existence of consciousness, for consciousness is itself this fact; and to doubt the existence of consciousness is absolutely impossible; for as such a doubt could not exist, except in and through consciousness, it would, consequently,

The facts of consciousness to be considered in two points of view: either as evidencing their own ideal existence, or the objective existence of something beyond them.

How far doubt is possible in regard to a fact of Consciousness. Illustrated in the case of Perception.

a See *Reid's Works*, Note A, p. 743 *et seq.*—ED.

annihilate itself. We should doubt that we doubted. As contained,—as given, in an act of consciousness, the contrast of mind knowing and matter known cannot be denied.

But the whole phænomenon as given in consciousness may be admitted, and yet its inference disputed. It may be said, consciousness gives the mental subject as perceiving an external object, contradistinguished from it as perceived : all this we do not, and cannot, deny. But consciousness is only a phænomenon ;— the contrast between the subject and object may be only apparent, not real ; the object given as an external reality, may only be a mental representation, which the mind is, by an unknown law, determined unconsciously to produce, and to mistake for something different from itself. All this may be said and believed, without self-contradiction,—nay, all this has, by the immense majority of modern philosophers, been actually said and believed.

In like manner, in an act of Memory consciousness connects a present existence with a past. I cannot deny the actual phænomenon, because my denial would be suicidal, but I can, without self-contradiction, assert that consciousness may be a false witness in regard to any former existence ; and I may maintain, if I please, that the memory of the past, in consciousness, is nothing but a phænomenon, which has no reality beyond the present. There are many other facts of consciousness which we cannot but admit as ideal phænomena, but may discredit as guaranteeing aught beyond their phænomenal existence itself. The legality of this doubt I do not at present consider, but only its possibility; all that I have now in view being to show that we must not confound, as has been done, the double im-

port of the facts, and the two degrees of evidence for
their reality. This mistake has, among others, been
made by Mr Stewart.[a] "The belief," he says, "which
accompanies consciousness, as to the present existence
of its appropriate phænomena, has been commonly considered as much less obnoxious to cavil, than any of
the other principles which philosophers are accustomed
to assume as self-evident, in the formation of their
metaphysical systems. No doubts on this head have
yet been suggested by any philosopher how sceptical
soever, even by those who have called in question
the existence both of mind and of matter. And yet
the fact is, that it rests on no foundation more solid
than our belief of the existence of external objects ; or
our belief, that other men possess intellectual powers
and faculties similar to those of which we are conscious
in ourselves. In all these cases, the only account that
can be given of our belief is, that it forms a necessary
part of our constitution; against which metaphysicians
may easily argue so as to perplex the judgment, but
of which it is impossible for us to divest ourselves for
a moment, when we are called on to employ our reason, either in the business of life, or in the pursuits
of science. While we are under the influence of our
appetites, passions, or affections, or even of a strong
speculative curiosity, all those difficulties which be-
wildered us in the solitude of the closet, vanish before
the essential principles of the human frame."

With all the respect to which the opinion of so distinguished a philosopher as Mr Stewart is justly entitled, I must be permitted to say, that I cannot but
regard his assertion,—that the present existence of
the phænomena of consciousness, and the reality of

a *Phil. Essays—Works*, vol. v. p. 57.

that to which these phænomena bear witness, rest on
a foundation equally solid,—as wholly untenable. The
second fact, the fact testified to, may be worthy of all
credit,—as I agree with Mr Stewart in thinking that
it is; but still it does not rest on a foundation equally
solid as the fact of the testimony itself. Mr Stewart
confesses that of the former no doubt had ever been
suggested by the boldest sceptic; and the latter, in
so far as it assures us of our having an immediate
knowledge of the external world,—which is the case
alleged by Mr Stewart,—has been doubted, nay, denied,
not merely by sceptics, but by modern philosophers
almost to a man. This historical circumstance, therefore, of itself, would create a strong presumption, that
the two facts must stand on very different foundations;
and this presumption is confirmed when we investigate what these foundations themselves are.

The one fact,—the fact of the testimony, is an act
of consciousness itself; it cannot, therefore, be invali-
dated without self-contradiction. For, as we have frequently observed, to doubt of the reality of that of
which we are conscious is impossible; for as we can
only doubt through consciousness, to doubt of consciousness is to doubt of consciousness by consciousness. If, on the one hand, we affirm the reality of the
doubt, we thereby explicitly affirm the reality of consciousness, and contradict our doubt; if, on the other
hand, we deny the reality of consciousness, we implicitly deny the reality of our denial itself. Thus, in
the act of perception, consciousness gives us a conjunct
fact, an ego or mind, and a non-ego or matter, known
together, and contradistinguished from each other.
Now, as a present phænomenon, this double fact cannot possibly be denied. I cannot, therefore, refuse the

fact, that, in perception, I am conscious of a phænomenon, which I am compelled to regard as the attribute
of something different from my mind or self.   This
I must perforce admit, or run into self-contradiction.
But admitting this, may I not still, without self-contradiction, maintain that what I am compelled to view
as the phænomenon of something different from me is
nevertheless (unknown to me) only a modification of
my mind ?   In this I admit the fact of the testimony
of consciousness as given, but deny the truth of its
report.   Whether this denial of the truth of consciousness as a witness, is or is not legitimate, we are not,
at this moment, to consider : all I have in view at
present is, as I said, to show that we must distinguish
in consciousness two kinds of facts,—the fact of consciousness testifying, and the fact of which consciousness testifies ; and that we must not, as Mr Stewart
has done, hold that we can as little doubt of the fact
of the existence of an external world, as of the fact
that consciousness gives, in mutual contrast, the phænomenon of self, in contrast to the phænomenon of
not-self.[a]

Under this first law, let it, therefore, be laid down, Results of
in the first place, that by a fact of consciousness, pro- Parcimony
perly so called, is meant a primary and universal fact
of our intellectual being ; and, in the second, that
such facts are of two kinds,—1°, The facts given in
the act of consciousness itself ; and, 2°, The facts
which consciousness does not at once give, but to the

a The only philosopher whom I   an external world is not self-contra-
have met with, touching on the ques-   dictory ; by no means,—he is only
tion, is Father Buffier, and he seems   mad."—*Traité des Premières Vérités*,
to strike the nail upon the head. He   c. xi. § 89. [See *Reid's Works*, p.
says, as I recollect,—" He who gain-   787.—ED.]
says the evidence of consciousness of

LECT.
XV.

reality of which it only bears evidence. And as sim-
plification is always a matter of importance, we may
throw out of account altogether the former class of
these facts; for of such no doubt can be, or has been,
entertained. It is only the authority of these facts as
evidence of something beyond themselves,—that is,
only the second class of facts,—which becomes matter
of discussion; it is not the reality of consciousness that
we have to prove, but its veracity.[a]

II. The
Law of
Integrity.

The second rule is,—That the whole facts of consciousness be taken without reserve or hesitation,
whether given as constituent, or as regulative, data.
This rule is too manifest to require much elucidation.
As philosophy is only a development of the phænomena and laws of consciousness, it is evident that
philosophy can only be complete, as it comprehends,
in one harmonious system, all the constituent, and all
the regulative, facts of consciousness. If any phænomenon or constituent fact of consciousness be omitted,
the system is not complete; if any law or regulative
fact is excluded, the system is not legitimate.

III. The
Law of
Harmony.

The violation of this second rule is, in general, connected with a violation of the third, and we shall accordingly illustrate them together. The third is,—
That nothing but the facts of consciousness be taken,
or if inferences of reasoning be admitted, that these
at least be recognised as legitimate only as deduced
from, and only in subordination to, the immediate
data of consciousness, and that every position be rejected as illegitimate which is contradictory of these.

These illus-
trated in
conjunc-
tion.

The truth and necessity of this rule are not less
evident than the truth and necessity of the preceding.
Philosophy is only a systematic evolution of the con-

[a] See *Reid's Works*, p. 743 *et seq.*—ED.

tents of consciousness, by the instrumentality of con-sciousness; it, therefore, necessarily supposes, in both respects, the veracity of consciousness.

But, though this be too evident to admit of doubt, and though no philosopher has ever openly thrown off allegiance to the authority of consciousness, we find, nevertheless, that its testimony has been silently overlooked, and systems established upon principles in direct hostility to the primary data of intelligence. It is only such a violation of the integrity of con-sciousness, by the dogmatist, that affords, to the sceptic, the foundation on which he can establish his proof of the nullity of philosophy. The sceptic cannot assail the truth of the facts of consciousness in them-selves. In attempting this he would run at once into self-contradiction. In the first place, he would enact the part of a dogmatist, that is, he would positively, —dogmatically, establish his doubt. In the second, waiving this, how can he accomplish what he thus proposes? For why? He must attack conscious-ness either from a higher ground, or from conscious-ness itself. Higher ground than consciousness there is none; he must, therefore, invalidate the facts of consciousness from the ground of consciousness itself. On this ground, he cannot, as we have seen, deny the facts of consciousness as given; he can only attempt to invalidate their testimony. But this again can be done only by showing that consciousness tells dif-ferent tales,—that its evidence is contradictory,—that its data are repugnant. But this no sceptic has ever yet been able to do. Neither does the sceptic or negative philosopher himself assume his principles; he only accepts those on which the dogmatist or positive philosopher attempts to establish his doc-

How scep-
ticism
arises out
of partial
dogmatic
systems.

trine ; and this doctrine he reduces to zero, by showing that its principles are, either mutually repugnant, or repugnant to facts of consciousness on
which, though it may not expressly found, still, as
facts of consciousness, it cannot refuse to recognise
without denying the possibility of philosophy in
general.

Violations
of the Se-
cond and
Third laws
in the writ-
ings of Dr
Thomas
Brown.

I shall illustrate the violation of this rule by examples taken from the writings of the late ingenious
Dr Thomas Brown.—I must, however, premise that
this philosopher, so far from being singular in his
easy way of appealing to, or overlooking, the facts of
consciousness, as he finds them convenient or inconvenient for his purpose, supplies only a specimen of

Brown's
doctrine of
External
Perception
involves an
inconsist-
ency.

the too ordinary style of philosophising. Now, you
must know, that Dr Brown maintains the common
doctrine of the philosophers, that we have no immediate knowledge of anything beyond the states or
modifications of our own minds,—that we are only
conscious of the ego,—the non-ego, as known, being
only a modification of self, which mankind at large
are illusively determined to view as external and different from self. This doctrine is contradictory of the
fact to which consciousness testifies,—that the object
of which we are conscious in perception, is the external
reality as existing, and not merely its representation
in the percipient mind. That this is the fact testified
to by consciousness, and believed by the common-
sense of mankind, is admitted even by those philosophers who reject the truth of the testimony and the
belief. It is of no consequence to us at present what
are the grounds on which the principle is founded,
that the mind can have no knowledge of aught besides itself ; it is sufficient to observe that, this prin-

ciple being contradictory of the testimony of consciousness, Dr Brown, by adopting it, virtually accuses consciousness of falsehood. But if consciousness be false
in its testimony to one fact, we can have no confidence
in its testimony to any other ; and Brown, having
himself belied the veracity of consciousness, cannot,
therefore, again appeal to this veracity as to a credible
authority. But he is not thus consistent. Although
he does not allow that we have any knowledge of the
existence of an outer world, the existence of that
world he still maintains. And on what grounds ? He
admits the reasoning of the idealist, that is, of the
philosopher who denies the reality of the material
universe,—he admits this to be invincible. How,
then, is his conclusion avoided ? Simply by appealing
to the universal belief of mankind in favour of the
existence of external things,[a]—that is, to the authority of a fact of consciousness. But to him this appeal
is incompetent. For, in the first place, having already
virtually given up, or rather positively rejected, the
testimony of consciousness, when consciousness de-
posed to our immediate knowledge of external things,
—how can he even found upon the veracity of that
mendacious principle, when bearing evidence to the
unknown existence of external things ? I cannot but
believe that the material reality exists ; therefore,
it does exist, for consciousness does not deceive us,—
this reasoning Dr Brown employs when defending his
assertion of an outer world. I cannot but believe
that the material reality is the object immediately
known in perception ; therefore, it is immediately
known, for consciousness does not deceive us,—this

a *Philosophy of the Human Mind,* See this argument further pursued in
Lecture xxviii., p. 175-177, ed. 1830. the Author's *Discussions,* p. 92.—ED.

reasoning Dr Brown rejects when establishing the foundation of his system. In the one case he maintains,—this belief, because irresistible, is true; in the other case, he maintains,—this belief, though irresistible, is false. Consciousness is veracious in the former belief, mendacious in the latter. I approbate the one, I reprobate the other. The inconsistency of this is apparent. It becomes more palpable when we consider, in the second place, that the belief which Dr Brown assumes as true rests on,—is, in fact, only the reflex of,—the belief which he repudiates as false. Why do mankind believe in the existence of an outer world? They do not believe in it as in something unknown; but, on the contrary, they believe it to exist, only because they believe that they immediately know it to exist. The former belief is only as it is founded on the latter. Of all absurdities, therefore, the greatest is to assert,—on the one hand, that consciousness deceives us in the belief that we know any material object to exist; and, on the other, that the material object exists, because, though on false grounds, we believe it to exist.

The same is true of Brown's proof of our Personal Identity.

I may give you another instance, from the same author, of the wild work that the application of this rule makes, among philosophical systems not legitimately established. Dr Brown, with other philosophers, rests the proof of our Personal Identity, and of our Mental Individuality, on the ground of beliefs, which, as "intuitive, universal, immediate, and irresistible," he, not unjustly, regards as the "internal and never-ceasing voice of our Creator,—revelations from on high, omnipotent, [and veracious], as their Author."[a]

a *Philosophy of the Human Mind,* also Sir W. Hamilton's *Discussions,* Lecture xiii., p. 79, ed. 1830. See p. 96.—ED.

To him this argument is, however, incompetent, as
contradictory.

What we know of self or person, we know only as
a fact of consciousness. In our perceptive consciousness, there is revealed, in contrast to each, a self and
a not-self. This contrast is either true or false. If
true, then am I conscious of an object different from
me,—that is, I have an immediate perception of the
external reality. If false, then am I not conscious
of anything different from me, but what I am constrained to regard as not-me is only a modification
of me, which, by an illusion of my nature, I mistake, and must mistake, for something different from
me.

Now, will it be credited that Dr Brown—and be it
remembered that I adduce him only as the representative of a great majority of philosophers—affirms or
denies, just as he finds it convenient or inconvenient,
this fact, this distinction, of consciousness? In his
doctrine of perception, he explicitly denies its truth,
in denying that mind is conscious of aught beyond
itself. But, in other parts of his philosophy, this false
fact, this illusive distinction, and the deceitful belief
founded thereupon, are appealed to, (I quote his expressions), as "revelations from on high, — as the
never-ceasing voice of our Creator," &c.

Thus, on the veracity of this mendacious belief, Dr
Brown establishes his proof of our personal identity.
Touching the object of perception, when its evidence
is inconvenient, this belief is quietly passed over, as
incompetent to distinguish not-self from self; in the
question regarding our personal identity, where its
testimony is convenient, it is clamorously cited as an
inspired witness, exclusively competent to distinguish

self from not-self. Yet why, if, in the one case, it
mistook self for not-self, it may not, in the other,
mistake not-self for self, would appear a problem not
of the easiest solution.

The same belief, with the same inconsistency, is
called in to prove the Individuality of mind.[a] But if
we are fallaciously determined, in our perceptive consciousness, to regard mind both as mind and as matter,
—for, on Brown's hypothesis, in perception, the object
perceived is only a mode of the percipient subject,—if,
I say, in this act, I must view what is supposed one
and indivisible, as plural, and different, and opposed,
—how is it possible to appeal to the authority of a testimony so treacherous as consciousness for an evidence
of the real simplicity of the thinking principle ? How,
says the materialist to Brown,—how can you appeal
against me to the testimony of consciousness, which
you yourself reject when against your own opinions,
and how can you, on the authority of that testimony,
maintain the unity of self to be more than an illusive
appearance, when self and not-self, as known to consciousness, are, on your own hypothesis, confessedly
only modifications of the same percipient subject?
If, on your doctrine, consciousness can split what you
hold to be one and indivisible into two, not only different but opposed, existences, — what absurdity is
there, on mine, that consciousness should exhibit as
phænomenally one, what we both hold to be really
manifold ? If you give the lie to consciousness in
favour of your hypothesis, you can have no reasonable
objection that I should give it the lie in favour of
mine. If you can maintain that not-self is only an
illusive phænomenon,—being, in fact, only self in dis-

a Lecture xii., p. 74, ed. 1830.—Ed.

guise ; I may also maintain, *a contra*, that self itself is only an illusive phænomenon, and that the apparent unity of the ego is only the result of an organic harmony of action between the particles of matter.

From these examples, the truth of the position I maintain is manifest,—that a fact of consciousness can only be rejected on the supposition of falsity, and that, the falsity of one fact of consciousness being admitted, the truth of no other fact of consciousness can be maintained. The legal brocard, *Falsus in uno, falsus in omnibus,* is a rule not more applicable to other witnesses than to consciousness. Thus, every system of philosophy which implies the negation of any fact of consciousness, is not only necessarily unable, without self-contradiction, to establish its own truth by any appeal to consciousness ; it is also unable, without self-contradiction, to appeal to consciousness against the falsehood of any other system. If the absolute and universal veracity of consciousness be once surrendered, every system is equally true, or rather all are equally false ; philosophy is impossible, for it has now no instrument by which truth can be discovered,—no standard by which it can be tried ; the root of our nature is a lie. But though it is thus manifestly the common interest of every scheme of philosophy to preserve intact the integrity of consciousness, almost every scheme of philosophy is only another mode in which this integrity has been violated. If, therefore, I am able to prove the fact of this various violation, and to show that the facts of consciousness have never, or hardly ever, been fairly evolved, it will follow, as I said, that no reproach can be justly addressed to consciousness as an ill-informed, or vacillating, or perfidious witness, but to those only who were too

The abso-
lute and
universal
veracity of
conscious-
ness must
be main-
tained.

LECT.
XV.

proud, or too negligent, to accept its testimony, to employ its materials, and to obey its laws. And on this supposition, so far should we be from despairing of the future advance of philosophy from the experience of its past wanderings, that we ought, on the contrary, to anticipate for it a steady progress, the moment that philosophers can be persuaded to look to consciousness, and to consciousness alone, for their materials and their rules.

LECTURE XVI.

CONSCIOUSNESS,—VIOLATIONS OF ITS AUTHORITY.

ON the principle, which no one has yet been found bold enough formally to deny, and which, indeed, requires only to be understood to be acknowledged,— viz., that as all philosophy is evolved from consciousness, so, on the truth of consciousness, the possibility of all philosophy is dependent, — it is manifest, at once, and without further reasoning, that no philosophical theory can pretend to truth except that single theory which comprehends and develops the fact of consciousness on which it founds, without retrenchment, distortion, or addition.   Were a philosophical system to pretend that it culls out all that is correct in a fact of consciousness, and rejects only what is erroneous,—what would be the inevitable result ?   In the first place, this system admits, and must admit, that it is wholly dependent on consciousness for its constituent elements, and for the rules by which these are selected and arranged,—in short, that it is wholly dependent on consciousness for its knowledge of true and false.   But, in the second place, it pretends to select a part, and to reject a part, of a fact given and guaranteed by consciousness.   Now, by what criterion, by what standard, can it discriminate the true from the false in this fact ?   This criterion must be either con-

sciousness itself, or an instrument different from consciousness. If it be an instrument different from
consciousness, what is it? No such instrument has
ever yet been named,—has ever yet been heard of. If
it exist, and if it enable us to criticise the data of consciousness, it must be a higher source of knowledge
than consciousness, and thus it will replace consciousness as the first and generative principle of philosophy.
But of any principle of this character, different from
consciousness, philosophy is yet in ignorance. It remains unenounced and unknown. It may, therefore,
be safely assumed not to be. The standard, therefore,
by which any philosophical theory can profess to regulate its choice among the elements of any fact of consciousness, must be consciousness itself. Now, mark
the dilemma. The theory makes consciousness the
discriminator between what is true and what is false
in its own testimony. But if consciousness be assumed to be a mendacious witness in certain parts of
its evidence, how can it be presumed a veracious witness in others? This it cannot be. It must be held
as false in all, if false in any; and the philosophical
theory which starts from this hypothesis, starts from
a negation of itself in the negation of philosophy in
general. Again, on the hyphothesis that part of the
deliverance of consciousness is true, part false, how
can consciousness enable us to distinguish these? This
has never yet been shown; it is, in fact, inconceivable.
But, further, how is it discovered that any part of a
datum of consciousness is false, another true? This
can only be done if the datum involve a contradiction.
But if the facts of consciousness be contradictory,
then is consciousness a principle of falsehood; and the
greatest of conceivable follies would be an attempt

to employ such a principle in the discovery of truth. And such an act of folly is every philosophical theory which, departing from an admission that the data of consciousness are false, would still pretend to build out of them a system of truth. But, on the other hand, if the data of consciousness are not contradictory, and consciousness, therefore, not a self-convicted deceiver, how is the unapparent falsehood of its evidence to be evinced ? This is manifestly impossible ; for such falsehood is not to be presumed ; and, we have previously seen, there is no higher principle by which the testimony of consciousness can be canvassed and red-argued. Consciousness, therefore, is to be presumed veracious ; a philosophical theory which accepts one part of the harmonious data of consciousness and rejects another, is manifestly a mere caprice, a chimera not worthy of consideration, far less of articulate disproof. It is *ab initio* null.

I have been anxious thus again to inculcate upon you this view in regard to the relation of Philosophy to Consciousness, because it contains a preliminary refutation of all those proud and wayward systems which,—though they can only pretend to represent the truth, inasmuch as they fully and fairly develop the revelations vouchsafed to us through consciousness,—still do, one and all of them, depart from a false or partial acceptance of these revelations themselves ; and because it affords a clear and simple criterion of certainty in our own attempts at philosophical construction. If it be correct, it sweeps away at once a world of metaphysical speculation ; and if it curtail the dominions of human reason, it firmly establishes our authority over what remains.

In order still further to evince to you the importance

Violations
of the au-
thority of
conscious-
ness illus-
trated.

of the precept (viz., that we must look to consciousness
and to consciousness alone for the materials and rules
of philosophy), and to show articulately how all the
variations of philosophy have been determined by its
neglect, I will take those facts of consciousness which
lie at the very root of philosophy, and with which, consequently, all philosophical systems are necessarily and
primarily conversant; and point out how, besides the
one true doctrine which accepts and simply states the
fact as given, there are always as many various actual
theories as there are various possible modes of distort-

ing or mutilating this fact. I shall commence with
that great fact to which I have already alluded,—that
we are immediately conscious in perception of an ego
and a non-ego, known together, and known in contrast to each other. This is the fact of the Duality of
Consciousness. It is clear and manifest. When I concentrate my attention in the simplest act of perception, I return from my observation with the most irresistible conviction of two facts, or rather two branches
of the same fact;—that I am,—and that something
different from me exists. In this act, I am conscious
of myself as the perceiving subject, and of an external
reality as the object perceived; and I am conscious of
both existences in the same indivisible moment of intuition. The knowledge of the subject does not precede, nor follow, the knowledge of the object,—neither
determines, neither is determined by, the other.

The fact of
the testi-
mony of
conscious-
ness in Per-
ception al-
lowed by
those who
deny its
truth.

Such is the fact of perception revealed in consciousness, and as it determines mankind in general in their
almost equal assurance of the reality of an external
world, as of the existence of their own minds. Consciousness declares our knowledge of material qualities
to be intuitive or immediate,—not representative or

mediate. Nor is the fact, as given, denied even by those LECT.
who disallow its truth. So clear is the deliverance, XVI.
that even the philosophers who reject an intuitive perception, find it impossible not to admit, that their doctrine stands decidedly opposed to the voice of consciousness,—to the natural convictions of mankind.
I may give you some examples of the admission of this
fact, which it is of the utmost importance to place
beyond the possibility of doubt. I quote, of course,
only from those philosophers whose systems are in
contradiction of the testimony of consciousness, which
they are forced to admit. I might quote to you confessions to this effect from Descartes, *De Passionibus*,
article 23, and from Malebranche, *Recherche*, liv. iii.
c. 1. To these I only refer you.

The following is from Berkeley, towards the con- Berkeley.
clusion of the third and last dialogue, in which his
system of Idealism is established :—" When Hylas is
at last entirely converted, he observes to Philonous,—
'After all, the controversy about matter, in the strict
acceptation of it, lies altogether between you and the
philosophers, whose principles, I acknowledge, are not
near so natural, or so agreeable to the common sense
of mankind, and Holy Scripture, as yours.' Philonous
observes in the end,—'That he does not pretend to
be a setter-up of new notions; his endeavours tend
only to unite, and to place in a clearer light, that
truth which was before shared between the vulgar
and the philosophers; the former being of opinion,
that those things they immediately perceive are the
real things; and the latter, that the things immediately perceived are ideas which exist only in the
mind; which two things put together do, in effect,
constitute the substance of what he advances.' And

LECT.
XVI.

he concludes by observing,—'That those principles which at first view lead to scepticism, pursued to a certain point, bring men back to common sense.'" [a]

Here you will notice that Berkeley admits that the common belief of mankind is, that the things immediately perceived are not representative objects in the mind, but the external realities themselves. Hume, in like manner, makes the same confession; and the confession of that sceptical idealist, or sceptical nihilist, is of the utmost weight.

Hume.

"It seems evident that men are carried by a natural instinct or prepossession to repose faith in their senses; and that, without any reasoning, or even almost before the use of reason, we always suppose an external universe, which depends not on our perception, but would exist though we and every sensible creature were absent or annihilated. Even the animal creation are governed by a like opinion, and preserve this belief of external objects in all their thoughts, designs, and actions.

"It seems also evident that, when men follow this blind and powerful instinct of nature, they always suppose the very images presented by the senses, to be the external objects, and never entertain any suspicion that the one are nothing but representations of the other. This very table, which we see white, and which we feel hard, is believed to exist, independent of our perception, and to be something external to our mind, which perceives it. Our presence bestows not being on it,—our absence does not annihilate it. It preserves its existence uniform and entire, independent of the situation of intelligent beings, who perceive or contemplate it.

"But this universal and primary opinion of all men

a See *Reid's Works,* p. 284.—ED.

is soon destroyed by the slightest philosophy, which teaches us that nothing can ever be present to the mind but an image or perception, and that the senses are only the inlets through which these images are conveyed, without being able to produce any immediate intercourse between the mind and the object. The table, which we see, seems to diminish as we remove further from it; but the real table, which exists independent of us, suffers no alteration ; it was, there-' fore, nothing but its image which was present to the mind.   These are the obvious dictates of reason ; and no man who reflects, ever doubted that the existences which we consider, when we say, *this house* and *that tree*, are nothing but perceptions in the mind, and fleeting copies or representations of other existences, which remain uniform and independent. . . . .

" Do you follow the instincts and propensities of nature, may they say, in assenting to the veracity of sense ? But these lead you to believe that the very perception or sensible image is the external object. Do you disclaim this principle, in order to embrace a more rational opinion, that the perceptions are only representations of something external ? You here depart from your natural propensities and more obvious sentiments ; and yet are not able to satisfy your reason, which can never find any convincing argument from experience to prove that the perceptions are connected with any external objects." [a]

The fact that consciousness does testify to an immediate knowledge by mind of an object different from

a *Essays*, vol. ii. pp. 154-155, 156-157 (edit. 1788). Similar confessions are made by Hume in his *Treatise of Human Nature*, vol. i. pp. 330, 338, 353, 355, 361, 369, (original edit.);— in a word, you may read from 330 to 370 ; and the same thing is acknowledged by Kant, by Fichte, by Schelling, by Tennemann, by Jacobi. Several of these testimonies you will find extracted and translated in a note of my *Discussions on Philosophy*, p. 92

any modification of its own, is thus admitted even by those philosophers who still do not hesitate to deny the truth of the testimony; for to say that all men do naturally believe in such a knowledge, is only, in other words, to say that they believe it upon the authority of consciousness. A fact of consciousness, and a fact of the common sense of mankind, are only various expressions of the same import. We may, therefore, lay it down as an undisputed truth, that consciousness gives, as an ultimate fact, a primitive duality;—a knowledge of the ego in relation and contrast to the non-ego; and a knowledge of the non-ego in relation and contrast to the ego. The ego and non-ego are, thus, given in an original synthesis, as conjoined in the unity of knowledge, and, in an original antithesis, as opposed in the contrariety of existence. In other words, we are conscious of them in an indivisible act of knowledge together and at once,— but we are conscious of them as, in themselves, different and exclusive of each other.

Again, consciousness not only gives us a duality, but it gives its elements in equal counterpoise and independence. The ego and non-ego,—mind and matter, are not only given together, but in absolute coequality. The one does not precede, the other does not follow; and, in their mutual relation, each is equally dependent, equally independent. Such is the

fact as given in and by consciousness. Philosophers have not, however, been content to accept the fact in its integrity, but have been pleased to accept it only under such qualifications as it suited their systems to devise. In truth, there are just as many different philosophical systems originating in this fact, as it admits of various possible modifications. An enumeration of these modifications, accordingly, affords an enumeration of philosophical theories.

LECT. XVI.

In the first place, there is the grand division of philosophers into those who do, and those who do not, accept the fact in its integrity.[a] Of modern philosophers, almost all are comprehended under the latter category, while of the former, if we do not remount to the schoolmen and the ancients,—I am only aware of a single philosopher[β] before Reid, who did not reject, at least in part, the fact as consciousness affords it.

1. Those who do, and those who do not, accept in its integrity the fact of the Duality of Consciousness.

As it is always expedient to possess a precise name for a precise distinction, I would be inclined to denominate those who implicitly acquiesce in the primitive duality as given in consciousness, the Natural Realists or Natural Dualists, and their doctrine, Natural Realism or Natural Dualism.

The former called Natural Realists or Natural Dualists.

In the second place, the philosophers who do not accept the fact, and the whole fact, may be divided and subdivided into various classes by various principles of distribution.

The latter, variously subdivided.

The first subdivision will be taken from the total, or partial, rejections of the import of the fact. I have previously shown you, that to deny any fact of consciousness as an actual phænomenon is utterly impossible. But, though necessarily admitted as a present phænomenon, the import of this phænomenon,—all beyond our actual consciousness of its existence,—may be denied. We are able, without self-contradiction, to suppose, and, consequently, to assert, that all to which the phenomenon of which we are conscious refers, is a deception,—that, for example, the past, to

a See the Author's *Suppl. Dissert. to Reid's Works,* Note C.—ED.

β This philosopher is doubtless Peter Poiret. John Sergeant is subsequently referred to by Sir W. Hamilton, as holding a similar doctrine in a paradoxical form. See below, vol. ii. pp. 92, 124.—ED.

LECT.
XVI.

Into Real-
ists and
Nihilists.

which an act of memory refers, is only an illusion involved in our consciousness of the present,—that the unknown subject to which every phænomenon of which we are conscious involves a reference, has no reality beyond this reference itself,—in short, that all our knowledge of mind or matter, is only a consciousness of various bundles of baseless appearances. This doctrine, as refusing a substantial reality to the phænomenal existence of which we are conscious, is called Nihilism ; and, consequently, philosophers, as they affirm or deny the authority of consciousness in guaranteeing a substratum or substance to the manifestations of the ego and non-ego, are divided into Realists or Substantialists, and into Nihilists or non-Substantialists. Of positive or dogmatic Nihilism there is no example in modern philosophy, for Oken's deduction of the universe from the original nothing,[a]—the nothing being equivalent to the Absolute or God,—is only the paradoxical foundation of a system of realism ; and, in ancient philosophy, we know too little of the book of Gorgias the Sophist, entitled Περὶ τοῦ μὴ ὄντος, ἢ περὶ φύσεως,[b]—*Concerning Nature or the Non-Existent,*—to be able to affirm whether it were maintained by him as a dogmatic and *bona fide* doctrine. But as a sceptical conclusion from the premises of previous philosophers, we have an illustrious example of Nihilism in Hume ; and the celebrated Fichte admits that the speculative principles of his own idealism would, unless corrected by his practical, terminate in this result.[7]

<hr>

a *Lehrbuch der Naturphilosophie,* § 30-43, (ed. 1831). This work has been translated for the Ray Society by Tulk. On Oken's doctrine of Nihilism, see also *Discussions,* pp. 21, 22.—ED.

β See Sextus Empiricus, *Adv.*

Math., vii. 65.—ED.

γ See a remarkable passage in the *Bestimmung des Menschen,* p. 174, (*Werke,* vol. ii. p. 245), translated by Sir W. Hamilton, *Reid's Works,* p. 129.—ED.

The Realists or Substantialists, again, are divided into Dualists, and into Unitarians or Monists, according as they are, or are not, contented with the testimony of consciousness to the ultimate duplicity of subject and object in perception. The Dualists, of whom we are now first speaking, are distinguished from the Natural Dualists of whom we formerly spoke, in this, —that the latter establish the existence of the two worlds of mind and matter on the immediate knowledge we possess of both series of phænomena,—a knowledge of which consciousness assures us; whereas the former, surrendering the veracity of consciousness to our immediate knowledge of material phænomena, and, consequently, our immediate knowledge of the existence of matter, still endeavour, by various hypotheses and reasonings, to maintain the existence of an unknown external world. As we denominate those who maintain a dualism as involved in the fact of consciousness, Natural Dualists; so we may style those dualists who deny the evidence of consciousness to our immediate knowledge of aught beyond the sphere of mind, Hypothetical Dualists or Cosmothetic Idealists.

To the class of Cosmothetic Idealists, the great majority of modern philosophers are to be referred. Denying an immediate or intuitive knowledge of the external reality, whose existence they maintain, they, of course, hold a doctrine of mediate or representative perception; and, according to the various modifications of that doctrine, they are again subdivided into those who view, in the immediate object of perception, a representative entity present to the mind, but not a mere mental modification, and into those who hold that the immediate object is only a representative modification of the mind itself. It is not always easy to

Realists divided into Hypothetical Dualists and Monists.

The majority of modern philosophers belong to the former of these classes, and are subdivided according to their view of the representation in perception.

determine to which of these classes some philosophers belong. To the former, or class holding the cruder hypothesis of representation, certainly belong the followers of Democritus and Epicurus, those Aristotelians who held the vulgar doctrine of species, (Aristotle himself was probably a natural dualist[a]), and in recent times, among many others, Malebranche, Berkeley, Clarke, Newton, Abraham Tucker, &c. To these is also, but problematically, to be referred Locke. To the second, or class holding the finer hypothesis of representation, belong, without any doubt, many of the Platonists, Leibnitz, Arnauld, Crousaz, Condillac, Kant, &c.; and to this class is also probably to be referred Descartes.[β]

Monists, subdivided,

The philosophical Unitarians or Monists reject the testimony of consciousness to the ultimate duality of the subject and object in perception, but they arrive at the unity of these in different ways. Some admit the testimony of consciousness to the equipoise of the mental and material phænomena, and do not attempt to reduce either mind to matter, or matter to mind. They reject, however, the evidence of consciousness to their antithesis in existence, and maintain that mind and matter are only phænomenal modifications of the same common substance. This is the doctrine of Absolute Identity,—a doctrine of which the most illustrious representatives among recent philosophers are Schelling, Hegel, and Cousin. Others again deny the evidence of consciousness to the equipoise of the sub-

Into, 1.
Those who hold the doctrine of Absolute Identity;

a Aristotle's opinion is doubtful. In the *De Anima*, i. 5, he combats the theory of Empedocles, that like is known by like, and appears as a natural realist. But in the *Nicomachean Ethics*, vi. 1, he adopts the principle of similarity as the basis of all knowledge. See *Reid's Works*, pp. 300, n. *, 856; also (completed edition), p. 952 a, n. *; and M. St Hilaire's preface to his translation of the *De Anima*, p. xxii.—ED.

β See the Author's *Discussions*, p. 57 et seq.—ED.

ject and object as co-ordinate and co-original elements; LECT.
XVI.
and as the balance is inclined in favour of the one
relative or the other, two opposite schemes of psychology are determined. If the subject be taken as the
original and genetic, and the object evolved from it as 2. Ideal-
its product, the theory of Idealism is established. On ^{ists;}
the other hand, if the object be assumed as the original 3. Mate-
and genetic, and the subject evolved from it as its pro- ^{rialists.}
duct, the theory of Materialism is established.

In regard to these two opposite schemes of a one- How a phi-
sided philosophy, I would at present make an observa- losophical
tion to which it may be afterwards necessary to recur often pre-
—viz., that a philosophical system is often prevented vented
from falling into absolute idealism or absolute mate- absolute
rialism, and held in a kind of vacillating equilibrium, absolute
not in consequence of being based on the fact of con- ism.
sciousness, but from the circumstance that its mate-
rialistic tendency in one opinion happens to be counteracted by its idealistic tendency in another;—two
opposite errors, in short, co-operating to the same
result as one truth. On this ground is to be explained why the philosophy of Locke and Condillac
did not more easily slide into materialism. Deriving
our whole knowledge, mediately or immediately, from
the senses, this philosophy seemed destined to be fairly
analysed into a scheme of materialism ; but from this
it was for a long time preserved, in consequence of
involving a doctrine, which, on the other hand, if not
counteracted, would have naturally carried it over
into idealism. This was the doctrine of a representative perception. The legitimate issue of such a doctrine is now admitted, on all hands, to be absolute
idealism ; and the only ground on which it has been
latterly thought possible to avoid this conclusion,—

an appeal to the natural belief of mankind in the
existence of an external world,—is, as I showed you,
incompetent to the hypothetical dualist or cosmothetic
idealist. In his hands such an appeal is self-contradictory. For if this universal belief be fairly applied,
it only proves the existence of an outer world by dis-
proving the hypothesis of a representative perception.

Recapitula-
tion of fore-
going.
To recapitulate what I have now said :—The philosophical systems concerning the relation of mind
and matter, are coextensive with the various possible
modes in which the fact of the Duality of Consciousness may be accepted or refused. It may be accepted
either wholly and without reserve, or it may not.
The former alternative affords the class of Natural
Realists or Natural Dualists.

Those, again, who do not accept the fact in its
absolute integrity, are subdivided in various manners.
They are, first of all, distinguished into Realists or
Substantialists, and into Nihilists, as they do, or do
not, admit a subject, or subjects, to the two opposite
series of phænomena which consciousness reveals. The
former class is again distributed into Hypothetical
Dualists or Cosmothetic Idealists, and into Unitarians
or Monists.

The Hypothetical Dualists or Cosmothetic Idealists,
are divided, according to their different theories of the
representation in perception, into those who view in
the object immediately perceived a *tertium quid* different both from the external reality and from the
conscious mind, and into those who identify this
object with a modification of the mind itself.

The Unitarians or Monists fall into two classes, as
they do, or do not, preserve the equilibrium of subject and object. If, admitting the equilibrium of these,

they deny the reality of their opposition, the system of Absolute Identity emerges, which carries thought and extension, mind and matter, up into modes of the same common substance.

It would be turning aside from my present purpose, were I to attempt any articulate refutation of these various systems. What I have now in view is to exhibit to you how, the moment that the fact of consciousness in its absolute integrity is surrendered, philosophy at once falls from unity and truth into variety and error. In reality, by the very act of refusing any one datum of consciousness, philosophy invalidates the whole credibility of consciousness, and, consciousness ruined as an instrument, philosophy is extinct. Thus, the refusal of philosophers to accept the fact of the duality of consciousness, is virtually an act of philosophical suicide. Their various systems are now only so many empty spectres,—so many enchanted corpses, which the first exorcism of the sceptic reduces to their natural nothingness. The mutual polemic of these systems is like the warfare of shadows ; as the heroes in Valhalla, they hew each other into pieces, only in a twinkling to be reunited, and again to amuse themselves in other bloodless and indecisive contests.[a]

Having now given you a general view of the various systems of philosophy, in their mutual relations, as founded on the great fact of the Duality of Consciousness, I proceed, in subordination to this fact, to give you a brief account of certain famous hypotheses which it is necessary for you to know,—hypotheses proposed in solution of the problem of how inter-

Hypotheses proposed in regard to the mode of intercourse between Mind and Body.

a This simile is taken from Kant, (edit. 1799).—ED. *Kritik der reinen Vernunft*, p. 784,

LECT.
XVI.

course of substances so opposite as mind and body could be accomplished. These hypotheses, of course, belong exclusively to the doctrine of Dualism, for in the Unitarian system the difficulty is resolved by the annihilation of the opposition, and the reduction of the two substances to one. The hypotheses I allude to, are known under the names, 1°, Of the system of Assistance or of Occasional Causes ; 2°, Of the Pre-established Harmony ; 3°, Of the Plastic Medium ; and, 4°, Of Physical Influence. The first belongs to Descartes, De la Forge, Malebranche, and the Cartesians in general ; the second to Leibnitz and Wolf, though not universally adopted by their school ; the third was an ancient opinion revived in modern times by Cudworth and Leclerc;[a] the fourth is the common doctrine of the Schoolmen, and though not explicitly enounced, that generally prevalent at present;—among modern philosophers, it has been expounded with great perspicuity by Euler.[β] We shall take these in their order.

Four in number.

The hypothesis of Divine Assistance or of Occasional Causes, sets out from the apparent impossibility involved in Dualism of any actual communication between a spiritual and a material substance,—that is, between extended and non-extended existences ; and it terminates in the assertion, that the Deity, on occasion of the affections of matter—of the motions in the bodily organism, excites in the mind correspondent thoughts and representations ; and, on occasion of thoughts or representations arising in the

1. Occasional Causes.

[a] Cudworth, *Intellectual System of the Universe*, b. i. c. iii. § 37. Leclerc, *Bibliothèque Choisie*, vol. ii. p. 107 *et seq.* See also Leibnitz, *Considérations sur la Principe de Vie,—Opera,*

edit. Erdmann, p. 429.—ED.

[β] *Lettres à une Princesse d'Allemagne*, part. ii. let. 14, ed. Cournot. —ED.

mind, that He, in like manner, produces the correspondent movements in the body. But more explicitly:
—"God, according to the advocates of this scheme,
governs the universe, and its constituent existences,
by the laws according to which He has created them;
and as the world was originally called into being by
a mere fiat of the divine will, so it owes the continuance of its existence from moment to moment only
to the unremitted perseverance of the same volition.
Let the sustaining energy of the divine will cease but
for an instant, and the universe lapses into nothingness. The existence of created things is thus exclusively maintained by a creation, as it were, incessantly
renewed. God is, thus, the necessary cause of every
modification of body, and of every modification of
mind; and His efficiency is sufficient to afford an explanation of the union and intercourse of extended
and unextended substances.

"External objects determine certain movements in
our bodily organs of sense, and these movements are,
by the nerves and animal spirits, propagated to the
brain. The brain does not act immediately and really
upon the soul; the soul has no direct cognisance of
any modification of the brain; this is impossible. It
is God himself who, by a law which He has established
when movements are determined in the brain, produces analogous modifications in the conscious mind.
In like manner, suppose the mind has a volition to
move the arm; this volition is, of itself, inefficacious;
but God, in virtue of the same law, causes the answer-
ing motion in our limb. The body is not, therefore,
the real cause of the mental modifications; nor the
mind the real cause of the bodily movements. Nevertheless, as the soul would not be modified without the

LECT.
XVI.

antecedent changes in the body, nor the body moved
without the antecedent determination of the soul,—
these changes and determinations are in a certain sort
necessary. But this necessity is not absolute ; it is
only hypothetical or conditional. The organic changes,
and the mental determinations, are nothing but simple
conditions, and not real causes ; in short, they are
occasions, or occasional causes."[a] This doctrine of
Occasional Causes is called, likewise, the Hypothesis of
Assistance, as supposing the immediate co-operation
or intervention of the Deity. It is involved in the
Cartesian theory, and, therefore, belongs to Descartes ;[b]
but it was fully evolved by De la Forge,[c] Malebranche,[d]
and other followers of Descartes. It may, however, be
traced far higher. I find it first explicitly, and in all
its extent, maintained in the commencement of the
twelfth century by Algazel,[e] or Elgazali, of Bagdad,
surnamed the Imaun of the World ;—from him it
passed to the schools of the West, and many of the
most illustrious philosophers of the middle ages maintained that God is the only real agent in the universe.[f]

a [Laromiguière, Leçons de Philosophie, tom. ii. p. 255-6.]

β See Reid's Works, completed edition, p. 961 b, n. *.—ED.

γ [Tennemann (Gesch. der Phil.,
vol. x. p. 313) denies that De la
Forge is an advocate, far less the
first articulate expositor, of the system of Occasional Causes ; but erroneously. See Traité de l'Esprit de
l'Homme, c. xvi., and Sigwart's Leib-
niz'sche Lehre von der prästabilirten
Harmonie, p. 39 et seq.]

δ Recherche de la Vérité, lib. vi.
part ii. c. 3; Entretiens sur la Méta-
physique, Ent. vii.—ED.

e In his Destructio Philosophorum,
now only known through the refutation of it by Averroes, called Destructio Destructionis, preserved in

a barbarous Latin translation, in the
ninth volume of Aristotle's Works,
Venice, 1550. A full account of this
treatise is given in Tennemann's
Geschichte der Philosophie, vol. viii.
p. 387 et seq. See also Degerando.
Histoire Comparée, vol. iv. p. 226.—
ED.

ζ [For a history of the doctrine of
Occasional Causes before Descartes,
see Syrbius, Institutiones Philosophi-
cæ, (ed. Jenæ, 1726), p. 62, note.]
Averroes, l. c. p. 56 : " Agens com-
bustionis creavit nigredinem in stup-
pa et combustionem in partibus ejus,
et posuit eam combustam et cinerem,
et est Deus gloriosus mediantibus
angelis, aut immediate." See Tennemann, l. c. p. 405.—ED.

To this doctrine Dr Reid inclines,[a] and it is expressly maintained by Mr Stewart.[β]

This hypothesis did not satisfy Leibnitz. "He reproaches the Cartesians with converting the universe into a perpetual miracle, and of explaining the natural, by a supernatural, order. This would annihilate philosophy; for philosophy consists in the investigation and discovery of the second causes which produce the various phænomena of the universe.[γ] You degrade the Divinity, he subjoined;—you make Him act like a watchmaker, who, having constructed a timepiece, would still be obliged himself to turn the hands, to make it mark the hours. A skilful mechanist would so frame his clock that it would go for a certain period without assistance or interposition. So when God created man, He disposed his organs and faculties in such a manner that they are able of themselves to execute their functions and maintain their activity from birth to death."[δ]

Leibnitz thought he had devised a more philosophical scheme, in the hypothesis of the Pre-established or Predetermined Harmony, (*Systema Harmoniæ Præstabilitæ vel Prædeterminatæ*). This hypothesis denies all real connection, not only between spiritual and material substances, but between substances in general; and explains their apparent communion from a previously decreed coarrangement of the Supreme Being, in the following manner :—"God, before creating souls and bodies, knew all these souls and bodies; He knew also all possible souls and bodies.' Now, in

---

a See *Works*, pp. 257, 527.—Ed.

β See *Coll. Works*, vol. ii. pp. 97, 476-9; vol. iii. pp. 230, 248, 389-91.—Ed.

γ *Système Nouveau de la Nature*, § 13. *Opera*, ed. Erdmann, p. 137. Cf. *Théodicée*, § 61. *Opera*, p. 520.

—Ed.

δ [Laromiguière, *Leçons*, tom. ii. p. 256-7.] *Troisième Eclaircissement*. *Opera*, ed. Erdmann, p. 134.—Ed.

ε *Système Nouveau de la Nature*, § 14. *Théodicée*, § 62. These passages contain the substance of the

this infinite variety of possible souls and bodies, it was necessary that there should be souls whose series of perceptions and determinations would correspond to the series of movements which some of these possible bodies would execute; for in an infinite number of souls, and in an infinite number of bodies, there would be found all possible combinations. Now, suppose that, out of a soul whose series of modifications corresponded exactly to the series of modifications which a certain body was destined to perform, and of this body whose successive movements were correspondent to the successive modifications of this soul, God should make a man,—it is evident, that between the two substances which constitute this man, there would subsist the most perfect harmony. It is, thus, no longer necessary to devise theories to account for the reciprocal intercourse of the material and the spiritual substances. These have no communication, no mutual influence. The soul passes from one state, from one perception, to another, by virtue of its own nature. The body executes the series of its movements without any participation or interference of the soul in these. The soul and body are like two clocks accurately regulated, which point to the same hour and minute, although the spring which gives motion to the one is not the spring which gives motion to the other.[a] Thus the harmony which appears to combine the soul and body is, however, independent of any reciprocal action. This harmony was established before the creation of man; and hence it is called the pre-established or predetermined harmony."[β]

It is needless to attempt a refutation of this hypo-

remarks in the text, but not the   edit. Erdmann, p. 135.—ED.
words.—ED.   β [Laromiguière, Leçons, tom. ii.
   a Troisième Eclaircissement. Opera,   p. 257-8.]

thesis, which its author himself probably regarded more as a specimen of ingenuity than as a serious doctrine.

The third hypothesis is that of the Plastic Medium between soul and body. "This medium participates of the two natures; it is partly material, partly spiritual. As material, it can be acted on by the body; and as spiritual, it can act upon the mind. It is the middle term of a continuous proportion. It is a bridge thrown over the abyss which separates matter from spirit. This hypothesis is too absurd for refutation; it annihilates itself. Between an extended and unextended substance, there can be no middle existence, [these being not simply different in degree, but contradictory.] If the medium be neither body nor soul, it is a chimera; if it is at once body and soul, it is contradictory; or if, to avoid the contradiction, it is said to be, like us, the union of soul and body, it is itself in want of a medium."[a]

The fourth hypothesis is that of Physical Influence, (*Influxus Physicus*). "On this doctrine, external objects affect our senses, and the organic motion they determine is communicated to the brain. The brain acts upon the soul, and the soul has an idea,—a perception. The mind thus possessed of a perception or idea, is affected for good or ill. If it suffers, it seeks to be relieved of pain. It acts in its turn upon the brain, in which it causes a movement in the nervous system; the nervous system causes a muscular motion in the limbs,—a motion directed to remove or avoid the object which occasions the sensation of pain.

"The brain is the seat of the soul, and, on this hypothesis, the soul has been compared to a spider seated in the centre of its web. The moment the least agitation is caused at the extremity of this web, the

[margin note: 3. Plastic Medium.]

[margin note: 4. Physical Influence.]

a [Laromiguière, *Leçons*, tom. ii. p. 253-4.]

insect is advertised and put upon the watch. In like manner, the mind situated in the brain has a point on which all the nervous filaments converge; it is informed of what passes at the different parts of the body; and forthwith it takes its measures accordingly. The body thus acts with a real efficiency on the mind, and the mind acts with a real efficiency upon the body. This action or influence being real,—physical,—in the course of nature,—the body exerts a physical influence upon the soul, the soul a physical influence upon the body.

· "This system is simple, but it affords us no help in explaining the mysterious union of an extended and an unextended substance.

' Tangere enim et tangi nisi corpus nulla potest res.' *

Nothing can touch and be touched but what is extended; and if the soul be unextended, it can have no connection by touch with the body, and the physical influence is inconceivable or contradictory." β

Historical
order of
these hy-
potheses.
Physical
Influence,
first.
If we consider these hypotheses in relation to their historical manifestation, — the doctrine of Physical Influence would stand first; for this doctrine, which was only formally developed into system by the later Peripatetics, was that prevalent in the earlier schools of Greece. The Aristotelians, who held that the soul was the substantial form,—the vital principle, of the body, that the soul was all in the whole and all in every part of the body, naturally allowed a reciprocal influence of these. By influence, (in Latin *influxus*), you are to understand the relation of a cause to its effect; and the term, now adopted into every vulgar language of Europe, was brought into use principally

a Lucretius, i. 305.—ED.   p. 251-3.]
β [Laromiguière, *Leçons*, tom. ii.

by the authority of Suarez, a Spanish Jesuit, who
flourished at the close of the sixteenth and beginning
of the seventeenth centuries, and one of the most
illustrious metaphysicians of modern times. By him
a cause is defined, *Principium per se influens esse in
aliud.*[a] This definition, however, and the use of the
metaphysical term *influence*, (for it is nothing more),
are not, as is supposed, original with him. They are
to be found in the pseudo-Aristotelic treatise *De
Causis.* This is a translation from the Arabic, but
a translation made many centuries before Suarez.[β]
But this by the way.

The second hypothesis in chronological order is
that of the Plastic Medium. It is to be traced to
Plato. That philosopher, in illustrating the relation
of the two constituents of man, says that the soul is
in the body like a sailor in a ship; that the soul employs the body as its instrument; but that the energy,
or life and sense of the body, is the manifestation of
a different substance,—of a substance which holds a
kind of intermediate existence between mind and
matter. This conjecture, which Plato only obscurely
hinted at, was elaborated with peculiar partiality by
his followers of the Alexandrian school, and, in their
psychology, the ὄχος, or vehicle of the soul,—the medium through which it is united to the body,—is
a prominent element and distinctive principle.[γ]   To

*Plastic
Medium,
second.*

---

a *Disputationes Metaphysicæ*, Disp.
xii., § ii. 4.—ED.

β The *Libellus de Causis* is printed
in a Latin version made from a Hebrew one, in the seventh volume of
the Latin edition of Aristotle's Works,
Venice, 1550, f. 144. It has been
attributed to Aristotle, to Avempace,
to Alfarabi, and to Proclus. The
above definition does not occur in it
verbatim, though it may be gathered

in substance from Prop. 1.—ED.

γ The passage referred to in Plato
is probably *Timæus*, p. 69 : Οἱ δὲ
μιμούμενοι παραλαβόντες ἀρχὴν ψυχῆς
ἀθάνατον, τὸ μετὰ τοῦτο θνητὸν σῶμα
αὐτῇ περιετόρνευσαν ὄχημά τε πᾶν τὸ
σῶμα βόσαν κ.τ.λ. This passage, as
well as the simile of the chariot in
the *Phædrus*, p. 246, were interpreted
in this sense by the later Platonists.
See Ficinus, *Theologia Platonica*, lib.

this opinion St Austin,[a] among other Christian fathers, was inclined, and, in modern times, it has been revived and modified by Gassendi,[β] Cudworth,[γ] and Le Clerc.[δ]

Occasional
Causes,
third.

Descartes agrees with the Platonists in opposition to the Aristotelians, that the soul is not the substantial form of the body, but is connected with it only at a single point in the brain,—viz., the pineal gland. The pineal gland, he supposes, is the central point at which the organic movements of the body terminate, when conveying to the mind the determinations to voluntary motion.[ε] But Descartes did not allow, like the Platonists, any intermediate or connecting substance. The nature of the connection he himself does not very explicitly state ;—but his disciples have evolved the hypothesis, already explained, of Occasional Causes, in which God is the connecting principle,—an hypothesis at least implicitly contained in his philosophy.[ζ]

Finally, Leibnitz and Wolf agree with the Carte-

xviii. c. 4 : "Ex quo sequitur rationales animas tanquam medias tales esse debere, ut virtute quidem semper separabiles sint, ... actu autem sint semper conjunctæ, quia familiare corpus nanciscuntur ex æthere, quod servant per immortalitatem propriam immortale, quod Plato currum tam deorum tum animarum vocat in Phædro vehiculum in Timæo." The ship is more definitely expressed by Maximus Tyrius, Diss. xl. ε (referred to by Stallbaum, on the Timæus, l. c.) : Οὐχ ὁρᾷς καὶ τὸν ἐν τῇ θαλάττῃ πλοῦν, ἔνθα ὁ μὲν κυβερνήτης ἄρχει, ὡς ψυχὴ σώματος, ἡ δὲ ναῦς ἄρχεται, ὡς ὑπὸ ψυχῆς σῶμα. Cf. also Proclus, Inst. Theol., c. 206 et seq.; Cudworth, Intellectual System, b. i. c. v. § 3. Platner, Phil. Aphorismen, i. p. 627.—Ed.

a St Augustin seems to have adopt-
ed the ancient and Platonic dogma that matter (ὕλη) is incorporeal (ἀσώματος). He regarded matter as "quiddam inter formatum et nihil, nec formatum nec nihil, informe prope nihil." Confess., lib. xii. c. 6.—Ed.

β Gassendi, in his Physica, divides the human soul into two parts, the one rational and incorporeal, the other corporeal, including the nutritive and sensitive faculties. The latter he regards as the medium of connection between the rational soul and the body. See Opera, vol. ii. p. 256 (ed. 1658).—Ed.

γ See above, p. 300, note a.—Ed.

δ See above, p. 300, note a.—Ed.

ε De Pass. An., art. 31, 32; De Homine, art. 63. Cf. Reid's Works, (compl.ed.), pp. 234, n.*, 962 b.—Ed.

ζ See above, p. 302, note β.—Ed.

LECT.
XVI.

Pre-estab-
lished
Harmony,
fourth.

sians, that there is no real, but only an apparent intercourse between mind and body.   To explain this
apparent intercourse, they do not, however, resort to
the continual assistance or interposition of the Deity,
but have recourse to the supposition of a harmony between mind and body, established before the creation
of either.[a]

All these theories are unphilosophical, because they
all attempt to establish something beyond the sphere
of observation, and, consequently, beyond the sphere
of genuine philosophy ; and because they are either,
like the Cartesian and Leibnitian theories, contradictions of the fact of consciousness ; or, like the two
other hypotheses, at variance with the facts which
they suppose.   What St Austin so admirably says of
the substance, either of mind or of body,—" Materiam spiritumque cognoscendo ignorari et ignorando
cognosci,"[β]—I would exhort you to adopt as your
opinion in regard to the union of these two existences.
In short, in the words of Pascal,[γ] " Man is to himself the mightiest prodigy of nature ; for he is unable
to conceive what is body, still less what is mind, but
least of all is he able to conceive how a body can be
united to a mind ; yet this is his proper being."   A
contented ignorance is, indeed, wiser than a presumptuous knowledge ; but this is a lesson which seems
the last that philosophers are willing to learn.   In
the words of one of the acutest of modern thinkers[δ]—
" Magna immo maxima pars sapientiæ est, quædam
æquo animo nescire velle."

a [On these hypotheses in general,
see Zedler's *Lexicon*, v. *Seele*, p. 1098
et seq.]

β *Confess.*, lib. xii. c. 5.   See *ante*,
p. 139.—ED.

γ *Pensées*, partie i. art. vi., 26.
Vol. ii. p. 74, edit. Faugère.—ED.

δ Julius Cæsar Scaliger.   The passage is quoted more correctly in the
Author's *Discussions*, p. 640.—ED.

LECTURE XVII.

CONSCIOUSNESS,—GENERAL PHÆNOMENA,—ARE WE ALWAYS CONSCIOUSLY ACTIVE?

LECT.
XVII.
———
Activity
and Pas-
sivity of
Mind.

THE second General Fact of Consciousness which we shall consider, and out of which several questions of great interest arise, is the fact, or correlative facts, of the Activity and Passivity of Mind.

No pure
activity or
passivity
in creation.

There is no pure activity, no pure passivity in creation. All things in the universe of nature are reciprocally in a state of continual action and counter-action; they are always active and passive at once. God alone must be thought of as a being active without any mixture of passivity, as His activity is subjected to no limitation. But precisely because it is unlimited, is it for us wholly incomprehensible.

Activity
and Pas-
sivity al-
ways con-
joined in
the mani-
festations
of mind.

Activity and passivity are not, therefore, in the manifestations of mind, distinct and independent phæ-nomena. This is a great, though a common, error. They are always conjoined. There is no operation of mind which is purely active; no affection which is purely passive. In every mental modification, action and passion are the two necessary elements or factors of which it is composed. But though both are always present, each is not, however, always present in equal quantity. Sometimes the one constituent preponde-rates, sometimes the other; and it is from the pre-ponderance of the active element in some modifica-tions of the passive element in others, that we distinguish these modifications by different names, and consider them as activities or passivities according as they approximate to one or other of the two factors. Thus *faculty, operation, energy,* are words that we employ to designate the manifestations in which activity is predominant. *Faculty* denotes an active power; *action, operation, energy,* denote its present exertion. On the other hand, *capacity* expresses a passive power; *affection, passion,* express a present suffering. The terms *mode, modification, state,* may be used indifferently to signify both phænomena; but it must be acknowledged that these, especially the word *state,* are now closely associated with the passivity of mind, which they, therefore, tend rather to suggest. The passivity of mind is expressed by another term, *receptivity;* for passivity is only the condition, the necessary antecedent of activity,—only the property possessed by the mind of standing in relation to certain foreign causes,—of receiving from them impressions, determinations to act.

It is to be observed, that we are never directly conscious of passivity. Consciousness only commences with, is only cognisant of, the reaction consequent upon the foreign determination to act, and this reaction is not itself passive. In so far, therefore, as we are conscious, we are active; whether there may be a mental activity of which we are not conscious, is another question.[a]

There are certain arduous problems connected with the activity of mind, which will be more appropriately considered in a subsequent part of the course, when we come to speak of the inferences from the Phænomenology of Mind, or of Metaphysics Proper. At present, I shall only treat of those questions which

a See below, Lect. xviii. p. 338.—ED.

are conversant about the immediate phænomena of activity. Of these, the first that I shall consider is one of considerable interest, and which, though variously

The question, Are we always consciously active? raised.

determined by different philosophers, does not seem to lie beyond the sphere of observation. I allude to the question, Whether we are always consciously active?

Distinguished from other questions.

It is evident that this question is not convertible with the question, Have we always a memory of our consciousness?—for the latter problem must be at once answered in the negative. It is also evident, that we must exclude the consideration of those states in which the mind is apparently without consciousness, but in regard to which, in reality, we can obtain no information from experiment. Concerning these we must be contented to remain in ignorance; at least only to extend to them the analogical conclusions which our observations on those within the sphere of experiment warrant us inferring. Our question, as one of possible solution, must, therefore, be limited to the states of sleep and somnambulism, to the exclusion of those states of insensibility which we cannot terminate suddenly at will. It is hardly necessary to observe, that with the nature of sleep and somnambulism as psychological phænomena, we have at present nothing to do; our consideration is now strictly limited to the inquiry, Whether the mind, in as far as we can make it matter of observation, is always in a state of conscious activity. The general problem in

Treatment of the question by philosophers.

regard to the ceaseless activity of the mind has been one agitated from very ancient times, but it has also been one on which philosophers have pronounced less

Plato and the Platonists.

on grounds of experience than of theory. Plato and the Platonists were unanimous in maintaining the continual energy of intellect. The opinion of Aristotle appears doubtful, and passages may be quoted

from his works in favour of either alternative. The LECT.
XVII.
Aristotelians, in general, were opposed, but a considerable number were favourable, to the Platonic doctrine.
This doctrine was adopted by Cicero and St Augustin.
" Nunquam animus," says the former, " cogitatione et
motu vacuus esse potest." [a]   " Ad quid menti," says
the latter, " præceptum est, ut se ipsam cognoscat,
nisi ut semper vivat, et semper sit in actu." [β]   The
question, however, obtained its principal importance
in the philosophy of Descartes.   That philosopher
made the essence, the very existence, of the soul to
consist in actual thought,[7] under which he included
even the desires and feelings ; and *thought* he defined
all of which we are conscious.[8]   The assertion, therefore, of Descartes, that the mind always thinks, is, in
his employment of language, tantamount to the assertion that the mind is always conscious.

That the mind is always conscious, though a fundamental position of the Cartesian doctrine, was rather
assumed, than proved by an appeal to fact and experience.   All is theoretical in Descartes ; all is theoretical in his disciples.   Even Malebranche assumes our
consciousness in sleep, and explains our oblivion only
by a mechanical hypothesis.[e]   It was, therefore, easy
for Locke to deny the truth of the Cartesian opinion,

Marginal notes: Aristotle and the Aristotelians. — Cicero and St Augustin. — Descartes. — Malebranche. — Locke.

a *De Divinatione*, ii. 62: "Naturam eam dico, quæ nunquam animus
insistens, *agitatione* et motu esse vacuus potest."—ED.

β Eugenios, Ψυχολογία, p. 129.—
[Book iii. of his Στοιχεῖα τῆς Μετα-
φυσικῆς, (edit. 1805). The reference
in Eugenios is to *De Trinitate*, lib. x.
c. v., where a passage occurs, resembling in words the one quoted in the
text, but hardly supporting the doctrine in question. It is as follows :
" Ut quid ergo ei præceptum est, ut se
ipsam cognoscat ?   Credo ut se ipsam

cogitet, et secundum naturam suam
vivat."   But in the *De Anima et ejus
Origine*, lib. iv. c. vi. § 7,—*Opera*, t.
x. p. 391, (edit. Benedict.), occurs the
following explicit statement: "Sicut
motus non cessat in corde, unde se
pulsus diffundit usquequaque ven-
arum, ita non quiescimus aliquid co-
gitando vermare."—ED.]

7 *Principia*, pars i. § 53.—ED.

8 *Principia*, pars i. § 9.   Cf. *Reid's
Works*, (compl. ed.), p. 961 a, n. †.—ED.

e *Recherche de la Vérité*, liv. iii.
ch. 2.—ED.

and to give a strong semblance of probability to his
own doctrine by its apparent conformity with the
phænomena.   Omitting a good deal of what is either
irrelevant to the general question, or what is now
admitted to be false, as founded on his erroneous doctrine of personal identity, the following is the sum of
Locke's argument upon the point.   "It is an opinion,"
he says,[a] "that the soul always thinks, and that it
has the actual perception of ideas in itself constantly,
as long as it exists; and that actual thinking is as
inseparable from the soul, as actual extension is from
the body; which, if true, to inquire after the beginning of a man's ideas, is the same as to inquire after
the beginning of his soul.   For by this account, soul
and its ideas, as body and its extension, will begin to
exist both at the same time.

"But whether the soul be supposed to exist antecedent to, or coeval with, or some time after, the first
rudiments, or organisation, or the beginnings of life in
the body, I leave to be disputed by those who have
better thought of that matter.   I confess myself to
have one of those dull souls that doth not perceive
itself always to contemplate ideas; nor can conceive
it any more necessary for the soul always to think
than for the body always to move: the perception of
ideas being (as I conceive) to the soul, what motion
is to the body; not its essence, but one of its operations.   And, therefore, though thinking be supposed
ever so much the proper action of the soul, yet it is
not necessary to suppose that it should be always
thinking, always in action.   That perhaps is the privilege of the infinite Author and Preserver of things,
who never slumbers nor sleeps; but is not competent
to any finite being, at least not to the soul of man.

Locke's
argument
for the
negative.

a *Essay*, book ii. chap. i. §§ 9, 10, 14 *et seq.*

We know certainly by experience that we sometimes think, and thence draw this infallible consequence, that there is something in us that has a power to think : but whether that substance perpetually thinks or no, we can be no further assured than experience informs us. For to say that actual thinking is essential to the soul and inseparable from it, is to beg what is in question, and not to prove it by reason ; which is necessary to be done if it be not a self-evident proposition. But whether this, 'that the soul always thinks,' be a self - evident proposition, that everybody assents to at first hearing, I appeal to mankind. It is doubted whether I thought all last night or no ; the question being about a matter of fact, it is begging it to bring as a proof for it an hypothesis which is the very thing in dispute ; by which way one may prove anything; and it is but supposing that all watches, whilst the balance beats, think ; and it is sufficiently proved, and past doubt, that my watch thought all last night. But he that would not deceive himself, ought to build his hypothesis on matter of fact, and make it out by sensible experience, and not presume on matter of fact, because of his hypothesis ; that is, because he supposes it to be so ; which way of proving amounts to this, that I must necessarily think all last night because another supposes I always think, though I myself cannot perceive that I always do so." . . . . "It will perhaps be said that 'the soul thinks even in the soundest sleep, but the memory retains it not.' That the soul in a sleeping man should be this moment busy a-thinking, and the next moment in a waking man not remember nor be able to recollect one jot of all those thoughts, is very hard to be conceived, and would need some better proof than bare assertion to make it be believed. For who can, without any more ado but being barely told so, imagine that the greatest part of men do, during all their lives for several hours every day, think of something which, if they were asked even in the middle of these thoughts, they could remember nothing at all of? Most men, I think, pass a great part of their sleep without dreaming. I once knew a man that was bred a scholar and had no bad memory, who told me he had never dreamed in his life till he had that fever he was then newly recovered of, which was about the five or six and twentieth year of his age. I suppose the world affords more such instances; at least every one's acquaintance will furnish him with examples enough of such as pass most of their nights without dreaming." . . . . And again, "If they say that a man is always conscious to himself of thinking; I ask how they know it? 'Consciousness is the perception of what passes in a man's own mind. Can another man perceive that I am conscious of anything, when I perceive it not myself? No man's knowledge here can go beyond his experience. Wake a man out of a sound sleep, and ask him what he was that moment thinking on. If he himself be conscious of nothing he then thought on, he must be a notable diviner of thoughts that can assure him that he was thinking: may he not with more reason assure him he was not asleep? This is something beyond philosophy; and it cannot be less than revelation that discovers to another thoughts in my mind when I can find none there myself; and they must needs have a penetrating sight who can certainly see what I think when I cannot perceive it myself, and when I declare that I do not. This some may think to be a step beyond the Rosicrucians, it being easier to make one's self invisible to others, than to make

another's thoughts visible to one which are not visible
to himself.   But it is but defining the soul to be 'a
substance that always thinks,' and the business is
done.   If such definition be of any authority, I know
not what it can serve for, but to make many men
suspect that they have no souls at all, since they find
a good part of their lives pass away without thinking.
For no definitions that I know, no suppositions of any
sect, are of force enough to destroy constant experience; and perhaps it is the affectation of knowing
beyond what we perceive that makes so much useless
dispute and noise in the world."

This decision of Locke was rejected by Leibnitz in
the *New Essays on the Human Understanding*,[a] the
great work in which he canvassed from beginning to
end the Essay, under the same title, of the English
philosopher. He observes, in reply to the supposition
that continual consciousness is an attribute of Him
' who neither slumbereth nor sleepeth,' " that this affords no inference that in sleep we are wholly without
perception."   To the remark, " that it is difficult to
conceive, that a being can think and not be conscious
of thought," he replies, " that in this lies the whole
knot and difficulty of the matter.   But this is not insoluble."   " We must observe," he says, " that we think
of a multitude of things at once, but take heed only
of those thoughts that are the more prominent.   Nor
could it be otherwise.   For were we to take heed of
everything, it would be necessary to attend to an infinity of matters at the same moment, all of which make
an effectual impression on the senses.   Nay, I assert
that there remains always something of all our past
thoughts,—that none is ever entirely effaced.   Now,
when we sleep without dreaming, and when stunned

a Liv. ii. ch. 1.—Ed.

by a blow or other accident, there are formed in us an infinity of small confused perceptions." And again he remarks : "That even when we sleep without dreaming, there is always some feeble perception. The act of awakening, indeed, shows this : and the more easily we are roused, the clearer is the perception we have of what passes without, although this perception is not always strong enough to cause us to awake."

Now, in all this it will be observed, that Leibnitz does not precisely answer the question we have mooted. He maintains that the mind is never without perceptions, but, as he holds that perceptions exist without consciousness, he cannot, though he opposes Locke, be considered as affirming that the mind is never without consciousness during sleep,—in short, *Wolf.* does always dream. The doctrine of Wolf on this point is the same with that of his master,[a] though the *Nouveaux Essais* of Leibnitz were not published till long after the death of Wolf.

Kant. But if Leibnitz cannot be adduced as categorically asserting that there is no sleep without its dream, this cannot be said of Kant. That great thinker distinctly maintains that we always dream when asleep; that to cease to dream would be to cease to live ; and that those who fancy they have not dreamt have only forgotten their dream.[β] This is all that the manual of *Anthropology,* published by himself, contains upon the question ; but in a manuscript in my possession, which bears to be a work of Kant, but is probably only a compilation from notes taken at his lectures on Anthropology, it is further stated that we can dream more in a minute than we can act during a day, and that the great rapidity of the train of

a *Psychologia Rationalis,* § 59.—ED.   β *Anthropologie,* §§ 30, 36.—ED.

thought in sleep, is one of the principal causes why we do not always recollect what we dream.[a] He elsewhere also observes that the cessation of a force to act, is tantamount to its cessation to be.

Though the determination of this question is one that seems not extremely difficult, we find it dealt with by philosophers, on the one side and the other, rather by hypothesis than by experiment ; at least, we have, with one partial exception, which I am soon to quote to you, no observations sufficiently accurate and detailed to warrant us in establishing more than a very doubtful conclusion. I have myself at different times turned my attention to the point, and, as far as my observations go, they certainly tend to prove that, during sleep, the mind is never either inactive or wholly unconscious of its activity. As to the objection of Locke and others, that, as we have often no recollection of dreaming, we have, therefore, never dreamt, it is sufficient to say that the assumption in this argument,—that consciousness, and the recollection of consciousness, are convertible,—is disproved in the most emphatic manner by experience. You have all heard of the phænomenon of somnambulism. In this remarkable state, the various mental faculties are usually in a higher degree of power than in the natural. The patient has recollections of what he has wholly forgotten. He speaks languages of which, when awake, he remembers not a word. If he use a vulgar dialect when out of this state, in it he employs only a correct and elegant phraseology. The imagination, the sense of propriety, and the faculty of

The question dealt with by philosophers rather by hypothesis than by experiment.

Conclusion from experiments made by the Author.

Locke's assumption, that consciousness and the recollection of consciousness are convertible, disproved by the phænomena of somnambulism.

---

a The substance of this passage is published in the *Menschenkunde oder Philosophische Anthropologie*, edited by Starke in 1831, from Kant's Lectures. See p. 164.—ED.

reasoning, are all in general exalted.[a] The bodily
powers are in high activity, and under the complete
control of the will ; and, it is well known, persons in
this state have frequently performed feats, of which,
when out of it, they would not even have imagined
the possibility. And what is even more remarkable,
the difference of the faculties in the two states, seems
not confined merely to a difference in degree. For it
happens, for example, that a person who has no ear
for music when awake, shall, in his somnambulic crisis,
sing with the utmost correctness and with full enjoyment of his performance. Under this affection persons sometimes live half their lifetime, alternating
between the normal and abnormal states, and performing the ordinary functions of life indifferently in
both, with this distinction, that if the patient be dull
and doltish when he is said to be awake, he is comparatively alert and intelligent when nominally asleep.
I am in possession of three works, written during the
crisis by three different somnambulists.[b] Now it is
evident that consciousness, and an exalted consciousness, must be allowed in somnambulism. This cannot

Conscious-
ness with-
out mem-
ory the
character-
istic of
somnam-
bulism.
possibly be denied,—but mark what follows. It is
the peculiarity of somnambulism,—it is the differential
quality by which that state is contradistinguished
from the state of dreaming, that we have no recollection, when we awake, of what has occurred during
its continuance. Consciousness is thus cut in two ;
memory does not connect the train of consciousness
in the one state with the train of consciousness in the
other. When the patient again relapses into the state

---

a For some interesting illustrations of this state, see Abercrombie,
On the Intellectual Powers, part iii.
sect. iv. § 2.—Ed.

β Of these works we have failed to
discover any trace.—Ed.

of somnambulism, he again remembers all that had
occurred during every former alternative of that state ;
but he not only remembers this, he recalls also the
events of his normal existence : so that whereas the
patient in his somnambulic crisis, has a memory of
his whole life, in his waking intervals he has a memory only of half his life.

At the time of Locke, the phænomena of somnam- Dreaming
possible
bulism had been very little studied ; nay, so great is without
memory.
the ignorance that prevails in this country in regard
to its nature even now, that you will find this, its distinctive character, wholly unnoticed in the best works
upon the subject.[a]   But this distinction, you observe,
is incompetent always to discriminate the states of
dreaming and somnambulism.   It may be true that
if we recollect our visions during sleep, this recollection excludes somnambulism, but the want of memory
by no means proves that the visions we are known by
others to have had, were not common dreams.   The
phænomena, indeed, do not always enable us to discriminate the two states.   Somnambulism may exist
in many different degrees.   The sleep-walking from
which it takes its name is only one of its higher phænomena, and one comparatively rare.   In general, the
subject of this affection does not leave his bed, and
it is then frequently impossible to say whether the
manifestations exhibited, are the phænomena of somnambulism or of dreaming.   Talking during sleep, for
example, may be a symptom of either, and it is often
only from our general knowledge of the habits and
predispositions of the sleeper, that we are warranted
in referring this effect to the one and not to the other

a This deficiency has been ably *Principles of Human Physiology*, §
supplied by Dr Carpenter. See his   827, (4th edition).—En.

LECT.
XVII.
class of phænomena. We have, however, abundant evidence to prove that forgetfulness is not a decisive criterion of somnambulism. Persons whom there is no reason to suspect of this affection, often manifest during sleep the strongest indications of dreaming, and yet, when they awaken in the morning, retain no memory of what they may have done or said during the night. Locke's argument, that because we do not always remember our consciousness during sleep, we have not, therefore, been always conscious, is thus, on the ground of fact and analogy, disproved.

That the mind remains conscious during sleep established by experience.

But this is not all. We can not only show that the fact of the mind remaining conscious during sleep is possible, is even probable, we can also show, by an articulate experience, that this actually occurs. The following observations are the result of my personal experience, and similar experiments every one of you is competent to institute for himself.

Results of the Author's personal experience.

In the first place, when we compose ourselves to rest, we do not always fall at once asleep, but remain for a time in a state of incipient slumber,—in a state intermediate between sleep and waking. Now, if we are gently roused from this transition-state, we find ourselves conscious of being in the commencement of a dream ; we find ourselves occupied with a train of thought, and this train we are still able to follow out to a point when it connects itself with certain actual perceptions. We can still trace imagination to sense, and show how, departing from the last sensible impressions of real objects, the fancy proceeds in its work of distorting, falsifying, and perplexing these, in order to construct out of their ruins its own grotesque edifices.

In the second place, I have always observed, that

when suddenly awakened during sleep, (and to ascertain the fact I have caused myself to be roused at different seasons of the night), I have always been able to observe that I was in the middle of a dream. The recollection of this dream was not always equally vivid. On some occasions, I was able to trace it back until the train was gradually lost at a remote distance; on others, I was hardly aware of more than one or two of the latter links of the chain; and, sometimes, was scarcely certain of more than the fact, that I was not awakened from an unconscious state. Why we should not always be able to recollect our dreams, it is not difficult to explain. In our waking and our sleeping states, we are placed in two worlds of thought, not only different but contrasted, and contrasted both in the character and in the intensity of their representations. When snatched suddenly from the twilight of our sleeping imaginations, and placed in the meridian lustre of our waking perceptions, the necessary effect of the transition is at once to eclipse or obliterate the traces of our dreams. The act itself also of rousing us from sleep, by abruptly interrupting the current of our thoughts, throws us into confusion, disqualifies us for a time from recollection, and before we have recovered from our consternation, what we could at first have easily discerned is fled or flying.

A sudden and violent is, however, in one respect, more favourable than a gradual and spontaneous wakening to the observation of the phænomena of sleep. For in the former case, the images presented are fresh and prominent; while in the latter, before our attention is applied, the objects of observation have withdrawn darkling into the background of the soul. We may, therefore, I think, assert, in general, that whether

LECT.
XVII.
we recollect our dreams or not, we always dream. Something similar, indeed, to the rapid oblivion of our sleeping consciousness, happens to us occasionally even when awake. When our mind is not intently occupied with any subject, or more frequently when fatigued, a thought suggests itself. We turn it lazily over and fix our eyes in vacancy; interrupted by the question what we are thinking of, we attempt to answer, but the thought is gone; we cannot recall it, and say that we were thinking of nothing.[a]

General conclusions from foregoing.

The observations I have hitherto made tend only to establish the fact, that the mind is never wholly inactive, and that we are never wholly unconscious of its activity. Of the degree and character of that activity, I at present say nothing; this may form the subject of our future consideration. But in confirmation of the opinion I have now hazarded, and in proof of something more even than I have ventured to maintain, I have great pleasure in quoting to you the substance of a very remarkable essay on sleep by one of the most distinguished of the philosophers of France,—

Jouffroy quoted in confirmation of the Author's view, and in proof of sundry other conclusions.

living when the extract was made, but now unfortunately lost to the science of mind which he cultivated with most distinguished success.—I refer to M. Jouffroy, who, along with M. Royer Collard, was at the head of the pure school of Scottish Philosophy in France.[β]

The mind frequently awake when the senses asleep.

"I have never well understood those who admit that in sleep the mind is dormant. When we dream, we are assuredly asleep, and assuredly also our mind is not asleep, because it thinks; it is, therefore, manifest, that the mind frequently wakes when the senses are in slumber. But this does not prove that it never

a Cf. Kant, *Anthropologie*, § 30, ed. 1838, (§ 28, ed. 1810).—ED.   β *Mélanges*, p. 318, [p. 290, second edition.—ED.]

sleeps along with them. To sleep is for the mind not to dream; and it is impossible to establish the fact, that there are in sleep moments in which the mind does not dream. To have no recollection of our dreams, does not prove that we have not dreamt; for it can be often proved that we have dreamt, although the dream has left no trace on our memory.

"The fact, then, that the mind sometimes wakes Probable that the mind is always awake. while the senses are asleep, is thus established; whereas the fact, that it sometimes sleeps along with them, is not: the probability, therefore, is, that it wakes always. It would require contradictory facts to destroy the force of this induction, which, on the contrary, every fact seems to confirm. I shall proceed to analyse some of these which appear to me curious and striking. They manifestly imply this conclusion, that the mind, during sleep, is not in a peculiar state, but that its activity is carried on precisely as when awake.

"When an inhabitant of the province comes to Paris, Induction of facts in support of this conclusion. his sleep is at first disturbed, and continually broken, by the noise of the carriages passing under his window. He soon, however, becomes accustomed to the turmoil, and ends by sleeping at Paris as he slept in his village.

"The noise, however, remains the same, and makes an equal impression on his senses; how comes it that this noise at first hinders, and then, at length, does not hinder, him from sleeping?

"The state of waking presents analogous facts. Every one knows that it is difficult to fix our attention on a book, when surrounded by persons engaged in conversation; at length, however, we acquire this faculty. A man unaccustomed to the tumult of the streets of Paris is unable to think consecutively while walking through them; a Parisian finds no difficulty.

He meditates as tranquilly in the midst of the crowd and bustle of men and carriages, as he could in the centre of the forest. The analogy between these facts taken from the state of waking, and the fact which I mentioned at the commencement, taken from the state of sleep, is so close, that the explanation of the former should throw some light upon the latter. We shall attempt this explanation.

Analysis and explanation of these phenomena — Attention and Distraction.

" Attention is the voluntary application of the mind to an object. It is established, by experience, that we cannot give our attention to two different objects at the same time. Distraction (*être distrait*) is the removal of our attention from a matter with which we are engaged, and our bestowal of it on another which crosses us. In distraction, attention is only diverted because it is attracted by a new perception or idea, soliciting it more strongly than that with which it is occupied ; and this diversion diminishes exactly in proportion as the solicitation is weaker on the part of the intrusive idea. All experience proves this. The more strongly attention is applied to a subject, the less susceptible is it of distraction ; thus it is, that a book which awakens a lively curiosity, retains the attention captive ; a person occupied with a matter affecting his life, his reputation, or his fortune, is not easily distracted ; he sees nothing, he understands nothing of what passes around him ; we say that he is deeply preoccupied. In like manner, the greater our curiosity, or the more curious the things that are spoken of around us, the less able are we to rivet our attention on the book we read. In like manner, also, if we are waiting in expectation of any one, the slightest noises occasion distraction, as these noises may be the signal of the approach we anticipate. All these facts tend to prove that distraction results only when the intrusive idea solicits us more strongly than that with which we are occupied.

" Hence it is that the stranger in Paris cannot think in the bustle of the streets. The impressions which assail his eyes and ears on every side being for him the signs of things new or little known, when they reach his mind interest him more strongly than the matter even to which he would apply his thoughts. Each of these impressions announces a cause which may be beautiful, rare, curious, or terrific ; the intellect cannot refrain from turning out to verify the fact. It turns out, however, no longer when experience has made it familiar with all that can strike the senses on the streets of Paris; it remains within, and no longer allows itself to be deranged.

" The other admits of a similar explanation. To read without distraction in the midst of an unknown company, would be impossible. Curiosity would be too strong. This would also be the case if the subject of conversation were very interesting. But in a familiar circle, whose ordinary topics of conversation are well known, the ideas of the book make an easy conquest of our thoughts.

" The will, likewise, is of some avail in resisting distraction. Not that it is able to retain the attention when disquieted and curious; but it can recall, and not indulge it in protracted absences, and, by constantly remitting it to the object of its volition, the interest of this object becomes at last predominant. Rational considerations, and the necessity of remaining attentive, likewise exert an influence ; they come in aid of the idea, and lend it, so to speak, a helping hand in concentrating on it the attention.

LECT.
XVII.

Distraction
and Non-
distraction
matters of
intelli-
gence.

"But, howsoever it may be with all these petty influences, it remains evident that distraction and non-distraction are neither of them matters of sense, but both matters of intelligence. It is not the senses which become accustomed to hear the noises of the street and the sounds of conversation, and which end in being less affected by them; if we are at first vehemently affected by the noises of the street or drawing-room, and then little or not at all, it is because at first attention occupies itself with these impressions, and afterwards neglects them: when it neglects them it is not diverted from its object, and distraction does not take place; when, on the contrary, it accords them notice, it abandons its object, and is then distracted.

"We may observe, in support of this conclusion, that the habit of hearing the same sounds renders us sometimes highly sensible to these, as occurs in savages and in the blind; sometimes, again, almost insensible to them, as exemplified in the apathy of the Parisian for the noise of carriages. If the effect were physical,—if it depended on the body and not on the mind, there would be a contradiction, for the habit of hearing the same sounds either blunts the organ or sharpens it; it could not at once have two, and two contrary, effects,—it could have only one. The fact is, it neither blunts nor sharpens; the organ remains the same; the same sensations are determined: but when these sensations interest the mind, it applies itself to them, and becomes accustomed to their discrimination; when they do not interest it, it becomes accustomed to neglect, and does not discriminate them. This is the whole mystery; the phænomenon is psychological, not physiological.

"Let us now turn our attention to the state of sleep,

and consider whether analogy does not demand a
similar explanation of the fact which we stated at the
commencement. What takes place when a noise
hinders us from sleeping? The body fatigued begins
to slumber; then, of a sudden, the senses are struck,
and we awake; then fatigue regains the ascendant,
we relapse into drowsiness, which is soon again interrupted; and so on for a certain continuance. When,
on the contrary, we are accustomed to noise, the impressions it makes no longer disturb our first sleep;
the drowsiness is prolonged, and we fall asleep. That
the senses are more torpid in sleep than in our waking
state, is not a matter of doubt. But when I am once
asleep, they are then equally torpid on the first night
of my arrival in Paris as on the hundredth. The noise
being the same, they receive the same impressions,
which they transmit in equal vivacity to the mind.
Whence comes it, then, that on the first night I am
awakened, and not on the hundredth? The physical
facts are identical; the difference can originate only
in the mind, as in the case of distraction and of non-
distraction in the waking state. Let us suppose that
the soul has fallen asleep along with the body; on this
hypothesis, the slumber would be equally deep, in both
cases, for the mind and for the senses, and we should
be unable to see why, in the one case, it was aroused
more than in the other. It remains, therefore, certain
that it does not sleep like the body; and that, in the
one case, disquieted by unusual impressions, it awakens
the senses to inquire what is the matter; whilst in the
other, knowing by experience of what external fact
these impressions are the sign, it remains tranquil,
and does not disturb the senses to obtain a useless
explanation.

"For let us remark, that the mind has need of the senses to obtain a knowledge of external things. In sleep, the senses are some of them closed, as the eyes; the others half torpid, as touch and hearing. If the soul be disquieted by the impressions which reach it, it requires the senses to ascertain the cause, and to relieve its inquietude. This is the cause why we find ourselves in a disquieted state, when aroused by an extraordinary noise; and this could not have occurred had we not been occupied with this noise before we awoke.

"This is, also, the cause why we sometimes feel, during sleep, the efforts we make to awaken our senses, when an unusual noise or any painful sensation disturbs our rest. If we are in a profound sleep, we are for a long time agitated before we have it in our power to awake,—we say to ourselves, we must awake in order to get out of pain; but the sleep of the senses resists, and it is only by little and little that we are able to rouse them from torpidity. Sometimes, when the noise ceases before the issue of the struggle, the awakening does not take place, and, in the morning, we have a confused recollection of having been disturbed during our sleep,—a recollection which becomes distinct only when we learn from others that such and such an occurrence has taken place while we were asleep.

"I had given orders some time ago, that a parlour adjoining to my bedroom should be swept before I was called in the morning. For the first two days the noise awoke me; but, thereafter, I was not aware of it. Whence arose the difference? The noises are the same and at the same hour; I am in the same degree of slumber; the same sensations, consequently, take

place.   Whence comes it that I awoke, and do no
longer awake ?   For this, it appears to me, there is
but one explanation,—viz. that my mind which wakes,
and which is now aware of the cause of these sensations, is no longer disquieted, and no longer rouses my
senses.   It is true that I do not retain the recollection
of this reasoning ; but this oblivion is not more extraordinary than that of so many others which cross our
mind both when awake and when asleep.

" I add a single observation.   The noise of the brush
on the carpet of my parlour is as nothing compared
with that of the heavy waggons which pass under my
windows at the same hour, and which do not trouble
my repose in the least.   I was, therefore, awakened by
a sensation much feebler than a crowd of others, which
I received at the same time.   Can that hypothesis
afford the reason, which supposes that the awakening
is a necessary event ; that the sensations rouse the
senses, and that the senses rouse the mind ?   It is
evident that my mind alone, and its activity, can
explain why the fainter sensation awoke me; as
these alone can explain why, when I am reading in
my study, the small noise of a mouse playing in a
corner can distract my attention, while the thunder-
ing noise of a passing waggon does not affect me at
all.

" The same explanation fully accounts for what
occurs with those who sleep in attendance on the sick.
All noises foreign to the patient have no effect on
them ; but let the patient turn him on his bed, let
him utter a groan or sigh, or let his breathing become
painful or interrupted, forthwith the attendant awakes,
however little inured to the vocation, or interested in
the welfare of the patient.   Whence comes this dis-

LECT.
XVII.

crimination between the noises which deserve the attention of the attendant, and those which do not, if, whilst the senses are asleep, the mind does not remain observant,—does not act the sentinel, does not consider the sensations which the senses convey, and does not awaken the senses as it finds these sensations disquieting or not? It is by being strongly impressed, previous to going to sleep, with the duty of attending to the respiration, motions, complaints of the sufferer, that we come to waken at all such noises, and at no others. The habitual repetition of such an impression gives this faculty to professional sick-nurses; a lively interest in the health of the patient gives it equally to the members of his family.

Awaking at an appointed hour.

"It is in precisely the same manner that we waken at the appointed hour, when before going to sleep we have made a firm resolution of so doing. I have this power in perfection; but I notice that I lose it if I depend on any one calling me. In this latter case, my mind does not take the trouble of measuring the time or of listening to the clock. But in the former, it is necessary that it do so, otherwise the phænomenon is inexplicable. Every one has made, or can make, this experiment; when it fails it will be found, if I mistake not, either that we have not been sufficiently preoccupied with the intention, or were overfatigued; for when the senses are strongly benumbed, they convey to the mind, on the one hand, more obtuse sensations of the monitory sounds, and, on the other, they resist for a longer time the efforts the mind makes to awaken them, when these sounds have reached it.

"After a night passed in this effort, we have, in general, the recollection, in the morning, of having been constantly occupied during sleep with this thought. The mind, therefore, watched, and, full of its resolution, awaited the moment. It is thus that when we go to bed much interested with any subject, we remember, on wakening, that during sleep we have been continually haunted by it. On these occasions the slumber is light, for, the mind being untranquil, its agitation is continually disturbing the torpor of the senses. When the mind is calm, it does not sleep more, but it is less restless.

" It would be curious to ascertain, whether persons of a feeble memory, and of a volatile disposition, are not less capable than others of awakening at an appointed hour ; for these two circumstances ought to produce this effect, if the notion I have formed of the phænomenon be correct. A volatile disposition is unable strongly to preoccupy itself with the thought, and to form a determined resolution ; and, on the other hand, it is the memory which preserves a recollection of the resolution taken before falling asleep. I have not had an opportunity of making the experiment.

" It appears to me, that from the previous observations, it inevitably follows :— General conclusions.

1°, That in sleep the senses are torpid, but that the mind wakes.

2°, That certain of our senses continue to transmit to the mind the imperfect sensations they receive.

3°, That the mind judges these sensations, and that it is in virtue of its judgments that it awakens, or does not awaken, the senses.

4°, That the reason why the mind awakens the senses is, that sometimes the sensation disquiets it, being unusual or painful ; that sometimes the sensa-

tion warns it to rouse the senses, as being an indication of the moment when it ought to do so.

5°, That the mind possesses the power of awakening the senses, but that it only accomplishes this by
its own activity overcoming their torpor; that this
torpor is an obstacle,—an obstacle greater or less as
it is more or less profound.

" If these inferences are just, it follows that we can
waken ourselves at will and at appointed signals;
that the instrument called an alarum (*réveil-matin*)
does not act so much by the noise it makes as by the
association we have established in going to bed between the noise and the thought of wakening; that,
therefore, an instrument much less noisy, and emitting
only a feeble sound, would probably produce the same
effect. It follows, moreover, that we can inure ourselves to sleep profoundly in the midst of the loudest
noises; that to accomplish this it is perhaps sufficient,
on the first night, to impress it on our minds that
these sounds do not deserve attention, and ought not
to waken us; and that by this means, any one may
probably sleep as well in the mill as the miller himself. It follows, in fine, that the sleep of the strong
and courageous ought to be less easily disturbed, all
things equal, than the sleep of the weak and timid.
Some historical facts may be quoted in proof of this
last conclusion."

Jouffroy's
theory cor-
roborated
by the case
of the post-
man of
Halle.

I shall not quote to you the observations of M.
Jouffroy on Reverie,[a] which form a sequel, and a confirmation of those he has made upon sleep. Before
terminating this subject, I may, however, notice a
rather curious case which occurs to my recollection,
and which tends to corroborate the theory of the

a See *Mélanges*, p. 304 *et seq.*—ED.

French psychologist. I give it on the authority of
Junker,[a] a celebrated physician and professor of Halle,
who flourished during the first half of last century,
and he says that he took every pains to verify the
facts by frequent personal observation. I regret that
I am unable at the moment to find the book in which
the case is recorded, but of all its relevant circumstances I have a vivid remembrance. The object of
observation was the postman between Halle and a
town, I forget which, some eight miles distant. This
distance the postman was in the habit of traversing
daily. A considerable part of his way lay across a
district of unenclosed champaign meadow-land, and
in walking over this smooth surface the postman was
generally asleep. But at the termination of this part
of his road, there was a narrow foot-bridge over a
stream, and to reach this bridge it was necessary to
ascend some broken steps. Now, it was ascertained
as completely as any fact of the kind could be,—the
observers were shrewd, and the object of observation
was a man of undoubted probity,—I say, it was completely ascertained :—1°, That the postman was asleep
in passing over this level course ; 2°, That he held on
his way in this state without deflection towards the
bridge ; and, 3°, That just before arriving at the
bridge, he awoke. But this case is not only deserving
of all credit from the positive testimony by which it
is vouched ; it is also credible as only one of a class
of analogous cases which it may be adduced as representing. This case, besides showing that the mind
must be active though the body is asleep, shows also
that certain bodily functions may be dormant, while

a *Gedanken vom Schlafe*, Halle, *buch der Psychologie*, p. 28-9.—
1746, p. 7. See Tiedemann, *Hand-* En.

others are alert. The locomotive faculty was here in exercise, while the senses were in slumber. This suggests to me another example of the same phænomenon.

Case of
Oporinus.

It is found in a story told by Erasmus [a] in one of his letters, concerning his learned friend Oporinus, the celebrated professor and printer of Basle. Oporinus was on a journey with a bookseller; and, on their road, they had fallen in with a manuscript. Tired with their day's travelling, — travelling was then almost exclusively performed on horseback,—they came at nightfall to their inn. They were, however, curious to ascertain the contents of their manuscript, and Oporinus undertook the task of reading it aloud. This he continued for some time, when the bookseller found it necessary to put a question concerning a word which he had not rightly understood. It was now discovered that Oporinus was asleep, and being awakened by his companion, he found that he had no recollection of what for a considerable time he had been reading. Most of you, I daresay, have known or heard of similar occurrences, and I do not quote the anecdote as anything remarkable. But, still, it is a case concurring with a thousand others to prove, 1°, That one bodily sense or function may be asleep while another is awake; and, 2°, That the mind may be in a certain state of activity during sleep, and no memory of that activity remain after the sleep has ceased. The first is evident; for Oporinus, while reading, must have had his eyes and the muscles of his tongue and fauces awake, though his ears and other senses were asleep; and the second is no less so, for the act of reading

a This story is told by Felix Platerus (*Observationes*, lib. i. p. 11). The person to whom Oporinus read, was the father of the narrator, Thomas Platerus. See Bohn, *Noctambulatio;* (Haller, *Disputationes ad Morborum Hist. et Curat.*, t. vii. p. 443.)—ED.

supposed a very complex series of mental energies. I
may notice, by the way, that physiologists have observed, that our bodily senses and powers do not fall
asleep simultaneously, but in a certain succession.
We all know that the first symptom of slumber is the
relaxation of the eyelids; whereas, hearing continues
alert for a season after the power of vision has been
dormant. In the case last alluded to, this order was,
however, violated; and the sight was forcibly kept
awake while the hearing had lapsed into torpidity.

In the case of sleep, therefore, so far is it from
being proved that the mind is at any moment unconscious, that the result of observation would incline us
to the opposite conclusion.

LECTURE XVIII.

CONSCIOUSNESS,—GENERAL PHÆNOMENA,—IS THE MIND EVER UNCONSCIOUSLY MODIFIED?

LECT.
XVIII.
———
Is the mind
ever uncon-
sciously
modified ?

I pass now to a question in some respects of still more proximate interest to the psychologist than that discussed in the preceding Lecture; for it is one which, according as it is decided, will determine the character of our explanation of many of the most important phænomena in the philosophy of mind, and, in particular, the great phænomena of Memory and Association. The question I refer to is, Whether the mind exerts energies, and is the subject of modifications, of neither of which it is conscious. This is the most general expression of a problem which has hardly been mentioned, far less mooted, in this country; and when it has attracted a passing notice, the supposition of an unconscious action or passion of the mind has been treated as something either unintelligible, or absurd. In Germany, on the contrary, it has not only been canvassed, but the alternative which the philosophers of this country have lightly considered as ridiculous, has been gravely established as a conclusion which the phænomena not only warrant, but enforce. The French philosophers, for a long time, viewed the question in the same light as the British. Condillac, indeed, set the latter the ex-

ample ;[a] but of late a revolution is apparent, and two
recent French psychologists[β] have marvellously propounded the doctrine, long and generally established
in Germany, as something new and unheard of before
their own assertion of the paradox.

This question is one not only of importance, but of
difficulty; I shall endeavour to make you understand
its purport by arguing it upon broader grounds than
has hitherto been done, and shall prepare you, by some
preliminary information, for its discussion.   I shall
first of all adduce some proof of the fact, that the
mind may, and does, contain far more latent furni-
ture than consciousness informs us it possesses.   To *Three degrees of mental latency.*
simplify the discussion, I shall distinguish three de-
grees of this mental latency.

In the first place, it is to be remembered that the *The first.*
riches,—the possessions, of our mind, are not to be
measured by its present momentary activities, but by
the amount of its acquired habits.   I know a science,
or language, not merely while I make a temporary
use of it, but inasmuch as I can apply it when and
how I will.   Thus the infinitely greater part of our
spiritual treasures, lies always beyond the sphere of
consciousness, hid in the obscure recesses of the mind.
This is the first degree of latency.   In regard to this,
there is no difficulty, or dispute; and I only take it
into account in order to obviate misconception, and
because it affords a transition towards the other two
degrees which it conduces to illustrate.

The second degree of latency exists when the mind *The second.*
contains certain systems of knowledge, or certain

a *Essai sur l'Origine des Connois-*   β Cardaillac and Damiron.   See
sances Humaines, Sect. ii. ch. 1, §   below, p. 363.—ED.
4-13.—ED.

habits of action, which it is wholly unconscious of
possessing in its ordinary state, but which are revealed
to consciousness in certain extraordinary exaltations
of its powers. The evidence on this point shows that
the mind frequently contains whole systems of knowledge, which, though in our normal state they have
faded into absolute oblivion, may, in certain abnormal
states, as madness, febrile delirium, somnambulism,
catalepsy, &c., flash out into luminous consciousness,
and even throw into the shade of unconsciousness
those other systems by which they had, for a long
period, been eclipsed and even extinguished. For
example, there are cases in which the extinct memory of whole languages was suddenly restored, and,
what is even still more remarkable, in which the
faculty was exhibited of accurately repeating, in
known or unknown tongues, passages which were
never within the grasp of conscious memory in the
normal state. This degree,—this phænomenon, of
latency, is one of the most marvellous in the whole
compass of philosophy, and the proof of its reality
will prepare us for an enlightened consideration, of
the third, of which the evidence, though not less
certain, is not equally obtrusive. But, however remarkable and important, this phænomenon has been
almost wholly neglected by psychologists,* and the
cases which I adduce in illustration of its reality have
never been previously collected and applied. That in
madness, in fever, in somnambulism, and other abnormal states, the mind should betray capacities and extensive systems of knowledge, of which it was at other

a These remarks were probably *Powers*. He collects some very
written before the publication of curious instances, see p. 314, 10th
Abercrombie, *On the Intellectual* edition.—ED.

times wholly unconscious, is a fact so remarkable that it may well demand the highest evidence to establish its truth. But of such a character is the evidence which I am now to give you. It consists of cases reported by the most intelligent and trustworthy observers,—by observers wholly ignorant of each other's testimony; and the phænomena observed were of so palpable and unambiguous a nature that they could not possibly have been mistaken or misinterpreted.

The first, and least interesting, evidence I shall adduce, is derived from cases of madness; it is given by a celebrated American physician, Dr Rush.

" The records of the wit and cunning of madmen," says Dr Rush, " are numerous in every country. Talents for eloquence, poetry, music, and painting, and uncommon ingenuity in several of the mechanical arts, are often evolved in this state of madness. A gentleman, whom I attended in an hospital in the year 1810, often delighted as well as astonished the patients and officers of our hospital by his displays of oratory, in preaching from a table in the hospital yard every Sunday. A female patient of mine who became insane, after parturition, in the year 1807, sang hymns and songs of her own composition during the latter stage of her illness, with a tone of voice so soft and pleasant that I hung upon it with delight every time I visited her. She had never discovered a talent for poetry or music, in any previous part of her life. Two instances of a talent for drawing, evolved by madness, have occurred within my knowledge. And where is the hospital for mad people, in which elegant and completely rigged ships, and curious pieces of machinery, have not been exhibited by persons who never discovered the least turn for a mechanical art, previously to their derangement? Sometimes we observe in mad people an unexpected resuscitation of knowledge; hence we hear them describe past events, and speak in ancient or modern languages, or repeat long and interesting passages from books, none of which, we are sure, they were capable of recollecting in the natural and healthy state of their mind."[a]

From cases of fever.

The second class of cases are those of fever; and the first I shall adduce is given on the authority of the patient himself. This is Mr Flint, a very intelligent American clergyman. I take it from his *Recollections of the Valley of the Mississippi*. He was travelling in the State of Illinois, and suffered the common lot of visitants from other climates, in being taken down with a bilious fever.—" I am aware," he remarks, " that every sufferer in this way is apt to think his own case extraordinary. My physicians agreed with all who saw me that my case was so. As very few live to record the issue of a sickness like mine, and as you have requested me, and as I have promised, to be particular, I will relate some of the circumstances of this disease. And it is in my view desirable, in the bitter agony of such diseases, that more of the symptoms, sensations, and sufferings should have been recorded than have been; and that others in similar predicaments may know, that some before them have had sufferings like theirs, and have survived them. I had had a fever before, and had risen, and been dressed every day. But in this, with the first day I was prostrated to infantine weakness, and felt, with its first attack, that it was a thing very different from what I had yet experienced. Paroxysms of derangement occurred the third day, and this was

a Beasley, *On the Mind*, p. 474.

to me a new state of mind. That state of disease in
which partial derangement is mixed with a consciousness generally sound, and a sensibility preternaturally
excited, I should suppose the most distressing of all
its forms. At the same time that I was unable to
recognise my friends, I was informed that my memory
was more than ordinarily exact and retentive, and
that I repeated whole passages in the different languages which I knew, with entire accuracy. I recited,
without losing or misplacing a word, a passage of
poetry which I could not so repeat after I recovered
my health."

The following more curious case is given by Lord Case of the
Comtesse
de Laval.
Monboddo in his *Antient Metaphysics.*ᵃ

" It was communicated in a letter from the late Mr
Hans Stanley, a gentleman well known both to the
learned and political world, who did me the honour
to correspond with me upon the subject of my first
volume of metaphysics. I will give it in the words
of that gentleman. He introduces it, by saying, that
it is an extraordinary fact in the history of mind,
which he believes stands single, and for which he does
not pretend to account. Then he goes on to narrate
it:—' About six-and-twenty years ago, when I was in
France, I had an intimacy in the family of the late
Maréchal de Montmorenci de Laval. His son, the
Comte de Laval, was married to Mademoiselle de
Maupeaux, the daughter of a Lieutenant-General of
that name, and the niece of the late Chancellor. This
gentleman was killed at the battle of Hastenbeck; his
widow survived him some years, but is since dead.

" ' The following fact comes from her own mouth.
She has told it me repeatedly. She was a woman of

ᵃ Vol. ii. p. 217.

perfect veracity, and very good sense. She appealed to her servants and family for the truth. Nor did she, indeed, seem to be sensible that the matter was so extraordinary as it appeared to me. I wrote it down at the time; and I have the memorandum among some of my papers.

"'The Comtesse de Laval, had been observed, by servants who sate up with her on account of some indisposition, to talk in her sleep a language that none of them understood; nor were they sure, or, indeed, herself able to guess, upon the sounds being repeated to her, whether it was or was not gibberish.

"'Upon her lying in of one of her children, she was attended by a nurse, who was of the province of Brittany, and who immediately knew the meaning of what she said, it being in the idiom of the natives of that country; but she herself, when awake, did not understand a single syllable of what she had uttered in her sleep, upon its being retold her.

"'She was born in that province, and had been nursed in a family where nothing but that language was spoken; so that, in her first infancy, she had known it, and no other; but, when she returned to her parents, she had no opportunity of keeping up the use of it; and, as I have before said, she did not understand a word of *Breton* when awake, though she spoke it in her sleep.

"'I need not say that the Comtesse de Laval never said or imagined that she used any words of the Breton idiom, more than were necessary to express those ideas that are within the compass of a child's knowledge of objects,'" &c.

Case given by Coleridge.

A highly interesting case is given by Mr Coleridge in his *Biographia Literaria*.[a]

a Vol. i. p. 117, (edit. 1847).

"It occurred," says Mr Coleridge, "in a Roman
Catholic town in Germany, a year or two before my
arrival at Göttingen, and had not then ceased to be a
frequent subject of conversation.   A young woman of
four or five and twenty, who could neither read nor
write, was seized with a nervous fever; during which,
according to the asseverations of all the priests and
monks of the neighbourhood, she became possessed,
and, as it appeared, by a very learned devil. She continued incessantly talking Latin, Greek, and Hebrew,
in very pompous tones, and with most distinct enunciation.   This possession was rendered more probable
by the known fact that she was or had been a heretic.
Voltaire humorously advises the devil to decline all
acquaintance with medical men; and it would have
been more to his reputation, if he had taken this advice in the present instance.   The case had attracted
the particular attention of a young physician, and by
his statement many eminent physiologists and psychologists visited the town, and cross-examined the
case on the spot. Sheets full of her ravings were
taken down from her own mouth, and were found to
consist of sentences, coherent and intelligible each for
itself, but with little or no connection with each other.
Of the Hebrew, a small portion only could be traced
to the Bible, the remainder seemed to be in the Rabbinical dialect.   All trick or conspiracy was out of the
question.   Not only had the young woman ever been
a harmless, simple creature; but she was evidently
labouring under a nervous fever.   In the town, in
which she had been resident for many years as a servant in different families, no solution presented itself.
The young physician, however, determined to trace
her past life step by step; for the patient herself was
incapable of returning a rational answer. He at length

succeeded in discovering the place where her parents
had lived : travelled thither, found them dead, but an
uncle surviving ; and from him learned that the patient had been charitably taken by an old Protestant
pastor at nine years old, and had remained with him
some years, even till the old man's death. Of this
pastor the uncle knew nothing, but that he was a very
good man. With great difficulty, and after much
search, our young medical philosopher discovered a
niece of the pastor's who had lived with him as his
housekeeper, and had inherited his effects. She remembered the girl ; related that her venerable uncle
had been too indulgent, and could not bear to hear the
girl scolded ; that she was willing to have kept her,
but that, after her patron's death, the girl herself refused to stay. Anxious inquiries were then, of course,
made concerning the pastor's habits ; and the solution of the phænomenon was soon obtained. For it
appeared that it had been the old man's custom, for
years, to walk up and down a passage of his house
into which the kitchen-door opened, and to read to
himself, with a loud voice, out of his favourite books.
A considerable number of these were still in the
niece's possession. She added, that he was a very
learned man, and a great Hebraist. Among the books
were found a collection of Rabbinical writings, together with several of the Greek and Latin fathers ;
and the physician succeeded in identifying so many
passages with those taken down at the young woman's
bedside, that no doubt could remain in any rational
mind concerning the true origin of the impressions
made on her nervous system."

These cases thus evince the general fact that a
mental modification is not proved not to be, merely

because consciousness affords us no evidence of its
existence. This general fact being established, I now
proceed to consider the question in relation to the
third class or degree of latent modifications,—a class
in relation to, and on the ground of which alone, it
has ever hitherto been argued by philosophers.

The problem, then, in regard to this class is,—Are
there, in ordinary, mental modifications,—*i.e.* mental
activities and passivities, of which we are unconscious,
but which manifest their existence by effects of which
we are conscious?

The prob-
lem in re-
gard to this
degree
stated.

I have thus stated the question, because this appears to me the most unambiguous form in which it
can be expressed; and in treating of it, I shall, in the
first place, consider it in itself, and, in the second
place, in its history. I adopt this order, because the
principal difficulties which affect the problem arise
from the equivocal and indeterminate language of
philosophers. These it is obviously necessary to avoid
in the first instance; but having obtained an insight
into the question itself, it will be easy, in a subsequent historical narrative, to show how it has been perplexed and darkened by the mode in which it has been
handled by philosophers. I request your attention to
this matter, as in the solution of this general problem
is contained the solution of several important questions, which will arise under our consideration of the
special faculties. It is impossible, however, at the
present stage of our progress, to exhibit all, or even
the strongest part of, the evidence for the alternative
which I adopt; and you must bear in mind that there
is much more to be said in favour of this opinion than
what I am able at present to adduce to you.

To be con-
sidered in
itself, and
in its his-
tory.

In the question proposed, I am not only strongly

LECT.
XVIII.

The affir-
mative of
this ques-
tion main-
tained.

inclined to the affirmative,—nay, I do not hesitate to
maintain, that what we are conscious of is constructed
out of what we are not conscious of,—that our whole
knowledge, in fact, is made up of the unknown and
the incognisable.

To the affir-
mative two
objections.

This at first sight may appear not only paradoxical, but contradictory. It may be objected, 1°, How
can we know that to exist which lies beyond the one
condition of all knowledge,—consciousness ? And 2°,
How can knowledge arise out of ignorance,—consciousness out of unconsciousness,—the cognisable out of
the incognisable,—that is, how can one opposite proceed out of the other ?

The first
objection
obviated.

In answer to the first objection,—How can we
know that of which we are unconscious, seeing that
consciousness is the condition of knowledge ?—it is
enough to allege, that there are many things which
we neither know nor can know in themselves,—that
is, in their direct and immediate relation to our faculties of knowledge, but which manifest their existence
indirectly through the medium of their effects. This

The mental
modifica-
tions in
question
manifest
their exist-
ence
through
their
effects.

is the case with the mental modifications in question ;
they are not in themselves revealed to consciousness,
but as certain facts of consciousness necessarily suppose them to exist and to exert an influence in the
mental processes, we are thus constrained to admit
as modifications of mind, what are not in themselves

Established
from the
nature of
conscious-
ness itself.

phænomena of consciousness. The truth of this will
be apparent, if, before descending to any special illustration, we consider that consciousness cannot exist
independently of some peculiar modification of mind ;
we are only conscious as we are conscious of a determinate state. To be conscious, we must be conscious of some particular perception, or remembrance,

or imagination, or feeling, &c.; we have no general consciousness. But as consciousness supposes a special mental modification as its object, it must be remembered, that this modification or state supposes a change,—a transition from some other state or modification. But as the modification must be present, before we have a consciousness of the modification, it is evident that we can have no consciousness of its rise or awakening; for its rise or awakening is also the rise or awakening of consciousness. LECT. XVIII.

But the illustration of this is contained in an answer to the second objection which asks,—How can knowledge come out of ignorance,—consciousness out of unconsciousness,—the known out of the unknown,—how can one opposite be made up of the other? *The second objection.*

In the removal of this objection, the proof of the thesis which I support is involved. And without dealing in any general speculation, I shall at once descend to the special evidence which appears to me, not merely to warrant, but to necessitate, the conclusion, that the sphere of our conscious modifications is only a small circle in the centre of a far wider sphere of action and passion, of which we are only conscious through its effects. *The special evidence for the affirmative of the general problem adduced.*

Let us take our first example from Perception,—the perception of external objects, and in that faculty, let us commence with the sense of sight. Now, you either already know, or can be at once informed, what it is that has obtained the name of *Minimum Visibile.* You are of course aware, in general, that vision is the result of the rays of light, reflected from the surface of objects to the eye; a greater number of rays is reflected from a larger surface; if the superficial extent of an object, and, consequently, the number of *I. External Perception. 1. The sense of Sight. Minimum Visibile.*

the rays which it reflects, be diminished beyond a
certain limit, the object becomes invisible; and the
minimum visibile is the smallest expanse which can
be seen,—which can consciously affect us,—which we
can be conscious of seeing. This being understood,
it is plain that if we divide this *minimum visibile* into
two parts, neither half can, by itself, be an object
of vision, or visual consciousness. They are, severally
and apart, to consciousness as zero. But it is evident
that each half must, by itself, have produced in us a
certain modification, real though unperceived; for as
the perceived whole is nothing but the union of the
unperceived halves, so the perception,—the perceived
affection itself of which we are conscious,—is only the
sum of two modifications, each of which severally
eludes our consciousness. When we look at a distant
forest, we perceive a certain expanse of green. Of
this as an affection of our organism, we are clearly
and distinctly conscious. Now, the expanse of which
we are conscious is evidently made up of parts of
which we are not conscious. No leaf, perhaps no
tree, may be separately visible. But the greenness of
the forest is made up of the greenness of the leaves;
that is, the total impression of which we are conscious,
is made up of an infinitude of small impressions of
which we are not conscious.

Take another example, from the sense of hearing.
In this sense, there is, in like manner, a *Minimum
Audibile*, that is, a sound the least which can come
into perception and consciousness. But this *minimum audibile* is made up of parts which severally
affect the sense, but of which affections, separately, we
are not conscious, though of their joint result we are.
We must, therefore, here likewise admit the reality of

modifications beyond the sphere of consciousness. To take a special example. When we hear the distant murmur of the sea, what are the constituents of the total perception of which we are conscious? This murmur is a sum made up of parts, and the sum would be as zero if the parts did not count as something. The noise of the sea is the complement of the noise of its several waves;—

ποντίων τε κυμάτων

'Ανήριθμον γέλασμα· [a]

and if the noise of each wave made no impression on our sense, the noise of the sea, as the result of these impressions, could not be realised. But the noise of each several wave, at the distance we suppose, is inaudible; we must, however, admit that they produce a certain modification, beyond consciousness, on the precipient subject; for this is necessarily involved in the reality of their result. The same is equally the case in the other senses: the taste or smell of a dish, be it agreeable or disagreeable, is composed of a multitude of severally imperceptible effects, which the stimulating particles of the viand cause on different points of the nervous expansion of the gustatory and olfactory organs; and the pleasant or painful feeling of softness or roughness is the result of an infinity of unfelt modifications, which the body handled determines on the countless papillæ of the nerves of touch.[β]

3. The other senses.

Let us now take an example from another mental process. We have not yet spoken of what is called the Association of Ideas; and it is enough for our

II. Association of Ideas.

---

a Æschylus, *Prometheus*, 1. 89.— Ed.

β See Leibnitz, *Nouveaux Essais*,

Avant-Propos, p. 8-9, (ed. Raspe); and lib. ii. c. i. § 9 *et seq.*—Ed.

present purpose that you should be aware, that one
thought suggests another in conformity to certain
determinate laws,—laws to which the succession of
our whole mental states are subjected. Now it sometimes happens, that we find one thought rising immediately after another in consciousness, but whose
consecution we can reduce to no law of association.
Now in these cases we can generally discover by an
attentive observation, that these two thoughts, though
not themselves associated, are each associated with
certain other thoughts ; so that the whole consecution
would have been regular, had these intermediate
thoughts come into consciousness, between the two
which are not immediately associated. Suppose, for
instance, that A, B, C, are three thoughts,—that A and
C cannot immediately suggest each other, but that each
is associated with B, so that A will naturally suggest
B, and B naturally suggest C. Now it may happen,
that we are conscious of A, and immediately thereafter
of C. How is the anomaly to be explained ? It can
only be explained on the principle of latent modifications. A suggests C, not immediately, but through
B ; but as B, like the half of the *minimum visibile* or
minimum audibile, does not rise into consciousness, we
are apt to consider it as non-existent. You are probably aware of the following fact in mechanics. If a
number of billiard balls be placed in a straight row
and touching each other, and if a ball be made to
strike, in the line of the row, the ball at one end of
the series, what will happen ? The motion of the im-
pinging ball is not divided among the whole row ;
this, which we might *a priori* have expected, does not
happen, but the impetus is transmitted through the
intermediate balls which remain each in its place, to

the ball at the opposite end of the series, and this ball alone is impelled on. Something like this seems often to occur in the train of thought. One idea mediately suggests another into consciousness,—the suggestion passing through one or more ideas which do not themselves rise into consciousness. The awakening and awakened ideas here correspond to the ball striking and the ball struck off; while the intermediate ideas of which we are unconscious, but which carry on the suggestion, resemble the intermediate balls which remain moveless, but communicate the impulse. An instance of this occurs to me with which I was recently struck. Thinking of Ben Lomond, this thought was immediately followed by the thought of the Prussian system of education. Now, conceivable connection between these two ideas in themselves, there was none. A little reflection, however, explained the anomaly. On my last visit to the mountain, I had met upon its summit a German gentleman, and though I had no consciousness of the intermediate and unawakened links between Ben Lomond and the Prussian schools, they were undoubtedly these,—the German, — Germany, — Prussia, — and, these media being admitted, the connection between the extremes was manifest.

I should perhaps reserve for a future occasion, noticing Mr Stewart's explanation of this phænomenon. He admits that a perception or idea may pass through the mind without leaving any trace in the memory, and yet serve to introduce other ideas connected with it by the laws of association.[a] Mr Stewart can hardly be said to have contemplated the possibility of the existence and agency of mental

a *Elements*, part i. chap. ii.; *Works*, vol. ii. pp. 121, 122.

modifications of which we are unconscious. He grants the necessity of interpolating certain intermediate ideas, in order to account for the connection of thought, which could otherwise be explained by no theory of association ; and he admits that these intermediate ideas are not known by memory to have actually intervened. So far, there is no difference in the two doctrines. But now comes the separation. Mr Stewart supposes that the intermediate ideas are, for an instant, awakened into consciousness, but, in the same moment, utterly forgot ; whereas, the opinion I would prefer, holds that they are efficient without rising into consciousness. Mr Stewart's doctrine on this point is exposed to all the difficulties, and has none of the proofs in its favour, which concur in establishing the other.

Difficulties of Stewart's doctrine.

1. Assumes acts of consciousness of which there is no memory.
2. Violates the analogy of consciousness.

In the first place, to assume the existence of acts of consciousness of which there is no memory beyond the moment of existence, is at least as inconceivable an hypothesis as the other. But, in the second place, it violates the whole analogy of consciousness, which the other does not. Consciousness supposes memory ; and we are only conscious as we are able to connect and contrast one instance of our intellectual existence with another. Whereas, to suppose the existence and efficiency of modifications beyond consciousness, is not at variance with its conditions ; for consciousness, though it assures us of the reality of what is within its sphere, says nothing against the reality of what is without. In the third place, it is

3. Presumption in favour of latent acts in association.
demonstrated, that, in perception, there are modifications, efficient, though severally imperceptible ; why, therefore, in the other faculties, should there not likewise be modifications, efficient, though unapparent ?

In the fourth place, there must be some reason for the assumed fact, that there are perceptions or ideas of which we are conscious, but of which there is no memory. Now, the only reason that can possibly be assigned is that the consciousness was too faint to afford the condition of memory. But of consciousness, however faint, there must be some memory, however short. But this is at variance with the phænomenon, for the ideas A and C may precede and follow each other without any perceptible interval, and without any the feeblest memory of B. If there be no memory, there could have been no consciousness ; and, therefore, Mr Stewart's hypothesis, if strictly interrogated, must, even at last, take refuge in our doctrine ; for it can easily be shown, that the degree of memory is directly in proportion to the degree of consciousness, and, consequently, that an absolute ̄negation of memory is an absolute negation of consciousness.

Let us now turn to another class of phænomena, which in like manner are capable of an adequate explanation only on the theory I have advanced ;— I mean the operations resulting from our acquired Dexterities and Habits.

To explain these, three theories have been advanced. The first regards them as merely mechanical or automatic, and thus denying to the mind all active or voluntary intervention, consequently removes them beyond the sphere of consciousness. The second, again, allows to each several motion a separate act of conscious volition ; while the third, which I would maintain, holds a medium between these, constitutes the mind the agent, accords to it a conscious volition over the series, but denies to it a consciousness and

LECT.
XVIII.

4. Stewart's hypothesis must take refuge in the counter doctrine.

III. Our Acquired Dexterities and Habits.

To explain these, three theories advanced. The first.

The second.

The third.

LECT.
XVIII. deliberate volition in regard to each separate movement in the series which it determines.

The first or mechanical theory, maintained by Reid and Hartley.

The first of these has been maintained, among others, by two philosophers who, in other points, are not frequently at one,—by Reid and Hartley. "Habit," says Reid, "differs from instinct, not in its nature, but in its origin; the latter being natural, the former acquired. Both operate without will or intention, without thought, and therefore may be called mechanical principles."[a] In another passage, he expresses himself thus :—" I conceive it to be a part of our constitution, that what we have been accustomed to do, we acquire not only a facility but a proneness to do on like occasions; so that it requires a particular will or effort to forbear it, but to do it requires very often no will at all."[β]

The same doctrine is laid down still more explicitly by Dr Hartley. "Suppose," says he, " a person, who has a perfectly voluntary command over his fingers, to begin to learn to play on the harpsichord. The first step is to move his fingers, from key to key, with a slow motion, looking at the notes, and exerting an express act of volition in every motion. By degrees the motions cling to one another, and to the impressions of the notes, in the way of *association*, so often mentioned; the acts of volition growing less and less express all the time, till, at last, they become evanescent and imperceptible. For an expert performer will play from notes, or ideas laid up in the memory, and at the same time carry on a quite different train of thoughts in his mind; or even hold a conversation with another. Whence we conclude, that

a *Active Powers*, Essay iii., part i.   β *Ibid.*
chap. 3; *Works*, p. 550.

there is no intervention of the idea, or state of mind
called will." Cases of this sort Hartley calls "transitions of voluntary actions into automatic ones." [a]

The second theory is maintained against the first by
Mr Stewart; and I think his refutation valid, though
not his confirmation. " I cannot help thinking it,"
he says, "more philosophical to suppose that those
actions which are originally voluntary always continue
so, although, in the case of operations which are become habitual in consequence of long practice, we may
not be able to recollect every different volition. Thus,
in the case of a performer on the harpsichord, I apprehend that there is an act of the will preceding every
motion of every finger, although he may not be able
to recollect these volitions afterwards, and although
he may, during the time of his performance, be employed in carrying on a separate train of thought. For
it must be remarked, that the most rapid performer
can, when he pleases, play so slowly as to be able to
attend to, and to recollect, every separate act of his
will in the various movements of his fingers; and he
can gradually accelerate the rate of his execution till
he is unable to recollect these acts. Now, in this instance, one of two suppositions must be made. The
one is, that the operations in the two cases are carried
on precisely in the same manner, and differ only in the
degree of rapidity; and that when this rapidity exceeds a certain rate, the acts of the will are too momentary to leave any impression on the memory. The
other is, that when the rapidity exceeds a certain rate,
the operation is taken entirely out of our hands, and
is carried on by some unknown power, of the nature
of which we are as ignorant as of the cause of the cir-

The second
theory main-
tained,
validly as
against the
first, by
Stewart.

a Vol. i. pp. 108, 109. [*Observations on Man*, prop. xxi.—Ed.]

culation of the blood, or of the motion of the intestines. The last supposition seems to me to be somewhat similar to that of a man who should maintain that although a body projected with a moderate velocity is seen to pass through all the intermediate spaces in moving from one place to another, yet we are not entitled to conclude that this happens when the body moves so quickly as to become invisible to the eye. The former supposition is supported by the analogy of many other facts in our constitution. Of some of these I have already taken notice, and it would be easy to add to the number. An expert accountant, for example, can sum up, almost with a single glance of his eye, a long column of figures. He can tell the sum with unerring certainty, while at the same time he is unable to recollect any one of the figures of which that sum is composed; and yet nobody doubts that each of these figures has passed through his mind, or supposes that when the rapidity of the process becomes so great that he is unable to recollect the various steps of it, he obtains the result by a sort of inspiration. This last supposition would be perfectly analogous to Dr Hartley's doctrine concerning the nature of our habitual exertions.

" The only plausible objection which, I think, can be offered to the principles I have endeavoured to establish on this subject, is founded on the astonishing and almost incredible rapidity they necessarily suppose in our intellectual operations. When a person, for example, reads aloud, there must, according to this doctrine, be a separate volition preceding the articulation of every letter; and it has been found by actual trial, that it is possible to pronounce about two thousand letters in a minute. Is it reasonable to suppose

that the mind is capable of so many different acts in
an interval of time so very inconsiderable?

" With respect to this objection it may be observed,
in the first place, that all arguments against the foregoing doctrine with respect to our habitual exertions,
in so far as they are founded on the inconceivable
rapidity which they suppose in our intellectual operations, apply equally to the common doctrine concerning our perception of distance by the eye. But
this is not all. To what does the supposition amount
which is considered as so incredible? Only to this,
that the mind is so formed as to be able to carry on
certain intellectual processes in intervals of time too
short to be estimated by our faculties; a supposition
which, so far from being extravagant, is supported by
the analogy of many of our most certain conclusions
in natural philosophy. The discoveries made by the
microscope have laid open to our senses a world of
wonders, the existence of which hardly any man would
have admitted upon inferior evidence; and have gradually prepared the way for those physical speculations, which explain some of the most extraordinary
phænomena of nature by means of modifications of
matter far too subtile for the examination of our organs.
Why, then, should it be considered as unphilosophical,
after having demonstrated the existence of various intellectual processes which escape our attention in consequence of their rapidity, to carry the supposition a
little farther, in order to bring under the known laws
of the human constitution a class of mental operations
which must otherwise remain perfectly inexplicable?
Surely our ideas of time are merely relative, as well
as our ideas of extension; nor is there any good reason
for doubting that, if our powers of attention and me-

mory were more perfect than they are, so as to give us the same advantage in examining rapid events, which the microscope gives for examining minute portions of extension, they would enlarge our views with respect to the intellectual world, no less than that instrument has with respect to the material." [a]

The principle of Stewart's theory already shown to involve contradictions.

This doctrine of Mr Stewart, — that our acts of knowledge are made up of an infinite number of acts of attention, that is, of various acts of concentrated consciousness, there being required a separate act of attention for every minimum possible of knowledge, — I have already shown you, by various examples, to

But here specially refuted.

involve contradictions. In the present instance, its admission would constrain our assent to the most monstrous conclusions. Take the case of a person reading. Now, all of you must have experienced, if ever under the necessity of reading aloud, that, if the matter be uninteresting, your thoughts, while you are going on in the performance of your task, are wholly abstracted from the book and its subject, and you are perhaps deeply occupied in a train of serious meditation. Here the process of reading is performed without interruption, and with the most punctual accuracy; and, at the same time, the process of meditation is carried on without distraction or fatigue. Now, this, on Mr Stewart's doctrine, would seem impossible, for what does his theory suppose? It supposes that separate acts of concentrated consciousness or attention, are bestowed on each least movement in either process. But be the velocity of the mental operations what it may, it is impossible to conceive how transitions between such contrary operations could be kept up for a continuance without fatigue and distraction, even if

a *Elements*, vol. i. chap. ii.; *Works*, vol. ii. p. 127-131.

we throw out of account the fact that the acts of attention to be effectual must be simultaneous, which on Mr Stewart's theory is not allowed.

We could easily give examples of far more complex operations; but this, with what has been previously said, I deem sufficient to show, that we must either resort to the first theory, which, as nothing but the assumption of an occult and incomprehensible principle, in fact explains nothing, or adopt the theory that there are acts of mind so rapid and minute as to elude the ken of consciousness.

I shall now say something of the history of this opinion. It is a curious fact that Locke, in the passage I read to you a few days ago, attributes this opinion to the Cartesians, and he thinks it was employed by them to support their doctrine of the ceaseless activity of mind.[a] In this, as in many other points of the Cartesian philosophy, he is, however, wholly wrong. On the contrary, the Cartesians made consciousness the essence of thought;[β] and their assertion that the mind always thinks, is, in their language, precisely tantamount to the assertion that the mind is always conscious.

But what was not maintained by the Cartesians, and even in opposition to their doctrine, was advanced by Leibnitz.[γ] To this great philosopher belongs the honour of having originated this opinion, and of having supplied some of the strongest arguments in its support. He was, however, unfortunate in the terms which he employed to propound his doctrine. The latent

LECT.
XVIII.

History of the doctrine of unconscious mental modifications.

Leibnitz the first to proclaim this doctrine.

---

a *Essay on Human Understanding*, book ii. c. 1, §§ 18, 19. The Cartesians are intended, though not expressly mentioned.—ED.

β Descartes, *Principia*, pars i. §

9. See above, p. 313.—ED.

γ *Nouveaux Essais*, ii. 1; *Monadologie*, § 14. *Principes de la Nature et de la Grace*, § 4.—ED.

LECT.
XVIII.

Unfortu-
nate in the
terms he
employed
to desig-
nate it.

modifications,—the unconscious activities of mind, he
denominated *obscure ideas, obscure representations,
perceptions without apperception or consciousness,
insensible perceptions,* &c.　In this he violated the
universal usage of language.　For perception, and idea,
and representation, all properly involve the notion of
consciousness ; it being, in fact, contradictory to speak
of a representation not really represented,—a perception not really perceived,—an actual idea of whose
presence we are not aware.

Fate of the
doctrine in
France and
Britain.

The close affinity of mental modifications with perceptions, ideas, representations, and the consequent
commutation of these terms, have been undoubtedly
the reasons why the Leibnitian doctrine was not
more generally adopted, and why, in France and in
Britain, succeeding philosophers have almost admitted
as a self-evident truth, that there can be no modification of mind devoid of consciousness.　As to any
refutation of the Leibnitian doctrine, I know of none.

Condillac.

Condillac is, indeed, the only psychologist who can be
said to have formally proposed the question.　He,
like Mr Stewart, attempts to explain why it can be
supposed that the mind has modifications of which we
are not conscious, by asserting that we are in truth
conscious of the modification, but that it is imme-

The doc-
trine of
Leibnitz
adopted in
Germany.

diately forgotten.[a]　In Germany, the doctrine of
Leibnitz was almost universally adopted.　I am not
aware of a philosopher of the least note, by whom it
has been rejected.　In France, it has, I see, lately

De Cardail-
lac.

been broached by M. de Cardaillac,[b] as a theory of
his own, and this, his originality, is marvellously

a *Essai sur l'Origine des Connois-*　　β *Etudes Elémentaires de Philoso-*
sances Humaines, sect. ii. c. 1, § 4-　*phie,* t. ii. pp. 138, 139.
13.—Ed.

admitted by authors, like M. Damiron,[a] whom we
might reasonably expect to have been better informed.
It is hardly worth adding, that as the doctrine is
not new, so nothing new has been contributed to
its illustration. To British psychologists, the opinion
would hardly seem to have been known ; by none,
certainly, is it seriously considered.[b]

LECT.
XVIII.

Damiron.

a [*Ess. sur l'Hist. de Phil.*, Sup-
plément, p. 460 et seq., 5th edition.]
[In the *second* edition of Damiron's
Psychologie (t. i. p. 188), Leibnitz
is expressly cited. In the *first*
edition, however, though the doctrine of latency is stated, (t. i. p.
190), there is no reference to Leibnitz.—Ed.]

β Qualified exception; Kames'
*Essays on the Principles of Morality
and Natural Religion*, part ii. ess.
iv., *On Matter and Spirit*, p. 289 to
end, (3d edit.) [With Kames compare
F. A. Carus, *Psychologie*, ii. p. 185,
(edit. 1808). Tucker, *Light of Nature*, i. c. 10, § 4. Tralles, *De Animæ existentis Immaterialitate et Im-
mortalitate*, p. 39 et seq. On the
general subject of acts of mind beyond the sphere of consciousness,

compare Kant, *Anthropologie*, § 5.
Reinhold, *Theorie des menschlichen
Erkenntnissvermögens und Metaphysik*, i. p. 279 et seq. Fries, *Anthropologie*, i. p. 77, (edit. 1820). Schulze,
Philosophische Wissenschaften, i. p.
16-17. H. Schmid, *Versuch einer Metaphysik der inneren Natur*, pp. 23,
232 et seq. Damiron, *Cours de Philosophie*, i. p. 190, (edit. 1834).
Maass, *Einbildungskraft*, § 24, p. 65
et seq., (edit. 1797). Sulzer, *Ver-
mischte Schriften*, i. pp. 99 et seq.,
100, (edit. 1808). Denzinger, *Institutiones Logicæ*, § 260, t. i. p. 226, (ed.
1824). Beneke, *Lehrbuch der Psychologie*, § 96 et seq., p. 72, (edit. 1833).
Platner, *Philosophische Aphorismen*,
i. p. 70.] [See further, *Reid's Works*,
(completed edition), p. 938-939.—
Ed.]

LECTURE XIX.

CONSCIOUSNESS.—GENERAL PHÆNOMENA.—DIFFICULTIES AND FACILITIES OF PSYCHOLOGICAL STUDY.

LECT.
XIX.
———
Recapitulation.

In our last Lecture we were occupied with the last and principal part of the question, Are there mental agencies beyond the sphere of Consciousness?—in other words, Are there modifications of mind unknown in themselves, but the existence of which we must admit as the necessary causes of known effects? In dealing with this question, I showed, first of all, that there is indisputable evidence for the general fact, that even extensive systems of knowledge may, in our ordinary state, lie latent in the mind, beyond the sphere of consciousness and will; but which, in certain extraordinary states of organism, may again come forward into light, and even engross the mind to the exclusion of its everyday possessions. The establishment of the fact, that there are in the mind latent capacities, latent riches, which may occasionally exert a powerful and obtrusive agency, prepared us for

Are there,
in ordinary, latent
modifications of
mind, concurring to
the production of
manifest
effects?

the question, Are there, in ordinary, latent modifications of mind,—agencies unknown themselves as phænomena, but secretly concurring to the production of manifest effects? This problem, I endeavoured to show you, must be answered in the affirmative. I took for the medium of proof various operations of mind, analysed these, and found as a residuum a

certain constituent beyond the sphere of consciousness, and the reality of which cannot be disallowed,
as necessary for the realisation of the allowed effect.
My first examples were taken from the faculty of
External Perception. I showed you, in relation to all
the senses, that there is an ultimate perceptible minimum; that is, that there is no consciousness, no perception, of the modification determined by its object
in any sense, unless that object determines in the
sense a certain quantum of excitement. Now, this
quantum, though the minimum that can be consciously perceived, is still a whole composed even of
an infinity of lesser parts. Conceiving it, however,
only divided into two, each of these halves is unperceived,—neither is an object of consciousness; the
whole is a percept made up of the unperceived halves.
The halves must, however, have each produced its
effect towards the perception of the whole; and,
therefore, the smallest modification of which consciousness can take account, necessarily supposes, as
its constituents, smaller modifications, real, but elud-
ing the ken of consciousness. Could we magnify the
discerning power of consciousness, as we can magnify
the power of vision by the microscope, we might
enable consciousness to extend its cognisance to modifications twice, ten times, ten thousand times less,
than it is now competent to apprehend; but still
there must be some limit. And as every mental
modification is a quantity, and as no quantity can be
conceived not divisible *ad infinitum*, we must, even
on this hypothesis, allow (unless we assert that the
ken of consciousness is also infinite), that there are
modifications of mind unknown in themselves, but
the necessary coefficients of known results. On the

ground of perception, it is thus demonstratively
proved that latent agencies,—modifications of which
we are unconscious,—must be admitted as a ground-
work of the Phænomenology of Mind.

The fact of the existence of such latent agencies
being proved in reference to one faculty, the presumption is established that they exert an influence in all.
And this presumption holds, even if, in regard to
some others, we should be unable to demonstrate, in
so direct and exclusive a manner, the absolute necessity of their admission. This is shown in regard to
the Association of Ideas. In order to explain this, I
stated to you that the laws, which govern the train
or consecution of thought, are sometimes apparently
violated; and that philosophers are perforce obliged,
in order to explain the seeming anomaly, to interpolate, hypothetically, between the ostensibly suggest-
ing and the ostensibly suggested thought, certain
connecting links of which we have no knowledge.
Now, the necessity of such interpolation being admitted, as admitted it must be, the question arises, How
have these connecting thoughts, the reality of which
is supposed, escaped our cognisance? In explanation
of this, there can possibly be only two theories. It may
be said, in the first place, that these intermediate ideas
did rise into consciousness, operated their suggestion,
and were then instantaneously forgotten. It may be
said, in the second place, that these intermediate ideas
never did rise into consciousness, but, remaining latent
themselves, still served to awaken into consciousness
the thought, and thus explain its suggestion.

The former of these theories, which is the only one
whose possibility is contemplated in this country, I
endeavoured to show you ought not to be admitted,

The fact of
the exist-
ence of la-
tent agen-
cies in one
faculty, a
presump-
tion that
they exert
an influ-
ence in all.

Association
of Ideas.
The laws of
Association
sometimes
apparently
violated.

being obnoxious to the most insurmountable objec- tions. It violates the whole analogy of consciousness; and must at last found upon a reason which would identify it with the second theory. At the same time it violates the law of philosophising, called the law of Parcimony, which prescribes that a greater number of causes are not to be assumed than are necessary to explain the phænomena. Now, in the present case, if the existence of unconscious modifications,— of latent agencies, be demonstratively proved by the phænomena of perception, which they alone are competent to explain, why postulate a second unknown cause to account for the phænomena of association, when these can be better explained by the one cause, which the phænomena of perception compel us to admit?

The fact of latent agencies being once established, and shown to be applicable, as a principle of psychological solution, I showed you, by other examples, that it enables us to account, in an easy and satisfactory manner, for some of the most perplexing phænomena of mind. In particular, I did this by reference to our Acquired Dexterities and Habits. In these the consecution of the various operations is extremely rapid; but it is allowed on all hands that, though we are conscious of the series of operations,—that is, of the mental state which they conjunctly constitute,— of the several operations themselves as acts of volition we are wholly incognisant. Now, this incognisance may be explained, as I stated to you, on three possible hypotheses. In the first place, we may state that the whole process is effected without either volition, or even any action of the thinking principle, it being merely automatic or mechanical. The incognisance to be

LECT. XIX.

The anomaly solved by the doctrine of latent agencies.

The same principle explains the operations of our Acquired Dexterities and Habits.

explained is thus involved in this hypothesis. In the second place, it may be said that each individual act of which the process is made up, is not only an act of mental agency, but a conscious act of volition; but that, there being no memory of these acts, they, consequently, are unknown to us when past. In the third place, it may be said that each individual act of the process is an act of mental agency, but not of consciousness and separate volition. The reason of the incognisance is thus apparent. The first opinion is unphilosophical, because, in the first place, it assumes an occult, an incomprehensible principle, to enable us to comprehend the effect. In the second place, admitting the agency of the mind in accomplishing the series of movements before the habit or dexterity is formed, it afterwards takes it out of the hands of the mind, in order to bestow it upon another agent. This hypothesis thus violates the two great laws of philosophising, — to assume no occult principle without necessity, — to assume no second principle without necessity. This doctrine was held by Reid, Hartley, and others.

The mechanical theory.

The theory of consciousness without Memory.

The second hypothesis, which Mr Stewart adopts, is at once complex and contradictory. It supposes a consciousness and no memory. In the first place, in this it is altogether hypothetical,—it cannot advance a shadow of proof in support of the fact which it assumes, that an act of consciousness does or can take place without any, the least, continuance in memory. In the second place, this assumption is disproved by the whole analogy of our intellectual nature. It is a law of mind, that the intensity of the present consciousness determines the vivacity of the future memory. Memory and consciousness are thus in the direct ratio

Consciousness and Memory in the direct ratio of each other.

of each other. On the one hand, looking from cause to effect,—vivid consciousness, long memory; faint consciousness, short memory; no consciousness, no memory: and, on the other, looking from effect to cause, —long memory, vivid consciousness; short memory, faint consciousness; no memory, no consciousness. Thus, the hypothesis which postulates consciousness without memory, violates the fundamental laws of our intellectual being. But, in the third place, this hypothesis is not only a psychological solecism, it is, likewise, a psychological pleonasm; it is at once illegitimate and superfluous. As we must admit, from the analogy of perception, that efficient modifications may exist without any consciousness of their existence, and as this admission affords a solution of the present problem, the hypothesis in question here again violates the law of parcimony, by assuming without necessity a plurality of principles to account for what one more easily suffices.

The third hypothesis, then,—that which employs the single principle of latent agencies to account for so numerous a class of mental phænomena,—how does it explain the phænomenon under consideration? Nothing can be more simple and analogical than its solution. As—to take an example from vision—in the external perception of a stationary object, a certain space,—an expanse of surface, is necessary to the *minimum visibile*, in other words, an object of sight cannot come into consciousness unless it be of a certain size; in like manner, in the internal perception of a series of mental operations, a certain time,—a certain duration, is necessary for the smallest section of continuous energy to which consciousness is competent. Some minimum of time must be admitted as the con-

dition of consciousness; and as time is divisible *ad
infinitum*, whatever minimum be taken, there must be
admitted to be, beyond the cognisance of consciousness, intervals of time, in which, if mental agencies
be performed, these will be latent to consciousness. If
we suppose that the minimum of time to which consciousness can descend, be an interval called six, and
that six different movements be performed in this
interval, these, it is evident, will appear to consciousness as a simple indivisible point of modified time;
precisely as the *minimum visibile* appears as an indivisible point of modified space. And, as in the extended parts of the *minimum visibile*, each must
determine a certain modification on the percipient
subject, seeing that the effect of the whole is only the
conjoined effect of its parts; in like manner, the pro-
tended parts of each conscious instant,—of each distinguishable minimum of time,—though themselves
beyond the ken of consciousness, must contribute to
give the character to the whole mental state which
that instant, that minimum, comprises. This being
understood, it is easy to see how we lose the consciousness of the several acts, in the rapid succession of many
of our habits and dexterities. At first, and before the
habit is acquired, every act is slow, and we are conscious of the effort of deliberation, choice, and volition;
by degrees the mind proceeds with less vacillation and
uncertainty; at length the acts become secure and
precise: in proportion as this takes place, the velocity
of the procedure is increased, and as this acceleration
rises, the individual acts drop one by one from consciousness, as we lose the leaves in retiring further and
further from the tree; and, at last, we are only aware
of the general state which results from these uncon-

scious operations, as we can at last only perceive the
greenness which results from the unperceived leaves.

I have thus endeavoured to recapitulate and vary
the illustration of this important principle. At present, I can only attempt to offer you such evidence
of the fact as lies close to the surface. When we come
to a discussion of the special faculties, you will find
that this principle affords an explanation of many
interesting phænomena, and from them receives confirmation in return.

Before terminating the consideration of the general
phænomena of consciousness, there are Three Principal
Facts which it would be improper altogether to pass
over without notice, but the full discussion of which
I reserve for that part of the course which is conversant with Metaphysic Proper, and when we come to
establish upon their foundation our conclusions in
regard to the Immateriality and Immortality of Mind ;
—I mean the fact of our Mental Existence or Substantiality, the fact of our Mental Unity or Individuality, and the fact of our Mental Identity or Personality. In regard to these three facts, I shall, at present, only attempt to give you a very summary view
of what place they naturally occupy in our psychological system.

The first of these,—the fact of our own Existence,
—I have already incidentally touched on, in giving
you a view of the various possible modes in which the
fact of the Duality of Consciousness may be conditionally accepted.

The various modifications of which the thinking
subject, Ego, is conscious, are accompanied with the
feeling, or intuition, or belief,—or by whatever name

Three Prin-
cipal Facts
to be no-
ticed in
connection
with the
general
phæno-
mena of
conscious-
ness.

1. Self-Ex-
istence.

the conviction may be called,—that I, the thinking subject, exist. This feeling has been called by philosophers the apperception or consciousness of our own existence, but as it is a simple and ultimate fact of consciousness, though it be clearly given, it cannot be defined or described. And for the same reason that it cannot be defined, it cannot be deduced or demonstrated; and the apparent enthymeme of Descartes,— *Cogito ergo sum*,—if really intended for an inference, —if really intended to be more than a simple enunciation of the proposition, that the fact of our existence is given in the fact of our consciousness, is either tautological, or false. Tautological, because nothing is contained in the conclusion which was not explicitly given in the premise,—the premise, *Cogito, I think*, being only a grammatical equation of *Ego sum cogitans, I am*, or *exist, thinking*. False, inasmuch as there would, in the first place, be postulated the reality of thought as a quality or modification, and then, from the fact of this modification, inferred the fact of existence, and of the existence of a subject; whereas it is self-evident, that in the very possibility of a quality or modification, is supposed the reality of existence, and of an existing subject. Philosophers, in general, among whom may be particularly mentioned Locke and Leibnitz, have accordingly found the evidence in a clear and immediate belief in the simple datum of consciousness; and that this was likewise the opinion of Descartes himself, it would not be difficult to show.[a]

*Descartes'
Cogito ergo
sum.*

a That Descartes did not intend to prove the fact of existence from that of thought, but to state that personal existence consists in consciousness, is shown in M. Cousin's Dissertation, *Sur le vrai sens du cogito ergo sum*; printed in the earlier editions of the *Fragments Philosophiques*, and in vol. i. p. 27 of the collected edition of his works.—ED.

The second fact,—our Mental Unity or Individuality, —is given with equal evidence as the first. As clearly as I am conscious of existing, so clearly am I conscious at every moment of my existence, (and never more so than when the most heterogeneous mental modifications are in a state of rapid succession,) that the conscious Ego is not itself a mere modification, nor a series of modifications of any other subject, but that it is itself something different from all its modifications, and a self-subsistent entity. This feeling, belief, datum, or fact of our mental individuality or unity, is not more capable of explanation than the feeling or fact of our existence, which it indeed always involves. The fact of the deliverance of consciousness to our mental unity has, of course, never been doubted ; but philosophers have been found to doubt its truth. According to Hume,[a] our thinking Ego is nothing but a bundle of individual impressions and ideas, out of whose union in the imagination, the notion of a whole, as of a subject of that which is felt and thought, is formed. According to Kant,[β] it cannot be properly determined whether we exist as substance or as accident, because the datum of individuality is a condition of the possibility of our having thoughts and feelings, —in other words, of the possibility of consciousness ; and, therefore, although consciousness gives,—cannot but give, the phænomenon of individuality, it does not follow that this phænomenon may not be only a necessary illusion. An articulate refutation of these opinions I cannot attempt at present; but their refutation is, in fact, involved in their statement. In regard to Hume, his sceptical conclusion is only an

LECT. XIX.

2. Mental Unity.

The truth of the testimony of consciousness to our mental unity, doubted.

Hume.

Kant.

---

a *Treatise of Human Nature*, part iv., sect. v., vi.—ED.

β *Kritik der reinen Vernunft*, Trans. Dial. b. ii. c. 1.—ED.

inference from the premises of the dogmatical philosophers, who founded their systems on a violation or
distortion of the facts of consciousness. His conclusion
is, therefore, refuted in the refutation of their premises, which is accomplished in the simple exposition
that they at once found on, and deny, the veracity of
consciousness. And by this objection the doctrine of
Kant is overset. For if he attempts to philosophise,
he must assert the possibility of philosophy. But the
possibility of philosophy supposes the veracity of consciousness as to the contents of its testimony; therefore, in disputing the testimony of consciousness to
our mental unity and substantiality, Kant disputes
the possibility of philosophy, and, consequently, reduces his own attempts at philosophising to absurdity.

3. Mental
Identity.

The third datum under consideration is the Identity of Mind or Person. This consists in the assurance
we have, from consciousness, that our thinking Ego,
notwithstanding the ceaseless changes of state or modification, of which it is the subject, is essentially the
same thing,—the same person, at every period of its
existence. On this subject, laying out of account
certain subordinate differences in the mode of stating
the fact, philosophers, in general, are agreed. Locke,[a]
in the *Essay on the Human Understanding*; Leibnitz,[β] in the *Nouveaux Essais*; Butler,[γ] and Reid,[δ]
are particularly worthy of attention. In regard to
this deliverance of consciousness, the truth of which
is of vital importance, affording, as it does, the basis
of moral responsibility and hope of immortality,—it
is, like the last, denied by Kant to afford a valid

a Book ii. c. 27, especially § 9 *et*
seq.—ED.
β Liv. ii. c. 27.—ED.
γ *Analogy*, Diss. i. Of Personal

Identity.—ED.
δ *Intell. Powers*, Essay iii. cc. 4,
6; *Works*, pp. 334-46, 350-53.—
ED.

ground of scientific certainty.   He maintains that
there is no cogent proof of the substantial permanence
of our thinking self, because the feeling of identity is
only the condition under which thought is possible.
Kant's doubt in regard to the present fact is refuted
in the same manner as his doubt in regard to the preceding, and there are also a number of special grounds
on which it can be shown to be untenable.   But of
these at another time.

We have now terminated the consideration of Consciousness as the general faculty of thought, and as
the only instrument and only source of Philosophy.
But before proceeding to treat of the Special Faculties, The peculiar diffi-
it may be proper here to premise some observations in culties and
facilities of
relation to the peculiar Difficulties and peculiar Facili- psychological investi-
ties which we may expect in the application of con- gation.
sciousness to the study of its own phænomena.   I shall
first speak of the difficulties.

The first difficulty in psychological observation arises I. Difficulties.
from this, that the conscious mind is at once the ob- 1. The conscious mind
serving subject and the object observed.   What are the at once the
observing
consequences of this ?   In the first place, the mental subject and
the object
energy, instead of being concentrated, is divided, and observed.
divided in two divergent directions.   The state of mind
observed, and the act of mind observing, are mutually in an inverse ratio ; each tends to annihilate the
other.   Is the state to be observed intense, all reflex
observation is rendered impossible ; the mind cannot
view as a spectator, it is wholly occupied as an agent
or patient.   On the other hand, exactly in proportion
as the mind concentrates its force in the act of reflective observation, in the same proportion must the direct phænomenon lose in vivacity, and consequently,

LECT.
XIX.

in the precision and individuality of its character. This
difficulty is manifestly insuperable in those states of
mind, which, of their very nature, as suppressing consciousness, exclude all contemporaneous and voluntary
observation, as in sleep and fainting. In states like
dreaming, which allow at least of a mediate, but, therefore, only of an imperfect, observation, through recollection, it is not altogether exclusive. In all states of
strong mental emotion, the passion is itself to a certain extent a negation of the tranquillity requisite for
observation, so that we are thus impaled on the awkward dilemma,—either we possess the necessary tranquillity for observation, with little or nothing to observe, or there is something to observe, but we have
not the necessary tranquillity for observation. All
this is completely opposite in our observation of the
external world. There the objects lie always ready
for our inspection ; and we have only to open our eyes
and guard ourselves from the use of hypotheses and
green spectacles, to carry our observations to an easy
and successful termination."

2. Want of
mutual co-
operation.
.In the second place, in the study of external nature,
several observers may associate themselves in the pursuit ; and it is well known how co-operation and mutual
sympathy preclude tedium and languor, and brace up
the faculties to their highest vigour. Hence the old
proverb, *unus homo, nullus homo.* " As iron," says
Solomon, "sharpeneth iron, so a man sharpeneth the
understanding of his friend."[β] " In my opinion," says
Plato,[γ] " it is well expressed by Homer,

' By mutual confidence and mutual aid
Great deeds are done, and great discoveries made ;'

a [Cf. Biunde, *Versuch einer syste-
matischen Behandlung der empirisch-
en Psychologie,* i. p. 55.]

β *Proverbs,* xxvii. 17. The authorised version is *countenance.*—ED.
γ *Protagoras,* p. 348.—ED.

for if we labour in company, we are always more
prompt and capable for the investigation of any hidden
matter.   But if a man works out anything by solitary
meditation, he forthwith goes about to find some one
with whom he may commune, nor does he think his
discovery assured until confirmed by the acquiescence
of others."   Aristotle,[a] in like manner, referring to
the same passage of Homer, gives the same solution.
"Social operation," he says, "renders us more energetic both in thought and action ;" a sentiment which
is beautifully illustrated by Ovid,[β]

> "Scilicet ingeniis aliqua est concordia junctis,
>   Et servat studii fœdera quisque sui.
>   Utque meis numeris tua dat facundia nervos,
>   Sic venit a nobis in tua verba nitor."

Of this advantage the student of Mind is in a great
measure deprived.   He who would study the internal
world must isolate himself in the solitude of his own
thought ; and for man, who, as Aristotle observes,[γ] is
more social by nature than any bee or ant, this isola-
tion is not only painful in itself, but, in place of
strengthening his powers, tends to rob them of what
maintains their vigour, and stimulates their exertion.

In the third place, " In the study of the material
universe, it is not necessary that each observer should
himself make every observation.   The phænomena
are here so palpable and so easily described, that the
experience of one observer suffices to make the facts
which he has witnessed intelligible and credible to all.
In point of fact, our knowledge of the external world
is taken chiefly upon trust.   The phænomena of the

3. No fact of consciousness can be accepted at second-hand.

a *Eth. Nic.* viii. 1.   Cf. *ibid.*, ix. 9.—Ed.
β *Epist. ex Ponto*, ii. v. 59, 69.—
γ *Polit.*, i. 2.—Ed.

internal world, on the contrary, are not thus capable of being described ; all that the first observer can do is to lead others to repeat his experience : in the science of mind, we can believe nothing upon authority, take nothing upon trust. In the physical sciences, a fact viewed in different aspects and in different circumstances, by one or more observers of acknowledged sagacity and good faith, is not only comprehended as clearly by those who have not seen it for themselves, but is also admitted without hesitation, independently of all personal verification. Instruction thus suffices to make it understood, and the authority of the testimony carries with it a certainty which almost precludes the possibility of doubt.

" But this is not the case in the philosophy of mind. On the contrary, we can here neither understand nor believe at second-hand. Testimony can impose nothing on its own authority ; and instruction is only instruction when it enables us to teach ourselves. A fact of consciousness, however well observed, however clearly expressed, and however great may be our confidence in its observer, is for us as nothing, until, by an experience of our own, we have observed and recognised it ourselves. Till this be done we cannot comprehend what it means, far less admit it to be true. Hence it follows that, in philosophy proper, instruction is limited to an indication of the position in which the pupil ought to place himself, in order by his own observation to verify for himself the facts which his instructor pronounces true." [a]

In the fourth place, the phænomena of consciousness are not arrested during observation,—they are in a ceaseless and rapid flow ; each state of mind is in-

[a] Cardaillac, *Etudes de Philosophie,* i. p. 6.

divisible, but for a moment, and there are not two states or two moments of whose precise identity we can be assured. Thus, before we can observe a modification, it is already altered ; nay, the very intention of observing it, suffices for the change. It hence results that the phænomenon can only be studied through its reminiscence ; but memory reproduces it often very imperfectly, and always in lower vivacity and precision. The objects of the external world, on the other hand, either remain unaltered during our observation, or can be renewed without change ; and we can leave off at will and recommence our investigation without detriment to its result.[a]

4. Phæno-mena of consciousness not arrested during observation, but only to be studied through memory.

In the fifth place, " The phænomena of the mental world are not, like those of the material, placed by the side of each other in space. They want that form by which external objects attract and fetter our attention ; they appear only in rows on the thread of time, occupying their fleeting moment, and then vanishing into oblivion ; whereas, external objects stand before us steadfast, and distinct, and simultaneous, in all the life and emphasis of extension, figure, and colour."[β]

5. Presented only in succession.

In the sixth place, the perceptions of the different qualities of external objects are decisively discriminated by different corporeal organs, so that colour, sound, solidity, odour, flavour,' are, in the sensations themselves, contrasted, without the possibility of confusion. In an individual sense, on the contrary, it is not always easy to draw the line of separation between its perceptions, as these are continually running into each other. Thus red and yellow are, in their

6. Naturally blend with each other, and are presented in complexity.

a [Ancillon, *Nouv. Mélanges*, t. ii. p. 102. Cardaillac, *Etudes de Phi-los.*, i. pp. 3, 4.   β [Biunde, *Psychologie*, vol. i. p. 56.]

extreme points, easily distinguished, but the transition point from one to the other is not precisely determined. Now, in our internal observation, the mental phænomena cannot be discriminated like the perceptions of one sense from the perceptions of another, but only like the perceptions of the same. Thus the phænomenon of feeling,—of pleasure or pain, and the phænomenon of desire, are, when considered in their remoter divergent aspects, manifestly marked out and contradistinguished as different original modifications ; whereas, when viewed on their approximating side, they are seen to slide so insensibly into each other, that it becomes impossible to draw between them any accurate line of demarcation. Thus the various qualities of our internal life can be alone discriminated by a mental process called Abstraction ; and abstraction is exposed to many liabilities of error. Nay, the various mental operations do not present themselves distinct and separate ; they are all bound up in the same unity of action ; and as they are only possible through each other, they cannot, even in thought, be dealt with as isolated and apart. In the perception of an external object, the qualities are, indeed, likewise presented by the different senses in connection, as, for example, vinegar is at once seen as yellow, felt as liquid, tasted as sour, and so on ; nevertheless, the qualities easily allow themselves in abstraction to be viewed as really separable, because they are all the properties of an extended and divisible body; whereas in the mind, thoughts, feelings, desires do not stand separate, though in juxtaposition, but every mental act contains at once all these qualities, as the constituents of its indivisible simplicity.

In the seventh place, the act of reflection on our

internal modifications is not accompanied with that frequent and varied sentiment of pleasure, which we experience from the impression of external things. Self-observation costs us a greater effort, and has less excitement than the contemplation of the material world; and the higher and more refined gratification which it supplies when its habit has been once formed, cannot be conceived by those who have not as yet been trained to its enjoyment.[a] "The first part of our life is fled before we possess the capacity of reflective observation; while the impressions which, from earliest infancy, we receive from material objects, the wants of our animal nature, and the prior development of our external senses, all contribute to concentrate, even from the first breath of life, our attention on the world without. The second passes without our caring to observe ourselves. The outer life is too agreeable to allow the soul to tear itself from its gratifications, and return frequently upon itself. And at the period when the material world has at length palled upon the senses, when the taste and the desire of reflection gradually become predominant, we then find ourselves, in a certain sort, already made up, and it is impossible for us to resume our life from its commencement, and to discover how we have become what we now are."[β] "Hitherto external objects have exclusively riveted our attention; our organs have acquired the flexibility requisite for this peculiar kind of observation; we have learned the method, acquired the habit, and feel the pleasure which results from performing what we perform with ease. But let us recoil upon ourselves; the scene changes; the charm is gone; diffi-

LECT. XIX.

7. The act of reflection not accompanied with the frequent and varied sentiment of pleasure, which we experience from the impression of external things.

a [Biunde, *Psychologie*, i. p. 56.]　p. 103.]
β [Ancillon, *Nouv. Mélanges*, t. ii.

culties accumulate, all that is done is done irksomely
and with effort; in a word, everything within repels,
everything without attracts; we reach the age of man-
hood without being taught another lesson than reading what takes place without and around us, whilst
we possess neither the habit nor the method of study-
ing the volume of our own thoughts"[a]   "For a long
time, we are too absorbed in life to be able to detach
ourselves from it in thought; and when the desires
and the feelings are at length weakened or tranquil-
lised,—when we are at length restored to ourselves,
we can no longer judge of the preceding state, because
we can no longer reproduce or replace it. Thus it is
that our life, in a philosophical sense, runs like water
through our fingers. We are carried along, lost,
whelmed in our life; we live, but rarely see ourselves
to live.

"The reflective Ego, which distinguishes self from
its transitory modifications, and which separates the
spectator from the spectacle of life, which it is continually representing to itself, is never developed in
the majority of mankind at all, and even in the
thoughtful and reflective few, it is formed only at a
mature period, and is even then only in activity by
starts and at intervals."[β]

But Philosophy has not only peculiar difficulties, it
has also peculiar facilities. There is indeed only one
external condition on which it is dependent, and that
is language; and when, in the progress of civilisation,
a language is once formed of a copiousness and
pliability capable of embodying its abstractions without figurative ambiguity, then a genuine philosophy

a [Cardaillac, *Etudes de Philoso-*
phie, t. i. p. 3.]

β [Ancillon, *Nouv. Mélanges,* t. ii.
pp. 103, 104, 105.]

may commence. With this one condition all is given;
the Philosopher requires for his discoveries no preliminary preparations,—no apparatus of instruments and materials. He has no new events to seek as the Historian ; no new combinations to form as the Mathematician. The Botanist, the Zoologist, the Mineralogist, can accumulate only by care, and trouble, and expense, an inadequate assortment of the objects necessary for their labours and observations. But that most important and interesting of all studies of which man himself is the object, has no need of anything external ; it is only necessary that the observer enter into his inner self in order to find there all he stands in need of, or rather it is only by doing this that he can hope to find anything at all. If he only effectively pursue the method of observation and analysis, he may even dispense with the study of philosophical systems. This is at best only useful as a mean towards a deeper and more varied study of himself, and is often only a tribute paid by philosophy to erudition.[a]

a [Cf. Fries, *Logik*, § 126, p. 587 *l'Etude de la Philosophie*, t. i., Disc. (edit. 1819). Thurot, *Introduction à* Prél. p. 35.]

APPENDIX.

I. A.—FRAGMENT ON ACADEMICAL HONOURS—(1836).
(See Vol. I. p. 18.)

BEFORE commencing the Lecture of to-day, I would occupy a few minutes with a matter in which I am confident you generally feel an interest ;—I refer to the Academical Honours to be awarded to those who approve their zeal and ability in the business of the Class. After what I formerly had occasion to say, I conceive it wholly unnecessary now to attempt any proof of the fact, that it is not by anything done by others for you, but by what alone you do for yourselves, that your intellectual improvement must be determined. Reading and listening to Lectures are only profitable, inasmuch as they afford you the means and the occasions of exerting your faculties ; for these faculties are only developed in proportion as they are exercised. This is a principle I take for granted.

A second fact, I am assured you will also allow me to assume, is, that although strenuous energy is the one condition of all improvement, yet this energy is, at first and for a long time, comparatively painful. It is painful, because it is imperfect. But as it is gradually perfected, it becomes gradually more pleasing, and when finally perfect, that is, when its power is fully developed, it is purely pleasurable ; for pleasure is nothing but the concomitant or reflex of the unforced and unimpeded energy of a faculty or habit,—the degree of pleasure being always in proportion to the degree of such energy. The great problem in education is, therefore, how to induce the pupil to undertake and go through with a course of exertion, in its result good and even agreeable, but immediately and in itself, irksome. There is no royal road to learning. "The gods," says Epicharmus,[a] "sell us everything for

[a] Xenophon, *Memorabilia*, ii. 1. 20.—ED.

toil ;" and the curse inherited from Adam,—that in the sweat of
his face man should eat his bread,—is true of every human acquisition. Hesiod, not less beautifully than philosophically, sings
of the painful commencement, and the pleasant consummation, of
virtue, in the passage of which the following is the commencement :—

> Τῆς δ' Ἀρετῆς ἱδρῶτα θεοὶ προπάροιθεν ἔθηκαν
> Ἀθάνατοι α

(a passage which, it will be recollected, Milton has not less beautifully imitated) ; β and the Latin poet has, likewise, well expressed
the principle, touching literary excellence in particular :—

> —— " Gaudent sudoribus artes
> Et sua difficilem reddunt ad limina cursum." γ

But as the pain is immediate, while the profit and the pleasure
are remote, you will grant, I presume, without difficulty, a third
fact, that the requisite degree and continuance of effort can only
be insured, by applying a stimulus to counteract and overcome the
repressive effect of the feeling with which the exertion is for a
season accompanied. A fourth fact will not be denied, that emulation and the love of honour constitute the appropriate stimulus
in education.   These affections are of course implanted in man
for the wisest purposes ; and, though they may be misdirected,
the inference from the possibility of their abuse to the absolute
inexpediency of their employment, is invalid.   However disguised, their influence is universal :—

> " Ad has se
> Romanus, Graiusque, et Barbarus induperator
> Erexit : causas discriminis atque laboris
> Inde habuit ;" δ

and Cicero shrewdly remarks, that the philosophers themselves
prefix their names to the very books they write on the contempt
of glory.ε These passions actuate most powerfully the noblest
minds. "Optimos mortalium,"ζ says the father of the Senate

α Opera et Dies, 287.—Ed.
β Sir W. Hamilton here probably
refers to the lines in Lycidas,—
" Fame is the spur that the clear spirit
   doth raise," &c.
—Ed.

γ B. Mantuanus, Carmen de suscepto
Theologico Magisterio, — Opera, Ant-
verpiæ, 1576, tom. i. p. 174.—Ed.
δ Juvenal, Sat., x. 138.—Ed.
ε Pro Archia, c. 11.—Ed.
ζ Tacitus, Ann., iv. 38.—Ed.

to Tiberius,—" Optimos mortalium altissima cupere : contemptu famœ contemni virtutes." " Naturâ," says Seneca,[a] " gloriosa est virtus, et anteire priores cupit ;" and Cicero,[β] in more proximate reference to our immediate object,—" Honor alit artes omnesqe incenduntur ad studia gloriâ." But, though their influence be universal, it is most powerfully conspicuous in the young, of whom Aristotle has noted it as one of the most discriminating characteristics, that they are lovers of honour, but still more lovers of victory.[γ] If, therefore, it could be but too justly proclaimed of man in general :—

> —— " Quis enim virtutem amplectitur ipsam,
> Præmia si tollas ?"[δ]

it was least of all to be expected that youth should do so. " In learning," says the wisdom of Bacon, " the flight will be [low and] slow without some feathers of ostentation."[ε] Nothing, therefore, could betray a greater ignorance of human nature, or a greater negligence in employing the most efficient mean within its grasp, than for any seminary of education to leave unapplied these great promoting principles of activity, and to take for granted that its pupils would act precisely as they ought, though left with every inducement strong against, and without any sufficient motive in favour of, exertion.

Now, I express, I believe, the universal sentiment, both within and without these walls, in saying, that this University has been unhappily all too remiss, in leaving the most powerful mean of academical education nearly, if not altogether, unemployed. You will observe I use the term *University* in contradiction to individual Professors, for many of these have done much in this respect, and all of them, I believe, are satisfied that a great deal more ought to be done. But it is not in the power of individual instructors to accomplish what can only be accomplished by the public institution. The rewards proposed to meritorious effort are not sufficiently honourable ; and the efforts to which they are frequently accorded, not of the kind or degree to be of any great or general advantage. I shall explain myself.

A distinction is sought after with a zeal proportioned to its

---

a *De Beneficiis*, iii. 36.—ED.
β *Tusc. Quæst.*, i. 2.—ED.
γ *Rhet.*, ii. 12.—ED.

δ Juvenal, *Sat.*, x. 141.—ED.
ε *Essay* liv. *Of Vain Glory.*—ED.

value; and its value is measured by the estimation which it holds in public opinion. Now, though there are prizes given in many of our classes, nothing has been done to give them proper value by raising them in public estimation. They are not conferred as matters of importance by any external solemnity; they are not conferred in any general meeting of the University; far less under circumstances which make their distribution a matter of public curiosity and interest. Compared to the publicity that might easily have been secured, they are left, so to speak, to be given in holes and corners; and while little thought of to-day, are wholly forgotten to-morrow; so that the wonder only is, that what the University has thus treated with such apparent contempt, should have awakened even the inadequate emulation that has been so laudably displayed. Of this great defect in our discipline, I may safely say that every Professor is aware, and it is now actually under the consideration of the Senatus, what are the most expedient measures to obtain a system of means of full efficiency for the encouragement and reward of academical merit. It will, of course, form the foundation of any such improvement, that the distribution of prizes be made an act of the University at large; and one of the most public and imposing character. By this means a far more powerful emulation will be roused; a spirit which will not be limited to a certain proportion of the students, but will more or less pervade the whole,—nay, not merely the students themselves, but their families; so that when this system is brought to its adequate perfection, it will be next to impossible for a young man of generous dispositions not to put forth every energy to raise himself as high as possible in the scale of so honourable a competition.

But besides those which can only be effected by an act of the whole University, important improvements may, I think, be accomplished in this respect in the several classes. In what I now say, I would not be supposed to express any opinion in regard to other classes; but confine my observations to one under the circumstances of our own.

In the first place, then, I am convinced that excitement and rewards are principally required to promote a general and continued diligence in the ordinary business of the class. I mean, therefore, that the prizes should with us be awarded for general eminence, as shown in the Examinations and Exercises; and I am averse on principle from proposing any premium during the course of the sessional labours for single and detached efforts. The effect of this would naturally be to distract attention from what ought to be the principal and constant object of occupation; and if honour is to be gained by an irregular and transient spirit of activity, less encouragement will necessarily be afforded to regular and sedulous application. Prizes for individual Essays, for Written Analyses of important books, and for Oral Examination on their contents, may, however, with great advantage, be proposed as occupation during the summer vacation; and this I shall do. But the honours of the Winter Session must belong to those who have regularly gone through its toils.

In the second place, the value of the prizes may be greatly enhanced by giving them greater and more permanent publicity. A very simple mode, and one which I mean to adopt, is to record upon a tablet each year, the names of the successful competitors; this tablet to be permanently affixed to the walls of the class-room, while a duplicate may, in like manner, be placed in the Common Reading-Room of the Library.

In the third place, the importance of the prizes for general eminence in the business of the class may be considerably raised, by making the competitors the judges of merit among themselves. This I am persuaded is a measure of the very highest efficiency. On theory I would argue this, and in practice it has been fully verified. On this head, I shall quote to you the experience of my venerated preceptor, the late Professor Jardine of Glasgow,—a man, I will make bold to say, who, in the chair of Logic of that University, did more for the intellectual improvement of his pupils than any other public instructor in this country within the memory of man. This he did not accomplish either by great erudition or great philosophical talent, —though he was both a learned and an able thinker,—but by the application of that primary principle of education, which, wherever employed, has been employed with success,—I mean the determination of the pupil to self-activity,—doing nothing for him which he is able to do for himself. This principle, which has been always inculcated by theorists on education, has, however, by few been carried fully into effect.

" One difficult and very important part," says Mr Jardine,[a] " in administering the system of prizes, still remains to be stated; and this is the method by which the different degrees of merit are determined, a point in which any error with regard to principle, or suspicion of practical mistake, would completely destroy all the good effects aimed at by the establishment in question. It has been already mentioned, that the qualifications which form the ground of competition for the class prizes, as they are sometimes called, and which are to be distinguished from the university prizes, are diligence, regularity of attendance, general eminence at the daily examinations, and in the execution of themes, propriety of academical conduct, and habitual good manners ; and, on these heads, it is very obvious, a judgment must be pronounced either by the professor, or by the students themselves, as no others have access to the requisite information.

" It may be imagined, at first view, that the office of judge would be best performed by the professor ; but, after long experience, and much attention to the subject in all its bearings, I am inclined to give a decided preference to the exercise of this right as vested in the students. Were the professor to take this duty upon himself, it would be impossible, even with the most perfect conviction, on the part of the students, that his judgment and candour were unimpeachable, to give satisfaction to all parties ; while, on the other hand, were there the slightest reason to suspect his impartiality in either of these points, or the remotest ground for insinuation that he gave undue advantage to any individuals, in bringing forward their claims to the prejudice of others, the charm of emulation would be dissolved at once, and every future effort among his pupils would be enfeebled.

*　　*　　*　　*　　*　　*

" The indispensable qualities of good judges, then, are a competent knowledge of the grounds upon which their judgment is to rest, and a firm resolution to determine on the matter before them with strict impartiality. It is presumed that the students, in these respects, are sufficiently qualified. They are every-day witnesses of the manner in which the business of the class goes on, and have, accordingly, the best opportunities of judging as to the merits of their fellow-students ; they have it in their power to observe the regularity of their attendance, and the general propriety of their conduct ; they hear the questions which are put, with the answers which are given ; their various themes are read aloud, and observations are made on them from the chair. They have, likewise, an opportunity of comparing the respective merits of all the competitors in the extemporaneous exercises of the class ; and they, no doubt, hear the performances of one another canvassed in conversation, and made the subject of a comparative estimate. Besides, as every individual is, himself, deeply interested, it is not possible but that he should pay the closest attention to what is going on around him ; whilst he cannot fail to be aware that he, in like manner, is constantly observed by others, and subjected to the ordeal of daily criticism. In truth, the character, the abilities, the diligence, and progress of students

---

a *Outlines of Philosophical Education*, &c., pp. 384, 385; 387, 389.

are as well known to one another, before the close of the session, as their faces. There cannot, therefore, be any deficiency as to means of information, to enable them to act the part of enlightened and upright judges.

"But they likewise possess the other requisite for an equitable decision ; for the great majority have really a desire to judge honourably and fairly on the merit of their fellows. The natural candour and generosity of youth, the sense of right and obligations of justice, are not yet so perverted, by bad example and the ways of the world, as to permit any deliberate intention of violating the integrity on which they profess to act, or any wish to conspire in supporting an unrighteous judgment. There is greater danger, perhaps, that young persons, in their circumstances, may allow themselves to be influenced by friendship or personal dislike, rather than by the pure and unbiassed sense of meritorious exertion, or good abilities ; but, on the other hand, when an individual considers of how little consequence his single vote will be among so many, it is not at all likely that he will be induced to sacrifice it either to friendship or to enmity. There are, however, no perfect judges in any department of human life. Prejudices and unperceived biasses make their way into the minds even of the most upright of our fellow-creatures ; and there can be no doubt that votes are sometimes thrown away, or injudiciously given, by young students in the Logic class. Still, these little aberrations are never found to disturb the operation of the general principle on which the scale of merit is determined, and the list of honours filled up."

Now, Gentlemen, from what I know of you, I think it almost needless to say, that, in confiding to you a function, on the intelligent and upright discharge of which the value and significance of the prizes will wholly depend, I do this without any anxiety for the result. I am sure at least that if aught be wanting, the defect will be found neither in your incompetency nor in your want of will.

And here I would conclude what I propose to say to you on this subject ; (this has extended to a far greater length than I anticipated) ; I would conclude with a most earnest exhortation to those who may be discouraged from coming forward as competitors for academical honours, from a feeling or a fancy of inferiority. In the first place, I would dissuade them from this, because they may be deceived in the estimate of their own powers. Many individuals do not become aware of their own talents, till placed in circumstances which compel them to make strenuous exertion. Then they and those around them discover the mistake. In the second place, even though some of you may now find yourselves somewhat inferior to others, do not for a moment despair of the future. . The most powerful minds are fre-

quently of a tardy development, and you may rest assured, that
the sooner and more vigorously you exercise your faculties, the
speedier and more complete will be their evolution.   In the
third place, I exhort you to remember that the distinctions now
to be gained, are on their own account principally valuable as
means towards an end,—as motives to induce you to cultivate
your powers by exercise.   All of you, even though nearly equal,
cannot obtain equal honours in the struggle; but all of you will
obtain advantage equally substantial, if you all—what is wholly
in your own power—equally put forth your energies to strive.
And though you should all endeavour to be first, let me remind
you in the words of Cicero, that :—" Prima sequentem, pulchrum
est in secundis, tertiisque consistere." [a]

B.—FRAGMENTS ON THE SCOTTISH PHILOSOPHY.

(a) PORTION OF INTRODUCTORY LECTURE (1836).

Before entering on the proposed subjects of consideration,
I must be allowed a brief preliminary digression.   In entering on a course of the Philosophy of Mind, — of Philosophy
Proper,—we ought not, as Scotsmen, to forget that on this is, and
always has been, principally founded the scientific reputation of
Scotland ; and, therefore, that independently of the higher claims
of this philosophy to attention, it would argue almost a want of
patriotism in us, were we to neglect a study with the successful
cultivation of which our country, and in particular this University, have been so honourably associated.

Whether it be that the characteristic genius of our nation,—the
præfervidum Scotorum ingenium,—was more capable of powerful effort than of persevering industry, and, therefore, carried us
more to studies of principle than studies of detail; or, (what is
more probable), that institutions and circumstances have been
here less favourable, than in other countries, for the promotion of
erudition and research ; certain it is that the reputation for intellectual capacity which Scotland has always sustained among the

a Orator., c: i.

nations of Europe, is founded far less on the achievements of her sons in learning and scholarship, than on what they have done, or shown themselves capable of doing, in Philosophy Proper and its dependent sciences.

In former ages, Scotland presented but few objects for scientific and literary ambition; and Scotsmen of intellectual enterprise usually sought in other countries, that education, patronage, and applause which were denied them in their own. It is, indeed, an honourable testimony to the natural vigour of Scottish talent, that, while Scotland afforded so little encouragement for its production, a complement so large in amount and of so high a quality should have been, as it were, spontaneously supplied. During the sixteenth and seventeenth centuries, there was hardly to be found a Continental University without a Scottish professor. It was, indeed, a common saying that a Scottish pedlar and a Scottish professor were everywhere to be met with. France, however, was long the great nursery of Scottish talent; and this even after the political and religious estrangement of Scotland from her ancient ally, by the establishment of the Reformation and the accession of the Scottish monarch to the English crown; and the extent of this foreign patronage may be estimated from the fact, that a single prelate,—the illustrious Cardinal du Perron,—is recorded to have found places in the seminaries of France for a greater number of literary Scotsmen than all the schools and universities of Scotland maintained at home.[a]

But this favour to our countrymen was not without its reasons ; and the ground of partiality was not their superior erudition. What principally obtained for them reputation and patronage abroad, was their dialectical and metaphysical acuteness; and this they were found so generally to possess, that philosophical talent became almost a proverbial attribute of the nation.[β]

During the ascendant of the Aristotelic philosophy, and so long as dexterity in disputation was considered the highest academical accomplishment, the logical subtlety of our countrymen was in high and general demand. But they were remarkable less as writers than as instructors ; for were we to consider them only in the former capacity, the works that now remain to us of these expatriated philosophers,—these *Scoti extra Scotiam agentes,*—

a See *Discussions*, p. 120.—ED.   β See *Discussions*, p. 119.—ED.

though neither few nor unimportant, would still never enable us to account for the high and peculiar reputation which the Scottish dialecticians so long enjoyed throughout Europe.

Such was the literary character of Scotland, before the establishment of her intellectual independence, and such has it continued to the present day. In illustration of this, I cannot now attempt a comparative survey of the contributions made by this country and others to the different departments of knowledge, nor is it necessary; for no one, I am assured, will deny that it is only in the Philosophy of Mind that a Scotsman has established an epoch, or that Scotland, by the consent of Europe, has bestowed her name upon a School.

The man who gave the whole philosophy of Europe a new impulse and direction, and to whom, mediately or immediately, must be referred every subsequent advance in philosophical speculation, was our countryman,—David Hume. In speaking of this illustrious thinker, I feel anxious to be distinctly understood. I would, therefore, earnestly request of you to bear in mind, that religious disbelief and philosophical scepticism are not merely not the same, but have no natural connection; and that while the one must ever be a matter of reprobation and regret, the other is in itself deserving of applause. Both were united in Hume ; and this union has unfortunately contributed to associate them together in popular opinion, and to involve them equally in one vague condemnation. They must, therefore, I repeat, be accurately distinguished; and thus, though decidedly opposed to one and all of Hume's theological conclusions, I have no hesitation in asserting of his philosophical scepticism, that this was not only beneficial in its results, but, in the circumstances of the period, even a necessary step in the progress of Philosophy towards truth. In the first place, it was requisite in order to arouse thought from its lethargy. Men had fallen asleep over their dogmatic systems. In Germany, the Rationalism of Leibnitz and Wolf; in England, the Sensualism of Locke, with all its melancholy results, had subsided almost into established faiths. The Scepticism of Hume, like an electric spark, sent life through the paralysed opinions ; philosophy awoke to renovated vigour, and its problems were again to be considered in other aspects, and subjected to a more searching analysis.

In the second place, it was necessary in order to manifest the inadequacy of the prevailing system. In this respect, scepticism is always highly advantageous; for scepticism is only the carrying out of erroneous philosophy to the absurdity which it always virtually involved. The sceptic, *qua* sceptic, cannot himself lay down his premises; he can only accept them from the dogmatist; if true, they can afford no foundation for the sceptical inference; if false, the sooner they are exposed in their real character the better. Accepting his principles from the dominant philosophies of Locke and Leibnitz, and deducing with irresistible evidence these principles to their legitimate results, Hume showed, by the extreme absurdity of these results themselves, either that Philosophy altogether was a delusion, or that the individual systems which afforded the premises, were erroneous or incomplete. He thus constrained philosophers to the alternative,—either of surrendering philosophy as null, or of ascending to higher principles, in order to re-establish it against the sceptical reduction. The dilemma of Hume constitutes, perhaps, the most memorable crisis in the history of philosophy; for out of it the whole subsequent Metaphysic of Europe has taken its rise.

To Hume we owe the Philosophy of Kant, and, therefore, also, in general, the latter philosophy of Germany. Kant explicitly acknowledges that it was by Hume's *reductio ad absurdum* of the previous doctrine of Causality, he was first roused from his dogmatic slumber. He saw the necessity that had arisen, of placing philosophy on a foundation beyond the reach of scepticism, or of surrendering it altogether; and this it was that led him to those researches into the conditions of thought, which, considered whether in themselves or in their consequences, whether in what they established or in what they subverted, are, perhaps, the most remarkable in the annals of speculation.

To Hume, in like manner, we owe the Philosophy of Reid, and, consequently, what is now distinctively known in Europe as the Philosophy of the Scottish School.

Unable to controvert the reasoning of Berkeley, as founded on the philosophy of Descartes and Locke, Reid had quietly resigned himself to Idealism; and he confesses that he would never have been led to question the legitimacy of the common doctrine of Perception, involving though it did the negation of an external world, had Hume not startled him into hesitation and inquiry, by showing that the same reasoning which disproved the Existence of Matter, disproved, when fairly carried out, also the Substantiality of Mind. Such was the origin of the philosophy founded by Reid,—illustrated and adorned by Stewart; and it is to this philosophy, and to the writings of these two illustrious thinkers, that Scotland is mainly indebted for the distinguished reputation which she at present enjoys, in every country where the study of Mind has not, as in England, been neglected for the study of Matter.

The Philosophy of Reid is at once our pride and our reproach. At home, mistaken and undervalued; abroad, understood and honoured. The assertion may be startling, yet is literally true, that the doctrines of the Scottish School have been nowhere less fairly appreciated than in Scotland itself. To explain how they have been misinterpreted, and, consequently, neglected, in the country of their birth, is more than I can now attempt; but as I believe that an equal ignorance prevails in regard to the high favour accorded to these speculations by those nations who are now in advance, as the most enlightened cultivators of philosophy, I shall endeavour, as briefly as possible, to show that it may be for our credit not rashly to disparage what other countries view as our chief national claim to scientific celebrity. In illustration of this, I shall only allude to the account in which our Scottish Philosophy is held in Germany and in France.

There is a strong general analogy between the philosophies of Reid and Kant; and Kant, I may observe by the way, was a Scotsman by proximate descent. Both originate in a recoil against the Scepticism of Hume;[a] both are equally opposed to the Sensualism of Locke; both vindicate with equal zeal the moral dignity of man; and both attempt to mete out and to define the legitimate sphere of our intellectual activity. There are, however, important differences between the doctrines, as might be anticipated from the very different characters of the men; and while Kant surpassed Reid in systematic power and comprehension, Reid excelled Kant in the caution and security of his proce-

a See the completed edition of *Reid's Works*, Memoranda for Preface, p. xv.—Ed.

dure. There is, however, one point of difference in which it is now acknowledged, even by the representatives of the Kantian philosophy, that Kant was wrong. I allude to the doctrine of Perception,—the doctrine which constitutes the very corner-stone of the philosophy of Reid. Though both philosophies were, in their origin, reactions against the scepticism of Hume, this reaction was not equally determined in each by the same obnoxious conclusion. For, as it was primarily to reconnect Effect and Cause that Kant was roused to speculation, so it was primarily to regain the worlds of Mind and Matter that Reid was awakened to activity. Accordingly Kant, admitting, without question, the previous doctrine of philosophers, that the mind has no immediate knowledge of any existence external to itself, adopted it without hesitation as a principle,—that the mind is cognisant of nothing beyond its own modifications, and that what our natural consciousness mistakes for an external world, is only an internal phænomenon, only a mental representation of the unknown and inconceivable. Reid, on the contrary, was fortunately led to question the grounds on which philosophers had given the lie to the natural beliefs of mankind; and his inquiry terminated in the conclusion, that there exists no valid ground for the hypothesis, universally admitted by the learned, that an immediate knowledge of material objects is impossible. The attempt of Kant, if the attempt were serious, to demonstrate the existence of an external and unknown world was, as is universally admitted, a signal failure ; and his Hypothetical Realism was soon analysed by an illustrious disciple,—Fichte,— into an Absolute Idealism, with a logical rigour that did not admit of refutation.[a] In the meanwhile, Reid's doctrine of Perception had attracted the attention of an acute opponent of the critical philosophy in Germany;[β] and that doctrine, divested of those superficial errors which have led some ingenious reasoners in this country to view and represent Reid as holding an opinion on this point identical with Kant's, was, in Kant's own country, placed in opposition against his opinion, fortified as that was by the authority of all modern philosophers.

[a] Some fragmentary criticisms of the Kantian philosophy in this respect, will be found appended to this dissertation.—See below, p. 401 et seq.— ED.

[β] Schulze, in his Æneidemus, published in 1792 ; and again in his Kritik der theoretischen Philosophie, 1801. See Reid's Works, p. 797.—ED.

And with what result? Simply this; — that the most distinguished representatives of the Kantian school now acknowledge Kant's doctrine of Perception to be erroneous, and one
analogous to that of Reid they have adopted in its stead. Thus,
while, in Scotland, the fundamental position of Reid's philosophy
has been misunderstood, his criticism of the ideal theory treated
as a blunder, and his peculiar doctrine of perception represented
as essentially the same with that of the philosophers whom he
assailed; in Germany, and by his own disciples, Kant's theory
of perception is admitted to be false, and the doctrine of Reid,
on this point, appreciated at its just value, and recognised as
one of the most important and original contributions ever made
to philosophy.

But in France, I may add Italy, the triumph of the Scottish
school has been even more signal than in Germany. The philosophy of Locke, first recommended to his countrymen by the bril-
liant fancy of Voltaire, was, by the lucid subtlety of Condillac,
reduced to a simplicity which not only obtained an ascendant over
the philosophy of Descartes, but rendered it in France the object
of all but universal admiration. Locke had deduced all knowledge
from Experience, but Condillac analysed every faculty into Sense.
Though its author was no materialist, the system of transformed
sensation is only a disguised materialism; and the import of the
doctrine soon became but too apparent in its effects. Melancholy,
however, as it was, this theory obtained an authority in France
unparalleled for its universality and continuance. For seventy
years, not a single work of an opposite tendency made the smallest impression on the public mind; all discussion of principles
had ceased; it remained only to develop the remoter consequences
of the system: philosophy seemed accomplished.

Such was the state of opinion in France until the downfall of
the Empire. In the period of tranquillity that followed the Resto-
ration, the minds of men were again turned with interest towards
metaphysical speculation; and it was then that the doctrines of
the Scottish Philosophy were, for the first time, heard in the
public schools of France. Recommended by the powerful talent
and high authority of Royer-Collard, these doctrines made converts of some of the loftiest intellects of France. A vigorous
assault, in which the prowess of Cousin was remarkable, was

made against the prevalent opinions, and with a success so decisive, that, after a controversy of twenty years, the school of Condillac is now, in its own country, considered as extinct; while our Scottish philosophy not only obtained an ascendant in public opinion, but, through the influence of my illustrious friend M. Cousin, forms the basis of philosophical instruction in the various Colleges connected with the University of France. It must not, however, be supposed, that the French have servilely adopted the opinions of our countrymen. On the contrary, what they have borrowed they have so ably amplified, strengthened, simplified, and improved, that the common doctrines of Reid and Stewart, of Royer-Collard and Jouffroy, (for Cousin falls under another category), ought in justice to be denominated the *Scoto-Gallican Philosophy,*—a name, indeed, already bestowed upon them by recent historians of philosophy in Germany.

* * * * * * * *

(*b.*) M. JOUFFROY'S CRITICISM OF THE SCOTTISH SCHOOL.[a]

(Probably 1837, or a little later. See *Œuvres de Reid*, vol. i., Préface, p. clxxxvi.-cxcix.—ED.)

* * * * I must be allowed to make an observation in reference to the criticism of M. Jouffroy.

Dr Reid and Mr Stewart not only denounce as absurd the attempt to demonstrate that the original data of Consciousness are for us the rule of what *we* ought to believe, that is, the criteria of a relative—human—subjective truth; but interdict as unphilosophical all question in regard to their validity, as the vehicles of an absolute or objective truth.

M. Jouffroy, of course, coincides with the Scottish philosophers in regard to the former; but, as to the latter, he maintains, with Kant, that the doubt is legitimate, and, though he admits it to be insoluble, he thinks it ought to be entertained. Nor, on the ground on which they and he consider the question, am I disposed to dissent from his conclusion. But on that on which I have now placed it,[β] I cannot but view the inquiry as incompetent. For what is the question in plain terms? Simply,—

a Published in a fuller form, in the completed edition of *Reid's Works,* Memoranda for Preface, p. xvii.—ED. β See *Reid's Works,* p. 746.—ED.

Whether what our nature compels us to believe as true and
real, be true and real, or only a consistent illusion ? Now this
question cannot be philosophically entertained, for two reasons.
1°, Because there exists a presumption in favour of the veracity
of our nature, which either precludes or peremptorily repels
a gratuitous supposition of its mendacity.   2°, Because we have
no mean out of Consciousness of testing Consciousness. If its
data are found concordant, they must be presumed trustworthy ;
if repugnant, they are already proved unworthy of credit. Unless,
therefore, the mutual collation of the primary data of Consciousness be held such an inquiry, it is, I think, manifestly incompetent.   It is only in the case of one or more of these original facts
being rejected as false, that the question can emerge in regard to
the truth of the others. But, in reality, on this hypothesis, the problem is already decided ; their character for truth is gone ; and all
subsequent canvassing of their probability is profitless speculation.

Kant started, like the philosophers in general, with the non-
acceptance of the deliverance of Consciousness,—that we are
immediately cognisant of extended objects. This first step
decided the destiny of his philosophy. The external world, as
known, was therefore only a phænomenon of the internal ; and
our knowledge in general only of self, the objective only subjective ; and truth only the harmony of thought with thought,
not of thought with things ;—reality only a necessary illusion.

It was quite in order, that Kant should canvass the veracity
of all our primary beliefs, having founded his philosophy on the
presumed falsehood of one ; and an inquiry followed out with
such consistency and talent could not, from such a commencement, terminate in a different result.

(c.) General Characteristics of the Scottish School.

(Written in connection with proposed MEMOIR OF MR DUGALD STEWART.
On Desk, May 1856 ; written Autumn 1855.—ED.)

The Scottish School of Philosophy is distinctively characterised
by its opposition to all the destructive schemes of speculation ;—
in particular, to Scepticism, or the uncertainty of knowledge; to
Idealism, or the non-existence of the material world; to Fatalism,

ERROR

or the denial of a moral universe. Reid has the merit of originating this movement, and Stewart the honour of continuing, and promoting, and extending it.

In the philosophy which prevailed before Descartes, in whose doctrines it may be affirmed that modern speculation took its rise, we find all these schemes, indeed, but all marked and modified in a peculiar manner. In antiquity, we have the scepticism of Pyrrho and Ænesidemus; but this, however ingenious its object, never became popular or dangerous, and, without a formal or decisive refutation, gradually died out.

In the scholastic ages, Idealism was [countenanced] by the dominant psychology, and would perhaps have taken root, but for the check it encountered from the Church, to the dogmas of which all philosophy was then voluntarily subjected. The doctrine of Representative Perception, in its cruder form, was generally accepted, and the question often mooted, "Could not God maintain the species in the sensory, the object (external reality) being annihilated?" This problem, as philosophy affirmed, theology denied. It was possible, nay probable, according to the former; impossible, because heretical, according to the latter.[a]

Finally, on the other hand, the Absolute Decrees of God might, at the first view, be thought, not only to favour, but to establish, a doctrine of unconditioned Fatalism. But this inference was disavowed by the most strenuous advocates of Prescience and Predestination; and the Freewill of man asserted no less vehemently than the Free Grace of God.

(d.) KANT AND REID.

(Written in connection with proposed MEMOIR OF MR STEWART. On Desk, May 1856; written Autumn 1855.—ED.)

* * * * * *

In like manner, Kant assailed Scepticism, and the scepticism of Hume; but with a very different result. For, if in one conclusion he controverted scepticism, he himself introduced and

[a] See *Discussions*, p. 198, second edition,—why Idealism and the doctrine of Transubstantiation were incompatible.

patronised the most unexclusive doubt. He showed, indeed, that Hume's rejection of the notion of Causality was groundless. He proved that, although this notion was not, and could not be, constructed from experience, still Causality was a real and efficient principle, native and necessary in human intelligence ; and that although experience did not explain its genesis, experience always supposes its operation. So far so good. But Kant did not stop here. He endeavoured to evince that pure Reason, —that Intelligence, is naturally, is necessarily, repugnant with itself, and that speculation ends in a series of insoluble anti-logies. In its highest potence, in its very essence, thought is thus infected with contradiction ; and the worst and most pervading scepticism is the melancholy result. If I have done anything meritorious in philosophy, it is in the attempt to explain the phænomena of these contradictions ; in showing that they arise only when intelligence transcends the limits to which its legitimate exercise is restricted ; and that within those bounds, (the Conditioned), natural thought is neither fallible nor mendacious—

"Neque decipitur, nec decipit unquam."

If this view be correct, Kant's antinomies, with their consequent scepticism, are solved ; and the human mind, however weak, is shown not to be the work of a treacherous Creator.

Reid, on the contrary, did not subvert the trustworthiness of the one witness, on whose absolute veracity he relied. In his hands natural (and, therefore, necessary) thought—Consciousness—Common Sense—are always held out as entitled to our implicit and thorough-going confidence. The fact of the testimony sufficiently guarantees the truth of what the testimony avouches. The testimony, if delivered, is to be deemed *pro tanto* impeccable.

* * * * * *

(e.) Kant's Doctrine of Space and Time.

(Fragments from early Papers. Probably before 1836.—Ed.)

Kant, 1°, Made our actual world one merely of illusion. Time and Space, under which we must perceive and think, he reduced

to mere subjective spectral forms, which have no real archetype
in the noumenal or real universe. We can infer nothing from
this to that. Cause and Effect govern thing and thought in the
world of Space and Time ; the relation will not subsist where
Time and Space have no reality. (Lines from Fracastorius).[a]
Corresponds with the Platonic, but more thorough-going. Kant,
2°, Made Reason, Intelligence, contradict itself in its legitimate
exercise. Antilogy,—antinomy, part and parcel of its nature ;
not only " reasoning, but to err," but reason itself.

Thus, the conviction that we live in a world of unreality and
illusion, and that our very faculty of knowledge is only given us
to mislead, is the result of our criticism ;—Scepticism.

On the contrary, my doctrine holds, 1°, That Space and Time,
as given, are real forms of thought and conditions of things ;
2°, That Intelligence,—Reason,—within its legitimate limits, is
legitimate ; within this sphere it never deceives ; and it is only
when transcending that sphere, when founding on its illegitimate as on its legitimate exercise, that it affords a contradictory
result ;—" Ne sapiamus ultra facultates." The dogmatic assertion of necessity,—of Fatalism, and the dogmatic assertion of
Liberty, are the counter and equally inconceivable conclusions
from reliance on the illegitimate and one-sided.

*　　*　　*　　*　　*　　*

Kant holds the subjectivity of Space (and Time), and, if he .
does not deny, will not affirm the existence of a real space,
external to our minds ; because it is a mere form of our perceptive faculty. He holds that we have no knowledge of any
external thing as really existing, and that all our perceptions
are merely appearances, i.e., subjective representations,—subjective modifications,—which the mind is determined to exhibit,
as an apparently objective opposition to itself,—its pure and
real subjective modifications. Yet, while he gives up the external existence of space, as beyond the sphere of consciousness,
he holds the reality of external material existences, (things in
themselves), which are equally beyond the sphere of consciousness. It was incumbent on him to render a reason for this
seeming inconsistency, and to explain how his system was not,
in its legitimate conclusions, an universal Idealism ; and he

a See below, Lect. xxi., vol. ii. p. 83.—Ed.

has accordingly attempted to establish, by necessary inference, what his philosophy could not accept as an immediate fact of consciousness.

In the second edition of his *Kritik der reinen Vernunft*, he has accordingly given what he calls a "*strict, and, as he is convinced, the only possible*, demonstration for the objective reality of our external perceptions;" and, at the same time, he declares that it would be the eternal scandal of Philosophy, and of the general reason of mankind, if we were compelled to yield our assent to the existence of an external world, only as an article of Faith, and were unable to oppose a satisfactory refutation to any sceptical objections that might be suggested touching their reality (Vorrede, p. xxxix). The demonstration which is thus exclusively and confidently proposed, attempts to prove that the existence of an external world is involved in the very consciousness of self,—that without a *Thou*, there could be no *I*, and that the *Cogito ergo sum* is not more certain than the *Cogito ergo es*.

*　　*　　*　　*　　*　　*

II.—PHYSIOLOGICAL. (See Vol. I., p. 264.)

(a.) PHRENOLOGY.

*　　*　　*　　*　　*　　*

Such is a very general view of that system [the Nervous] and its relations, which physiologists and philosophers in general have held to be the proximate organ of the thinking principle, and many to be even the thinking principle itself. That the mind, in its lower energies and affections, is immediately dependent on the conditions of the nervous system, and that, in general, the development of the brain in the different species of animals is correspondent to their intelligence,—these are conclusions established upon an induction too extensive and too certain to admit of doubt. But when we attempt to proceed a step farther, and to connect the mind or its faculties with particular parts of the nervous system, we find ourselves at once checked. Observation and experiment seem to fail; they afford only obscure and varying reports; and if, in this uncertainty, we hazard a conclusion, this is only a theory established upon some arbitrary hypothesis, in which fictions stand in place of facts. The uncertainty of such conclusions is shown by the unexampled diversity of opinion that has always reigned among those who, discontented with a prudent ignorance, have attempted to explain the phænomena of mind by the phænomena of organisation.

In the first place, some, (and their opinion is not, certainly, the least philosophical), hold that, in relation to the body, the soul is less contained than containing,—that it is all in the whole, and all in every part. This is the common doctrine of many of the Fathers, and of the scholastic Aristotelians.[a]

In the second place, others have attempted to connect the conscious principle in general with a particular part of the organism, but by very different relations. Some place it there, as in a local seat ; others make it dependent on that part, as on its organ ; while others hold that the mind stands in a more immediate relation to this part, only because it is the point of convergence where all the bodily sensations meet. I shall not attempt to enumerate the hundred and one conjectures in regard to the point in the corporeal organism, in proximate connection with the mind. It would occupy more than our hour to give you even a summary account of the hypotheses on this subject.

In the third place, no opinion has been more generally prevalent than that different faculties and dispositions of the mind are dependent on different parts of the bodily organism, and more especially on different parts of the nervous system. Under this head, I shall state to you one or two of the more famous opinions. The most celebrated doctrine,—that which was more universally adopted, and for a longer period, than any other,— was that which, with certain modifications, assigned different places in the Encephalos to Memory, Imagination, Sense, and the Locomotive Faculty,—Reason or Intelligence being left inorganic. This opinion we trace upward, through the Latin and Arabian schools,[β] to St Austin,[γ] Nemesius,[δ] the Greek

a See below, Lect. xx., vol. ii. p. 7.   Venice, 1560.]
—Ed.   γ De Genesi ad Literam, lib. vii. cc.
β [See Gassendi, Physica, Sect. iii.,   17, 18.—Ed.   [See Tennemann, t. vii.
Memb. Post., lib. viii.; Opera, t. ii. pp.   p. 241.]
400, 401. Averroes, Destruct. Destruc-   δ De Natura Hominis, c. xiii., p. 204,
tionum, Arist. Opera, t. x. p. 340.   edit. Matthæi.—Ed.

physician Aetius, and even to the anatomists Rufus and Posidonius. Memory, on this hypothesis, was placed in the substance of the cerebellum, or in the subjacent ventricle ; and as the phrenologists now attempt to prove that the seat of this faculty lies above the eyebrows, by the alleged fact, that when a man wishes to stimulate his recollection, he rubs the lower part of his forehead,—so, of old, the same conclusion was established on the more plausible assertion, that a man in such circumstances naturally scratches the back of his head. The one indication is at least as good as the other.

Among modern physiologists, Willis was the first who attempted a new attribution of mental functions to different parts of the nervous system. He placed Perception and Sensation in the *corpus callosum*, Imagination and Appetite in the *corpora striata*, Memory in the cerebral convolutions, Involuntary Motion in the cerebellum, &c.; and to Willis is to be traced the determination so conspicuous among subsequent physiologists, of attributing different mental uses to different parts of the brain.

It would be bootless to state to you the many various and contradictory conjectures in regard to these uses. To psychologists they are, with one exception, all comparatively uninteresting, as, were they even ascertained to be something better than conjectures, still, as the physical condition is in all of them occult, it could not be applied as an instrument of psychological discovery. The exception which I make is, the celebrated doctrine of Gall. If true, that doctrine would not only afford us a new instrument, but would in a great measure supersede the old. In fact, the psychology of consciousness, and the psychology founded on Gall's organology, are mere foolishness to each other. They arrive at conclusions the most contradictory; insomuch that the establishment of the one necessarily supposes the subversion of the other.

In these circumstances, no one interested in the philosophy of man can be indifferent to an inquiry into the truth or falsehood of the new doctrine. This doctrine cannot be passed over with contempt. It is maintained not only by too many, but by too able advocates, to be summarily rejected. That its results are repugnant to those previously admitted, is but a sorry reason for not inquiring into their foundation. This doctrine professes to have discovered new principles, and to arrive at new conclusions; and the truth or falsehood of these cannot, therefore, be estimated merely by their conformity or disconformity with those old results which the new professedly refute. To do so would be mere prejudice,—a mere assumption of the point at issue. At the same time, this doctrine professes to be founded on sensible facts. Sensible facts must be shown to be false, not by reasoning, but by experiment; for, as old Fernelius has well expressed it,—" Desipientis arrogantiæ est argumentationis necessitatem sensuum auctoritati anteponere." To oppose such a doctrine in such a manner is not to refute, but to recommend; and yet, unfortunately, this has been the usual mode in which the organology of Gall and his followers has been assailed. Such an opinion must be taken on its own ground. We must join issue with it upon the facts and inferences it embraces. If the facts are true, and if the inferences necessarily follow, the opinion must be admitted; the sooner, therefore, that we candidly inquire into these the better, for it is only thus that we shall be enabled to form a correct estimate of the evidence on which such a doctrine rests.

With these views, I many years ago undertook an investigation of the fundamental facts on which the phrenological doctrine, as it is unfortunately called, is established. By a fundamental fact I mean a fact by the truth of which the hypothesis could be proved, and, consequently, by the falsehood of which it could be disproved. Now, what are such facts? The one condition of such a fact is that it should be general. The phrenological theory is, that there is a correspondence between the volume of certain parts of the brain, and the intensity of certain qualities of mind and character;—the former they call development, the latter manifestation. Now, individual cases of alleged conformity of development and manifestation could prove little in favour of the doctrine, as individual cases of alleged disconformity could prove little against it; because, 1°, The phrenologists had no standard by which the proportion of cerebral development could be measured by themselves or their opponents; 2°, Because the mental manifestation was vague and indeterminate; 3°, Because they had introduced, as subsidiary hypotheses, the occult qualities of temperament and activity, so that, in individual cases, any given head could always be explained in harmony with any given character. Individual cases were thus ambiguous; they were worthless either to establish or to refute the theory. But where the phrenologists had proclaimed a general fact, by that fact their doctrine could be tried. For example, when they asserted as the most illustrious discovery of Gall, and as the surest inference of their doctrine, that the cerebellum is the organ of the sexual appetite, and established this inference as the basis of certain general facts which, as common to the whole animal kingdom, could easily be made matter of precise experiment;—by these facts the truth of their doctrine could be brought to the test, and this on ground the most favourable for them. For the general probability of their doctrine was thus estimated by the truth of its best - established element. But, on the other hand, if such general facts were found false, their disproval afforded the most satisfactory refutation of the whole system. For the phrenologists themselves readily admit that their theory is exploded, if their doctrine of the function of the cerebellum is disproved. Because, therefore, an examination of the general facts of Phrenology was at once decisive and comparatively easy, I determined, on this ground, to try the truth of the opinion. I shall state to you very generally a few results of the investigation, of which I may, without boasting, affirm that no inquiry of the kind was ever conducted with greater care or more scrupulous accuracy.

I shall commence with the phrenological doctrine of the cerebellum, on which you will see the propriety of dwelling as briefly as I can. I may mention that the extent of my experiments on this organ is wholly unconnected with phrenology. My attention was, indeed, originally turned to the relation of the after-brain to the other parts of the nervous system, when testing the accuracy of the phrenological doctrine on this point; but that end was very soon accomplished, and it was certain discoveries which I made in regard to the laws of development and the function of this organ, and the desire of establishing these by an induction from as many of the species as possible of the animal kingdom, that led me into a more extensive inquiry than has hitherto been instituted by any professional physiologist.

When I publish its results, they will disprove a hundred times over all the phrenological assertions in regard to the cerebellum ; but this will be only an accidental circumstance, and of comparatively little importance. I may add, that my tables extend to above 1000 brains of above 50 species of animals accurately weighed by a delicate balance ; and you will remark that the phrenologists have not a single observation of any accuracy to which they can appeal. The only evidence in the shape of precise experiment on which they can found, is a table of Serres, who is no phrenologist, affording the general averages of certain weighings, said to have been made by him, of the brain and cerebellum in the human subject. I shall prove that table an imaginary fabrication in support of a now exploded hypothesis of the author.

The alleged facts on which Gall and his followers establish their conclusion in regard to the function of the cerebellum are the following :—

The first is, that, in all animals, females have this organ, on an average, greatly smaller, in proportion to the brain proper, than males. Now, so far is this assertion from being correct, it is the very reverse of truth ; and I have ascertained, by an immense induction, that in no species of animal has the female a proportionally smaller cerebellum than the male, but that in most species, and this according to a certain law, she has a considerably larger. In no animal is this difference more determinate than in man. Women have on an average a cerebellum to the brain proper, as 1 : 7 ; men as 1 : 8. This is a general fact which I have completely established.[a]

The second alleged fact is, that in impuberal animals the cerebellum is in proportion to the brain proper greatly less than in adults. This is equally erroneous. In all animals, long previous to puberty, has the cerebellum attained its maximum proportion. And here, also, I am indebted to the phrenologists for having led me to make the discovery of another curious law, and to establish the real function of the cerebellum. Physiologists have hitherto believed that the cerebella of all animals, indifferently, were, for a certain period subsequent to birth, greatly less, in

a See below, (b), On Weight of Brain, p. 419.—ED.

proportion to the brain proper, than in adults; and have taken no note of the differences in this respect between different classes. Thus, completely wrong in regard to the fact, they have necessarily overlooked the law by which it is governed. In those animals that have from the first the full power of voluntary motion, and which depend immediately on their own exertions, and on their own power of assimilation for nutriment, the proportion of the cerebellum is as large, nay larger, than in the adult. In the chicken of the common fowl, pheasant, partridge, &c., this is the case; and most remarkably after the first week or ten days, when the yolk, (corresponding in a certain sort to the milk in quadrupeds), has been absorbed. In the calf, kid, lamb, and probably in the colt, the proportion of the cerebellum at birth is very little less than in the adult. In those birds that do not possess at once the full power of voluntary motion, but which are in a rapid state of growth, the cerebellum, within a few days at least after being hatched, and by the time the yolk is absorbed, is not less or larger than in the adult; the pigeon, sparrow, &c. &c., are examples. In the young of those quadrupeds that for some time wholly depend for support on the milk of the mother, as on half-assimilated food, and which have at first feeble powers of regulated motion, the proportion of the cerebellum to the brain proper is at birth very small; but by the end of the full period of lactation, it has with them as with other animals, (nor is man properly an exception), reached the full proportion of the adult.[a] This, for example, is seen in the young rabbit, kitten, whelp, &c.; in them the cerebellum is to the brain proper at birth about as 1 to 14; at six and eight weeks old about as 1 to 6. Pigs, &c., as possessing immediately the power of regulated motion, but wholly dependent on the milk of the mother during at least the first month after birth, exhibit a medium between the two classes. At birth the proportion is in them about 1 to 9, in the adult as 1 to 6. This analogy, at which I now only hint, has never been suspected; it points at the new and important conclusion, (corroborated by many other facts), that the cerebellum is the intracranial organ of the nutritive faculty, that term being taken in its broadest signification; and it confirms also an old

a This may, perhaps, explain the apparent exception to Berkeley's theory noticed by Adam Smith. See below, vol. ii. p. 182.—ED.

opinion, recently revived, that it is the condition of voluntary or systematic motion.[a]

The third alleged fact is, that the proportion of the cerebellum to the brain proper in different species, is in proportion to the *energy* of the phrenological function attributed to it. This assertion is groundless as the others. There are many other fictions in regard to this organ; but these, I think, are a sufficient specimen of the truth of the doctrine in regard to the function of the cerebellum ; and the cerebellum, you will recollect, is the citadel of Phrenology.

I shall, however, give you the sample of another general fact. The organ of Veneration rises in the middle on the coronal surface of the head. Women, it is universally admitted, manifest religious feeling more strongly and generally than men ; and the phrenologists accordingly assert that the female cranium is higher in proportion in that region than the male. This I found to be the very reverse of truth, by a comparative average of nearly two hundred skulls of either sex. In man, the female encephalos is considerably smaller than that of the male, and in shape the crania of the sexes are different. By what dimension is the female skull less than the male? The female skull is longer, it is nearly as broad, but it is much lower than the male. This is only one of several curious sexual differences of the head.

I do not know whether it be worth while mentioning, that, by a comparison of all the crania of murderers preserved in the Anatomical Museum of this University, with about nearly two hundred ordinary skulls indifferently taken, I found that these criminals exhibited a development of the phrenological organs of Destructiveness and other evil propensities smaller, and a development of the higher moral and intellectual qualities larger, than the average. Nay, more, the same result was obtained when the murderers' skulls were compared, not merely with a common average, but with the individual crania of Robert Bruce, George Buchanan, and Dr David Gregory.

I omit all notice of many other decisive facts subversive of the hypothesis in question ; but I cannot leave the subject without alluding to one which disproves, at one blow, a multitude of

a From a communication by the *tomy of the Brain*, pp. 6, 7. See be-
Author, printed in Dr Munro's *Ana-* low, (b), *On Weight of Brain.*—ED.

organs, affords a significant example of their accuracy of statement, and shows how easily manifestation can, by the phrenologists, be accommodated to any development, real or supposed.
I refer to the Frontal Sinuses. These are cavities between the
tables of the frontal bone, in consequence of a divergence from
each other. They are found in all puberal crania, and are of
variable and, [from without], wholly inappreciable extent and
depth. Where they exist, they of course interpose an insuperable bar to any estimate of the cerebral development; and their
extent being undiscoverable, they completely baffle all certain
observation. Now, the phrenologists have fortunately, or unfortunately, concentrated the whole of their very smallest organs
over the region of the sinus; which thus, independently of other
impediments, renders all phrenological observation more or less
uncertain in regard to sixteen of their organs. Of these cavities
the anatomists in general seem to have known not much, and
the phrenologists absolutely nothing. At least, the former are
wrong in many of their positions, the latter wrong in all. I
shall give you a sample of the knowledge and consistency of
the phrenologists on this point.

Gall first of all answered the objection of the sinus, by asserting that even when it existed, the plates of the frontal bone
were still parallel. The truth is, that the cavity is only formed
by their divergence from parallelism, and thus it is now described
by the phrenologists themselves. In his latest works, Gall asserted that the sinus is frequently absent in men, and seldom or
never found in women. But Spurzheim carried the negation to
its highest climax, for he avers, (I quote his words), " that children and young adult persons have no holes between the two
tables of the skull at the forehead, and that they occur only in
old persons, or after chronic insanity." He did not always, indeed, assert as much, and in some of his works he allows that
they throw some uncertainty over the organs of Individuality
and Size, but not much over that of Locality.

Now the fact is, as I have established by an inspection of
several hundred crania, that *no skull is without a sinus.* This is,
indeed, the common doctrine of the anatomists. But I have also
proved that the vulgar doctrine of their increasing in extent, in
proportion as the subject advances in life, is wholly erroneous.

The smallest sinus I ever saw was in the cranium of a woman of a hundred years of age.

The two facts,—the fact of the universal existence of the sinus, and its great and various and inappreciable extent, and the fact of the ignorance of the phrenologists in regard to every circumstance connected with it,—these two facts prove that these observers have been going on finding always manifestation and development in exact conformity; when, lo! it turns out, that in nearly half their organs, the protuberance or depression apparent on the external bone has no connection with any correspondent protuberance or depression in the brain. Now, what does this evince? Not merely that they were wrong in regard to these particular observations and the particular organs established upon the mistake. Of course, the whole organs lying over the sinuses are swept away. But this is not all; for the theory supposes, as its condition, that the amount of the two qualities of mental manifestation and cerebral development can be first accurately measured apart, and then compared together, and found either to be conformable or disconformable: and the doctrine, assuming this possibility, proves its truth only by showing that the two qualities thus severally estimated, are, in all cases, in proportion to each other. Now, if the possibility thus assumed by Phrenology were true, it would at once have discovered that the apparent amount of development over the sinus was not in harmony with the mental manifestation. But this it never did;—it always found the apparent or cranial development over the sinus conformable to the mental manifestation, though this bony development bore no more a proportion to the cerebral brain than if it had been looked for on the great toe; and thus it is at once evident, that manifestation and development in general are, in their hands, such factitious, such arbitrary quantities, that they can always, under any circumstances, be easily brought into unison. Phrenology is thus shown to be a mere leaden rule, which bends to whatever it is applied; and, therefore, all phrenological observation is poisoned, in regard even to those organs where a similar obstacle did not prevent the discovery of the cerebral development. Suppose a mathematician to propose a new method for the solution of algebraical equations. If we applied it, and found it gave a false result, would the inventor be listened to if he said,—"True, my method is wrong in these cases in which it has been tried, but it is not, therefore, proved false in those in which it has not been put to the test"? Now, this is precisely the plea I have heard from the phrenologists in relation to the sinus. "Well!" they say, "we admit that Gall and Spurzheim have been all wrong about the sinus, and we give up the organs above the eyes; but our system is untouched in the others which are situate beyond the reach of that obnoxious cavity." To such reasoning there was no answer.

I should have noticed, that, even supposing there had been no intervening caverns in the forehead, the small organs arranged, like peas in a pod, along the eyebrows could not have severally manifested any difference of development. If we suppose, (what I make bold to say was never yet observed in the brain), that a portion of it so small in extent as any one of the six phrenological organs of Form, Size, Weight, Colour, Order, and Number, which lie side by side upon the eyebrows, was ever prominent beyond the surrounding surface,—I say, supposing the protuberance of so small a spot upon the cerebral convolutions, it could never determine a corresponding eminence on the external table of the skull. What would be the effect of such a protrusion of brain upon the cranium? It would only make room for itself in the thickness of the bone which it would attenuate. This is shown by two examples. The first is taken from the convolutions themselves. I should, however, state, that convolution, and anfractuosity or furrow, are correlative terms, like hill and valley,—the former (convolutions) being applied to the windings of the cerebral surface as rising up,—the latter (anfractuosity, or furrow) being applied to them as sinking in. Convolutions are the winding eminences between the furrows; anfractuosities the winding depressions between the convolutions. This being understood, we find, on looking to the internal surface of the cranium, that the convolutions attenuate the bone, which is sometimes quite transparent,—diaphanous,—over them, whereas it remains comparatively thick over the anfractuosities; but they cause no inequality on the outer surface. Yet the convolutions, which thus make room for themselves in the bone without elevating it externally, are often broader, and of course always longer, than the little organs which the phrenologists have placed along the eyebrows. *A fortiori*, therefore, we must suppose that an organ like Size, or Weight, or Colour, if it did project beyond the surrounding brain, would only render the superincumbent bone thinner, without causing it to rise, unless we admit that nature complaisantly changes her laws in accommodation to the new doctrine.

But we have another parallel instance still more precisely in point. In many heads there are certain rounded eminences, (called *Glandulæ Pacchioni*), on the coronal surface of the brain, which nearly correspond in size with the little organs in question. Now, if the phrenological supposition were correct, that an elevation on the brain, of so limited an extent, would cause an elevation on the external table of the bone, these eminences would do so far more certainly than any similar projection over the eyebrows. For the frontal bone in the frontal region is under the continual action of muscles, and this action would tend powerfully to prevent any partial elevation ; whereas, on the upper part of the head, the bone is almost wholly exempt from such an agency. But do the glands, as they are called, of Pacchioni, (though they are no glands),—do they determine an elevation on the external surface of the skull corresponding to the elevation they form on the cerebral surface ? Not in the very least ; the cranium is there outwardly quite equable,—level, —uniform,—though probably attenuated to the thinness of paper to accommodate the internal rising.

The other facts which I have stated as subversive of what the phrenologists regard as the best-established constituents of their system,—I could only state to you on my own authority. But they are founded on observations made with the greatest accuracy, and on phænomena, which every one is capable of verifying. If the general facts I gave you in regard to the cerebellum, &c., are false, then am I a deliberate deceiver; for these are of such a nature that no one with the ordinary discourse of reason could commit an error in regard to them, if he actually made the observations. The maxim, however, which I have myself always followed, and which I would earnestly impress upon you, is to take nothing upon trust that can possibly admit of doubt, and which you are able to verify for yourselves ; and had I not been obliged to hurry on to more important subjects, I might have been tempted to show you by experiment what I have now been compelled to state to you upon authority alone.[a]

I am here reminded of a fact, of which I believe none of our present phrenologists are aware,—at least all their books confidently assert the very reverse. It is this,—that the new system is the result, not of experience but of conjecture, and that Gall, instead of deducing the faculties from the organs, and generalising both from particular observations, first of all excogitated a faculty *a priori*, and then looked about for an organ with which to connect it. In short, Phrenology was not discovered but invented.

You must know, then, that there are two faculties, or rather two modifications of various faculties, which cut a conspicuous figure in the psychologies of Wolf and other philosophers of the Empire:—these are called in German *Tiefsinn* and *Scharfsinn*, —literally, *deep sense* and *sharp sense*, but are now known in English phrenological language by the terms *Causality* and *Comparison*. Now what I wish you to observe is, that Gall found these two clumsy modifications of mind, ready shaped out in the previous theories of philosophy prevalent in his own country, and then in the language itself. Now, this being understood, you must also know that, in 1798, Gall published a letter to Retzer of Vienna, wherein he, for the first time, promulgates the nature of his doctrine, and we here catch him,—*reum confitentem*,—in the very act of conjecturing. In this letter he says: " I am not yet so far advanced in my researches as to have discovered special organs for Scharfsinn and Tiefsinn, (Comparison and Causality), for the principle of the Representative Faculty, (*Vorstellungsvermögen*, — another faculty in German philosophy), and for the different varieties of judgment, &c." In this sentence we see exhibited the real source and veritable derivation of the system.

In the *Darstellung* of Froriep, a favourite pupil of Gall, under whose eye the work was published in the year 1800, twenty-two organs are given, of which the greater proportion are now either translated to new localities, or altogether thrown out. We find also that the sought-for organs had, in the interval, been found for Scharfsinn, (Comparison), and Tiefsinn, (Causality); and what

a See below, (d), *On Frontal Sinus*, p. 424.—ED.

further exhibits the hypothetical genealogy of the doctrine, is, that a great number of organs are assumed, which lie wholly beyond the possible sphere of observation, at the base and towards the centre of the brain; as those of the External Senses, those of Desire, Jealousy, Envy, love of Power, love of Pleasure, love of Life, &c.

An organ of Sensibility is placed above that of Amativeness, between and below two organs of Philoprogenitiveness; an organ of Liberality, (its deficiency standing instead of an organ of Avarice or Acquisitiveness), is situated above the eyebrows, in the position now occupied by that of Time. An organ of Imagination is intimately connected with that of Theosophy or Veneration, towards the vertex of the head; and Veracity is problematically established above an organ of Parental Love. An organ of Vitality is not to be forgotten, situated in the *medulla oblongata*, the development of which is measured by the size of the *foramen magnum* and the thickness of the neck. These faculties and organs are all now cashiered; and who does not perceive that, like those of Causality and Comparison, which are still suffered to remain, they were first devised, and then quartered on some department of the brain?

We thus see that, in the first edition of the craniological hypothesis, there were several tiers or stories of organs,—some at the base, some about the centre, and others on the surface of the brain. Gall went to lecture through Germany, and among other places he lectured at Göttingen. Here an objection was stated to his system by the learned Meiners. Gall measured the development of an external organ by its prominence. "How," says Meiners, "do you know that this prominence of the outer organ indicates its real size? May it not merely be pressed out, though itself of inferior volume, by the large development of a subjacent organ?" This objection it was easily seen was checkmate. A new game must be commenced, the pieces arranged again. Accordingly, all the organs at the base and about the centre of the brain were withdrawn, and the whole organs were made to run very conveniently upwards and outwards from the lower part of the brain to its outer periphery.

It would be tiresome to follow the history of phrenological variation through the works of Leune and Villars to those of

Bischoff and Blöde,—which last represent the doctrine as it
flourished in 1805.   In these, the whole complement of organs
which Gall ever admitted is detailed, with the exception of
Ideality. But their position was still vacillating. For example,
in Froriep, Bischoff, and Blöde, the organ of Destructiveness is
exhibited as lying principally on the parietal bone, above and a
little anterior to the organ of Combativeness; while the region
of the temporal bone, above and before the opening of the ear,
in other words, its present situation, is marked as *terra adhuc
incognita.*

No circumstance, however, is more remarkable than the successive changes of shape in the organs. Nothing can be more
opposite than the present form of these as compared with those
which the great work of Gall exhibits. In Gall's plates they
are round or oval, in the modern casts and plates they are of
every variety of angular configuration; and I have been told
that almost every new edition of these varies from the preceding. We may, therefore, well apply to the phrenologist and his
organology the line of Horace [a]—

> "Diruit, ædificat, mutat quadrata rotundis,"

with this modification, that we must read in the latter part,
mutat rotunda quadratis.

So much for Phrenology,—for the doctrine which would substitute the callipers for consciousness in the philosophy of man;
and the result of my observation,—the result at which I would
wish you also to arrive,—I cannot better express than in the
language of the Roman poet [β]—

> "Materiæ ne quære modum, sed perspice vires
> Quas ratio, non pondus habet."

In what I have said in opposition to the phrenological doctrine, I should, however, regret if it could be ever supposed that
I entertain any feeling of disrespect for those who are converted
to this opinion. On the contrary, I am prompt to acknowledge
that the sect comprises a large proportion of individuals of great
talent; and I am happy to count among these some of my most
valued and respected friends. To the question, How comes it
that so many able individuals can be believers in a groundless

[a] *Epist.*, lib. i. ep. i. 100.—E[d].   [β] Manilius, iv. 929.—E[d].

opinion?—I answer, that the opinion is not wholly groundless; it contains much of truth,—of old truth it must be allowed; but it is assuredly no disparagement to any one that he should not refuse to admit facts so strenuously asserted, and which, if true, so necessarily infer the whole conclusions of the system. But as to the mere circumstance of numbers, that is of comparatively little weight,—*argumentum pessimi turba,*[a]—and the phrenological doctrines are of such a nature that they are secure of finding ready converts among the many. There have been also, and there are now, opinions far more universally prevalent than the one in question, which nevertheless we do not consider on that account to be undeniable.

(*b.*) An Account of Experiments on the Weights and Relative Proportions of the Brain, Cerebellum, and Tuber Annulare in Man and Animals, under the various circumstances of Age, Sex, Country, &c.

(Published in Dr Monro's *Anatomy of the Brain,* p. 4-8, Edinburgh, 1831.—Ed.)

The following, among other conclusions, are founded on an induction drawn from above sixty human brains, from nearly three hundred human skulls, of determined sex,—the capacity of which, by a method I devised, was taken in sand, and the original weight of the brain thus recovered,—and from more than seven hundred brains of different animals.

1. In man, the adult male Encephalos is heavier than the female; the former nearly averaging, in the Scot's head, 3 lb. 8 oz. troy, the latter, 3 lb. 4 oz.; the difference, 4 oz. In males of this country, about one brain in seven is found above 4 lb. troy; in females, hardly one in one hundred.

2. In man, the Encephalos reaches its full size about seven years of age. This was never before proved. It is commonly believed that the brain and the body attain their full development

---

a Seneca, *De Vita Beata,* c. 2.—Ed. *Mimi et aliorum Sententiæ,* ed. Orellii, [After Publius Syrus] [See *Publii Syri* p. 14.—Ed.]

together. The Wenzels rashly generalised from two cases the conclusion, that the brain reaches its full size about seven years of age; as Sömmering had in like manner, on a single case, erroneously assumed that it attains its last growth by three. Gall and Spurzheim, on the other hand, assert that the increase of the Encephalos is only terminated about forty. This result of my induction is deduced from an average of thirty-six brains and skulls of children, compared with an average of several hundred brains and skulls of adults. It is perhaps superfluous to observe, that it is the greater development of the bones, muscles, and hair, which renders the adult head considerably larger than that of the child of seven.

3. It is extremely doubtful whether the cranial contents usually diminish in old age. The vulgar opinion that they do, rests on no adequate evidence, and my induction would rather prove the negative.

4. The common doctrine, that the African brain, and in particular that of the Negro, is greatly smaller than the European, is false. By a comparison of the capacity of two Caffre skulls, male and female, and of thirteen negro crania (six male, five female, and two of doubtful sex), the encephalos of the African was found not inferior to the average size of the European.

5. In man, the Cerebellum, in relation to the Brain proper, comes to its full proportion about three years. This anti-phrenological fact is proved by a great induction.

6. It is extremely doubtful whether the Cerebellum usually diminishes in old age; probably only in cases of *atrophia senilis.*

7. The female Cerebellum is, in general, considerably larger in proportion to the Brain proper, than the male. In the human subject, (the Tuber excluded), the former is nearly as 1 to 7.6 : the latter nearly as 1 to 8.4 : and this sexual difference appears to be more determinate in man than in most other animals. Almost the whole difference of weight between the male and female encephali lies in the brain proper; the cerebella of the two sexes, absolutely, are nearly equal,—the preponderance rather in favour of the women. This observation is new; and the truth of the phrenological hypothesis implies the reverse. It confirms the theory of the function of the cerebellum noticed in the following paragraph.

8. The proportion of the Cerebellum to the Brain proper at birth varies greatly in different animals.[a]

9. Castration has no effect in diminishing the Cerebellum, either absolutely or in relation to the Brain proper.[β] The opposite doctrine is an idle fancy, though asserted by the phrenologists as their most incontrovertible fact. Proved by a large induction.

10. The universal opinion is false, that man, of all or almost all animals, has the smallest Cerebellum in proportion to the Brain proper. Many of the commonest quadrupeds and birds have a cerebellum, in this relation, proportionally smaller than man.

11. What has not been observed, the proportion of the Tuber Annulare to the Cerebellum, (and, *a majore*, to the Brain proper), is greatly less in children than in adults. In a girl of one year, (in my table of human brains), it is as 1 to 16.1; in another of two, as 1 to 14.8; in a boy of three, as 1 to 15.5; and the average of children under seven, exhibits the Pons,[γ] in proportion to the cerebellum, much smaller than in the average of adults, in whom it is only as 1 to 8, or 1 to 9.

12. In specific gravity, contrary to the current doctrine, the encephalos and its parts vary very little, if at all, from one age to another. A child of two, and a woman of a hundred years, are, in this respect, nearly equal, and the intermediate ages show hardly more than individual differences.

13. The specific gravity of the brain does not vary in madness, (if one case of chronic insanity is to be depended on), contrary to what has been alleged. In fever it often does, and remarkably.

14. The Cerebellum, (the converse of the received opinion), has a greater specific gravity than the Brain proper; and this difference is considerably more marked in birds than in man and quadrupeds. The opinion also of the ancients is probably true, that the Cerebellum is harder than the Brain proper.

a For the remainder of this section, see above, Appendix II. (a), p. 409, "Physiologists," &c., to p. 410, "motion."—ED.

β The effect is, in fact, to increase the cerebellum. See the experiments recorded by M. Leuret, cited by Sir Benjamin Brodie, *Psychological Inquiries*, Note H.—ED.

γ I.e., the *Pons Varolii*, a term used by some anatomists as synonymous with the *Tuber Annulare*; and so here; though others distinguish between the two.—ED.

15. The human brain does not, as asserted, possess a greater specific gravity than that of other animals.

(c.) REMARKS ON DR MORTON'S TABLES ON THE SIZE OF THE BRAIN.

(Communicated to the *Edinburgh New Philosophical Journal*, conducted by Professor JAMIESON. See Vol. XLVIII., p. 330 (1850). For Dr MORTON'S Tables, see the same Journal, Vol. XLVIII., p. 262.—ED.)

What first strikes me in Dr Morton's tables completely invalidates his conclusions,—he has not distinguished male from female crania. Now, as the female encephalos is, on an average, some four ounces troy less than the male, it is impossible to compare national skulls with national skulls, in respect of their capacity, unless we compare male with male, female with female heads, or, at least, know how many of either sex go to make up the national complement.

A blunder of this kind is made by Mr Sims, in his paper and valuable correlative table of the weight of 253 brains (*Medico-Chirurgical Transactions*, vol. xix.) He there attacks the result of my observation, (published by Dr Monro, *Anatomy of the Brain*, &c., 1831), *that the human encephalos, (brain proper and after-brain), reaches its full size by seven years of age*, perhaps somewhat earlier. In refutation of this paradox, he slumps the male and female brains together; and then, because he finds that the average weight of his adults, among whom the males are greatly the more numerous, is larger than the average weight of his impuberals, among whom the females preponderate, he jumps at once to the conclusion, that I am wrong, and that the encephalos continues to grow, to diminish, and to grow again (!), for,—I forget how long after the period of maturity. Fortunately, along with his crotchets, he has given the detail of his weighings; and his table, when properly arranged, confutes himself, and superfluously confirms me. That is, comparing the girls with the women, and the boys with the men, it appears, from his own induction, that the cranial contents do reach the average amount even before the age of seven.

Tiedemann, (*Das Hirn des Negers, &c.*, 1837, p. 4), notes the contradiction of Sims' result and mine ; but he does not solve it. The same is done, and not done, by Dr Bostock, in his *Physiology*. Tiedemann, however, remarks, that his own observations coincide with mine (p. 10); as is, indeed, evident from his Table, (p. 11), " Of the cranial capacity from birth to adolescence," though, unfortunately, in that table, but in that alone, he has not discriminated the sex.

Dr Morton's conclusion as to the comparative size of the Negro brain, is contrary to Tiedemann's larger, and to my smaller, induction, which concur in proving that the Negro encephalos is not less than the European, and greatly larger than the Hindoo, the Ceylonese, and sundry other Asiatic brains. But the vice, already noticed, of Dr Morton's induction, renders it, however extensive, of no cogency in the question.

Dr Morton's method of measuring the capacity of the cranium, is, certainly, no " invention " of his friend Mr Philips, being, in either form, only a clumsy and unsatisfactory modification of mine. Tiedemann's millet-seed affords, likewise, only an inaccurate approximation to the truth ; for seeds, as found by me, vary in weight according to the drought and moisture of the atmosphere, and are otherwise ill adapted to recover the size of the brain in the smaller animals. The physiologists who have latterly followed the method of filling the cranium, to ascertain the amount of the cranial contents, have adopted, not without perversion, one-half of my process and altogether omitted the other. After rejecting mustard-seed, which I first thought of employing, and for the reasons specified, I found that pure silicious sand was the best mean of accomplishing the purpose, from its suitable ponderosity, incompressibility, equality of weight in all weathers, and tenuity. Tiedemann, (p. 21), says, that he did not employ sand, " because, by its greater specific gravity, it might easily burst the cranial bones at the sutures." He would, by trial, have found that this objection is futile. The thinnest skull of the youngest infant can resist the pressure of sand, were it many times greater than it is ; even Morton's lead shot proved harmless in this respect. But, while nothing could answer the purpose better than sand, still this afforded only one, and that an inadequate, mean towards an end. Another was requisite. By

weighing the brain of a young and healthy convict, who was
hanged, and afterwards weighing the sand which his prepared
cranium contained, I determined the proportion of the specific
gravity of cerebral substance, (which in all ages and animals is
nearly equal), to the specific gravity of the sand which was employed. I thus obtained a formula by which to recover the original weight of the encephalos in all the crania which were filled ;
and hereby brought brains weighed and skulls gauged into a
universal relation.   On the contrary, the comparisons of Tiedemann and Morton, as they stand, are limited to their own Tables.
I have once and again tested the accuracy of this process, by
experiment, in the lower animals, and have thus perfect confidence in the certainty of its result, be the problem to recover the
weight of the encephalos from the cranium of a sparrow, or from
the cranium of an elephant.

I may conclude by saying, that I have now established, apart
from the proof by averages, *that the human encephalos does not
increase after the age of seven, at highest.*   This has been done, by
measuring the heads of the same young persons, from infancy to
adolescence and maturity ; for the slight increase in the size of
the head, after seven (or six) is exhausted by the development to
be allowed in the bones, muscles, integuments, and hair.

(The following is an unpublished Memorandum in reference
to preceding.—ED.)

March 23, 1850.

Found that the specific weight of the sand I had employed
for measuring the capacity of crania, was that the sand filling
32 cubic inches weighed 12,160 grains.

Found at the same time that the millet-seed occupying the
same number of cubic inches, weighed 5665 grains.

Thus the proportion of millet-seed to sand, in specific gravity,
is as 1 : 2.147.

One cubic inch thus contains 380 grains sand ; and 177 grains
millet-seed.

(d.) ORIGINAL RESEARCHES ON THE FRONTAL SINUSES, WITH OBSERVATIONS ON THEIR BEARINGS ON THE DOGMAS OF PHRENOLOGY.

(From *The Medical Times*, May 1845, Vol. XII. p. 159; June 7, 1845, Vol. XII. p. 177; August 1845, Vol. XII. p. 371.—ED.)

Before proceeding to state in detail the various facts and fictions relative to the Frontal Sinus,ª it will be proper to premise some necessary information touching the nature and relations of the sinuses themselves.

These *cruces phrenologorum* are two cavities, separated from each other by a perpendicular osseous partition, and formed between the tables of the frontal bone, in consequence of a divergence of these tables from their parallelism, as they descend to join the bones of the nose, and to build the orbits of the eye. They are not, however, mere inorganic vacuities, arising from the recession of the bony plates; they constitute a part of the olfactory apparatus; they are lined with a membrane, a continuation of the pituitary, and this, copiously supplied with blood, secretes a lubricating mucus which is discharged by an aperture into the nose.

Various theories have been proposed to explain the mode of their formation; but it is only the fact of their existence, frequency, and degree, with which we are at present interested. In the fœtus manifested only in rudiment, they are gradually, but in different subjects variously, developed, until the age of puberty; they appear to obtain their ultimate expansion towards

ª It is proper to observe, that the notes of which the following is an abstract, were written above sixteen years ago, and have not since been added to or even looked at. They were intended for part of a treatise to be entitled "*The Fictions of Phrenology and the Facts of Nature.*" My researches, however, particularly into the relations of the cerebellum, and the general growth of the brain, convinced me that the phrenological doctrine was wholly unworthy of a serious refutation; and should the detail of my observations on these points be ever published, it will not be done in a polemical form. My notes on the frontal sinuses having, however, been cast in relation to the phrenological hypothesis, I have not thought it necessary to take the labour of altering them,—especially as the phrenological fiction is, in truth, a complement of all possible errors on the subject of these cavities.

the age of twenty-five. They are exclusively occasioned by the elevation of the external table, which determines, in fact, the rise of the nose at the period of adolescence, by affording to the nasal bones their formation and support.

Sundry hypotheses have likewise been advanced to explain their uses, but it will be enough for us, from the universality of their appearance, to refute the singular fancy of the phrenologists, that these cavities are abnormal varieties, the product of old age or disease.

But though the sinuses are rarely if ever absent, their size in every dimension varies to infinity. Laying aside all rarer enormities, and speaking, of course, only of subjects healthy and in the prime of life, in superficial extent the sinus sometimes reaches hardly above the root of the nose, sometimes it covers nearly the whole forehead, penetrates to the bottom of the orbit, and, turning the external angle of the eyebrow, is terminated only at the junction of the frontal and parietal bones. Now, a sinus is small, or almost null upon one side,—on the other it is, perhaps, unusually large; while in no dimension are the two cavities, in general, strictly correspondent, even although the outer forehead present the most symmetrical appearance. In depth (or transverse distance between the tables) the sinus is equally inconstant, varying indeterminably in different heads, from a line or less to half an inch and more. Now, a sinus gradually disappears by a gradual convergence of its walls; now, these walls, after running nearly parallel, suddenly unite. Now, the depth of the cavity decreases from centre to circumference; now, the plates approximate in the middle and recede farther from each other, immediately before they ultimately unite. In one cranium, a sinus, collected within itself, is fairly rounded off; in another, it runs into meandering bays, or is subdivided into separate chambers, these varying without end in their relative capacity and extent. In depth, as well as in extent, the capacity of the sinus is thus wholly indeterminable; and no one can predict, from external observation, whether the cavity shall be a lodging scanty for a fly or roomy for a mouse.

It is an error of the grossest, that the extent of the sinus is indicated by a ridge, or crest, or blister, in the external bony plate. Such a protuberance has no certain or even probable relation to the extent, depth, or even existence, of any vacuity beneath. Over the largest cavities there is frequently no bony elevation; and women, in whose crania these protuberances are in general absent or very small, exhibit the sinuses as universally existent, and not, perhaps, proportionably less extensive than those of men. The external ridge, however prominent, is often merely a sudden outward thickening of the bony wall, which sometimes has a small, sometimes no cavity at all, beneath. Apart also from the vacuity, though over the region of the sinus, no quarter of the cranium presents greater differences in thickness, whether in different subjects or in the same head, than the plates and diploe of the frontal bone; and I have found that the bony walls themselves presented an impediment which varied inappreciably from three to thirteen lines:—"*fronti nulla fides.*"

But the "*fronti nulla fides,*" in a phrenological relation, is further illustrated by the accidents of its sinus, which all concur in manifesting the universality and possibly capacious size of that cavity. That cavity is sometimes occupied by stony concretions, and is the seat of ulcers, cancer, polypus, and sarcoma. When acutely inflamed the sensibility of its membrane becomes painfully intense; and every one has experienced its irritation when simply affected with catarrh. The mucosity of this membrane, the great extent and security of the caverns, joined with their patent openings into the nose, render the sinuses a convenient harbour for the nidulation, hatching, and nourishment of many parasitic animals; indeed, the motley multitude of its guests might almost tempt us to regard it as

> ———"The cistern for all creeping things
> To knot and gender in."[a]

"Chacun a son Vercoquin dans la teste"—"Quemque suus vellicat Vermis"—are adages which, from the vulgarity of the literal occurrence, would seem more than metaphorically true.[b] With a frequency sometimes epidemic,[y] flies and insects here

[a] "Or keep it as a cistern for *foul toads* To knot and gender in." *Othello,* act. iv. sc. 2.—ED.

[b] In the frontal sinuses worms and insects are *not unfrequently* found.—

Voigtel, *Handb. d. Pathol. Anat.,* 1804, vol. i. p. 292. I quote him, *instar omnium,* as one of the best and one of the most recent authorities.

[y] Forestus, *Obs. Med.,* lib. xxi. schol. 28.

ascend to spawn their eggs, and maggots (other than phrenological) are bred and fostered in these genial labyrinths. Worms,
in every loathsome diversity of slime and hair,—reptiles armed
with fangs,—crawlers of a hundred feet,—ejected by the score,
and varying from an inch to half an ell in length, cause by their
suction, burrowing, and erosion, excruciating headache, convul-
sions, delirium, and phrensy. With many a nameless or nonde-
script visitor, the leech, the lumbricus, the ascaris, the ascaris
lumbricoides, the fasciola, the eruca, the oniscus, the gordius, the
forficula, the scolopendra, the scorpiodes, and even the scorpion,ᵃ
are by a hundred observers recorded as finding in these "antres
vast,"—these "spelunci ferarum,"—a birthplace or an asylum.ᵝ
And the fact, sufficiently striking in itself, is not without signi-

ᵃ Hollerius, *De Morb. Int.*, lib. i. c.
1; Gesner, *Hist. Anat.*, lib. v.; Bo-
neti, *Sepul. Obs.*, 121; Ferretti.—I here
refer to the scorpion alone.

ᵝ Long before the sinns was anato-
mically described by Carpi, this patho-
logical fact had been well known to
physicians. The prescription of the
Delphic oracle to Demosthenes of
Athens for his epilepsy, shows that the
Greeks were aware of the existence of
worms in the frontal sinuses of the
goat. (Alex. Trallian, lib. i. c. 15.)
Among the Arabians, Avicenna (Fen-
estella, lib. iii. tr. 2, c. 3) tells us it
was well known to the Indian physicians, that worms, generated in the
forehead, immediately above the root
of the nose, were frequently the cause
of headaches; and Rhazes (Continet,
lib. i. c. 10) observes that this was the
opinion of Schare and others. Among
the moderns, my medical ignorance
suggests more authorities than I can
almost summon patience simply to
name. The curious reader may con-
sult, among others, Valescus de Tar-
anta, Nicolaus de Nicolis, Vega, Mar-
cellus Donatus, Trincavelli, Benedetti,
Hollerius, Duretus, Fabricius Hildan-
us, Zacuta Lusitanus, Hercules de Sax-
onia, Petrus Paulus Magnus, Angellin-
us, Alsarius, Cornelius Gemma, Gesner,
Benevenius, Fernelius, Riolanus, Forestus, Bartholinus, Ferretti, Rolfinck,
Olaus Wormius (who himself ejected a
worm from the nose—was it a family
affection?), Smetius (who also relates
his own case), Tulpius, Heurnius,
Rousswus, Monardia, Schenk, Senertus,
Montuus, Borelli, Bonetus, Hertodius,
Kerkringius, Joubert, Volkammer,
Wohlfarth, Nannoni, Stalpert, Vander
Wiel, Morgagni, Clericus, De Blegny,
Salzmann, Honold, Hill, Kilgour, Lit-
tré, Malœt, Sandifort, Nil Rosen, Raz-
oux, Schaarschmidt, Quelmatz, Wolf,
Blumenbach, Ploucquet, Baur, Ried-
lin, Zacharides, Lange, Boettcher,
Welge, Wrisberg, Trois, Voigtel, Ru-
dolphi, Bremser, &c. &c.; and of jour-
nals—*Ephem. Misc.; Acta et Nova
Acta Curios. Nat.; Commerc. Liter.*,
Nov. 2; *Breslauer Sammlung; Dun-
can's Med. Journ.; Edinb. Med. Es-
says; London Chronicle; Philadelphia
Transactions; Blumenbach's Med.
Bibl.*, &c., &c.

I may here mention that the nidulation of the œstrus ovinus (which occasionally infests the human sinus) forms
a frequent epidemic among sheep and
goats. The horse, the dog (and probably most other animals) are similarly
afflicted.

ficance in relation to the present inquiry, that these intruders principally infest the sinuses of women, and more especially before the period of full puberty.

Such is the great and inappreciable variation of the frontal sinus and its walls, that we may well laugh at every attempt to estimate, in that quarter, the development of any part of the subjacent hemispheres, were that part larger than the largest even of the pretended phrenological organs. But this is nothing. Behind these spacious caverns, in utter ignorance of the extent, frequency, and even existence of this impediment, the phrenologists have placed, not one large, but seventeen of their very smallest organs ; and have thus enabled an almost insurmountable obstacle to operate in disproof of their system in its highest intensity.

By concentrating all their organs of the smallest size within the limits of the sinus, they have, in the first place, carried all those organs whose range of development was least, behind the obstacle whose range of development was greatest. Where the cranium is thinner and comparatively more equal in thickness, they have placed all the organs, (those of the propensities and sentiments), which present the broadest surface, and, as they themselves assure us, varying in their development from the centre to circumference by an inch and upwards ; while all the organs, (those of the intellect), which have the narrowest expansion, and whose varying range of development from the centre is stated to be only a quarter of an inch, (less even than the fourth of the variation of the others),ᵃ—these have been accumulated behind an impediment whose ordinary differences are far more than sufficient to explain every gradation of the pretended development of the pretended organs from their smallest to their largest size.

In the second place, they have thus at once thrown one half of their whole organology beyond the verge of possible discovery and possible proof.

In the third place, by thus evincing that their observations on that one half had been only illusive fancies, they have afforded a

---

a Combe's *System*, &c., p. 31. "The difference in development between a large and a small organ of the propensities and some of the sentiments, amounts to an inch and upwards ; and to a quarter of an inch in the organs of intellect, which are naturally smaller than the others."

criterion of the credit to be fairly accorded to their observations
in relation to the other; they have shown in this, as in other
parts of their doctrine, that *manifestation* and *development* are
quantities which, be they what they may, can on their doctrine
always be brought to an equation.

Nay, in the fourth place, as if determined to transcend themselves—to find "a lower deep beneath the lowest deep," they
have even placed the least of their least organs at the very point
where this, the greatest obstacle, was in its highest potency, by
placing the organs of configuration, size, weight, and resistance,
&c., towards the internal angle of the eyebrow, the situation
where the sinus is almost uniformly deepest.[a]

Nor, in the fifth place, were they less unfortunate in the loca-
tion of the rest of their minutest organs. These they arranged
in a series along the upper edge of the orbit, where, independently even of the sinus, the bone varies more in thickness, from
one individual and from one nation to another, than in any other
part of the skull; and where these organs, hardly larger, are
packed together more closely than peas in a pod. These pretended organs, if they even severally protruded from the brain,
as they never do—if no sinus intervened—and if, instead of
lying under the thickest, they were situate under the thinnest
bone of the cranium; these petty organs could not, even in
these circumstances, reveal their development by determining
any elevation, far less any sudden elevation, of the incumbent
bone. That bone they could only attenuate at the point of
contact, by causing an indentation on its inner surface. This is
shown by what are called the glands of Pacchioni, though erroneously. These bodies, which are often found as large as, or larger
than, the organs in question, and which arise on the coronal surface of the encephalos, attenuate to the thinnest, but never ele-
vate in the slightest, the external bony plate, though there the
action of the muscles presents a smaller impediment to a partial
elevation than in the superciliary region. This I have frequently taken note of.

a Every one who has ever examined the sinus knows that what Schulze has observed is true: "In illo angulo qui ad nares est, cavitatis fundus est, et hoc in loco fere ossium laminæ *a se invicem maxime distant*."—(*De Cav. Cranii; Acta Phys. Med. Acad. Cæs.*, i. p. 508.)

As it is, these minute organs are expected to betray their distinct and relative developments through the obstacle of two thick bony walls, and a large intervening chamber; the varying difference of the impediment being often considerably greater than the whole diameter even of the organs themselves. The fact, however, is, that these organs are commonly, if not always, developed only in the bone, and may be cut out of the cranium, even in an impuberal skull destitute of the sinus, without trenching on the confines of the brain itself. At the external angle of the eyebrow at the organ of slumber, the bone, exclusive of any sinus, is sometimes found to exceed an inch in thickness.

How then have the phrenologists attempted to obviate the objection of the sinus?

The first organs which Gall excogitated, he placed in the region of the sinus; and it is manifest he was then in happy unacquaintance with everything connected with that obnoxious cavity. In ignorance, however, Gall was totally eclipsed by Spurzheim; who, while he seems even for a time unaware of its existence as a normal occurrence, has multiplied the number and diminished the size of the organs which the sinus regularly covers. By both the founders, their organology was published before they had discovered the formidable nature of the impediment, and then it was too late to retract. They have attempted, indeed, to elude the objection; but the manner in which they have floundered on from blunder to bunder,—blunders not more inconsistent with each other, than contrary to the fact,—shows that they have never dared to open their eyes on the reality, or never dared to acknowledge their conviction of its effect. The series of fictions in relation to the frontal sinus, is, out of Phrenology, in truth, unparalleled in the history of science. These fictions are substituted for facts the simplest and most palpable in nature; they are substituted for facts contradicted by none, and proclaimed by every anatomical authority; and they are substituted for facts which, as determining the competency of phrenological proof, ought not to have been rejected without a critical refutation by the founders of that theory themselves. But while it seemed possible for the phrenologists to find only truth, they have yet continued to find nothing but error—error always at

the greatest possible distance from the truth. But if they were
thus so curiously wrong in matters so easy, notorious, and fundamental, how far may we not presume them to have gone astray
where they were not, as it were, preserved from wandering ?

The fictions by which phrenologists would obviate the objection of the frontal sinus, may, with the opposing facts, be
divided into four classes ;—as they relate, 1°, to its *nature* and
effect ; 2°, to its *indication* ; 3°, to its *frequency* ; and 4°, to its
size.

I.—NATURE AND EFFECT OF THE SINUS.

Fact.—The frontal sinus only exists in consequence of the recession of the two cranial tables from their parallelism ; and as
this recession is inappreciable, consequently, no indication is
afforded by the external plate of the eminence or depression of
the brain, in contact with the internal.

. To this fact, Gall opposed the following

Fiction.—The frontal sinus interposes no impediment to the
observation of cerebral development ; for as the walls of this
cavity are exactly parallel, the effect of the brain upon the inner
table must consequently be expressed by the outer.

Authorities for the Fiction.—This fiction was originally advanced by Gall, in his Lectures, and, though never formally re-
tracted, has not been repeated by him or Spurzheim in their
works subsequently published. I therefore adduce it, not as an
opinion now actually held by the phrenologists, but as a part
only of that cycle of vacillation and absurdity which, in their
attempts to elude the objection of the sinus, they have fruit-
lessly accomplished. That it was so originally advanced, is
shown by the following authorities ; which, as beyond the
reach of readers in general, I shall not merely refer to, but
translate.

The first is Froriep ; and I quote from the 3d edition of his
Darstellung, &c., which appeared in 1802. This author was a
pupil and friend of Gall, on whose doctrine he delivered lectures,
and his work is referred to by Gall, in his *Apologetic Memorial*
to the Austrian Government, in that very year, as containing an
authentic exposition of his opinions.—" Although at this place

the frontal sinuses are found, and here constitute the vaulting of the forehead, nevertheless, Gall maintains that the brain, in consequence of the walls of the sinuses lying quite parallel (? !), is able to affect likewise the outer plate, and to determine its protuberance."—P. 61. The doubt and wonder are by the disciple himself.

The second authority is Bartel's, whose *Anthropologische Bemerkungen* appeared in 1806. "In regard to the important objection drawn from the frontal sinuses, Gall's oral reply is very conformable to nature. 'Here, notwithstanding the intervening cavity in the bones, there is found a parallelism between the external and internal plates of the cranium.'"—P. 125.

Proof of the Fact.—In refutation of a fiction so ridiculous, it is unnecessary to say a single word; even the phrenologists now define the sinus by " a divergence from parallelism between the two tables of the bone."[a]

It was only in abandoning this one fiction, and from the conviction that the sinus, when it existed, did present an insuperable obstacle to observation, that the phrenologists were obliged to resort to a plurality of fictions of far inferior efficacy; for what mattered it to them, whether these cavities were indiscoverable, frequent, and capacious, if, in effect, they interpose no obstacle to an observation of the brain?

II.—INDICATION OF THE SINUS.

Fact.—There is no correlation between the extent and existence of a sinus, and the existence and extent of any elevation, whether superciliary or glabellar; either may be present without the other, and when both are coexistent they hold no reciprocal proportion in dimension or figure. Neither is there any form whatever of cranial development which guarantees either the absence or the presence of a subjacent cavity.

To this fact the phrenologists are unanimous in opposing the following

Fiction.—The sinus, when present, betrays its existence and extent by an irregular elevation of a peculiar character, under the appearance of a bony ridge, or crest, or blister, and is dis-

a Combe, *System*, p. 32.

tinguished from the regular forms under which the phrenological organs are developed.

Authorities for the Fiction.—It is sufficient to adduce Gall[a] and Spurzheim,[β] followed by Combe,[γ] and the phrenologists in general. In support of their position, they adduce no testimony by anatomists,—no evidence from nature.

Proof of the fact.—All anatomical authority, as will be seen in the sequel, is opposed to the fiction, for every anatomist concurs in holding that the sinuses are rarely, if ever, absent; whereas the crests or blisters which the phrenologists regard as an index of these cavities, are of comparatively rare occurrence. It must be admitted, however, that some anatomists have rashly connected the extent of the internal sinus with the extent of the external elevation. The statement of the *fact* is the result of my own observation of above three hundred crania; and any person who would in like manner interrogate nature, will find that the largest sinuses are frequently in those foreheads which present no superciliary or glabellar elevations. I may notice, that of the fifty skulls whose phrenological development was marked under the direction of Spurzheim, and of which a table is appended, the one only head where the frontal sinuses are noted, from the ridge, as present, is the male cranium No. 19; and that cranium, it will be seen, has sinuses considerably beneath even the average extent.

III.—FREQUENCY OF THE SINUS.

Fact.—The sinuses are rarely, if ever, wanting in any healthy adult head of either sex.

To this fact, the phrenologists oppose the three following inconsistent fictions:—

Fiction I.—The sinuses are only to be found in some male heads, being frequently absent in men until a pretty advanced age.

Fiction II.—In women the sinuses are rarely found.

Fiction III.—The presence of the sinus is abnormal; young

a *Anat. et Phys.*, t. iv. p. 43 *et seq.;*   *Object.*, p. 79; *Phren.*, p. 115.
and, in the same terms, *Sur les Fonct.*   γ *Syst.*, pp. 21, 35, 308.
β *Phys. Syst.*, p. 236; *Exam. of*

and adult persons have no cavities between the tables of the frontal bone,—the real frontal sinuses occurring only in old persons, or after chronic insanity.

Authorities for fiction I.—This fiction is held in terms by Gall.[a] The other phrenologists, as we shall see, are much further in the wrong. But even for this fiction they have adduced no testimony of other observers, and detailed no observations of their own.

Proof of the fact, in opposition to this fiction.—All anatomists —there is not a single exception—concur in maintaining a doctrine diametrically opposed to the figment of the phrenologists that the sinuses are, even in men, frequently or generally absent. Some, however, assert that the sinus in a state of health is *never* wanting; while others insist that, though *very rarely*, cases do occur in which it is actually deficient.

Of the latter opinion, Fallopius[β] holds that they are present "in all adults," except occasionally in the case of simous foreheads, an exception which Riolanus[γ] and others have shown to be false. Schulze,[δ] Winslow,[ε] Buddeus,[ζ] "that they are *sometimes* absolutely wanting in cases where the cranium is *spongy* and *honeycombed.*" Palfyn,[η] "that they are sometimes, though *rarely*, absent. Wittich,[θ] "that they are *almost always* present, though it may be admitted, that in *some very rare cases* they are wanting;" and Stalpart Van der Weil[ι] relates, that "he had seen in Nuck's Museum, preserved as a *special rarity*, a cranium without a frontal sinus." Of more recent authorities, Hippolyte Cloquet[κ] observes, "that they are *seldom wanting;*" and the present Dr Monro[λ] found, in forty-five skulls, that while three only were without the sinus, in two of them, (as observed by Schulze, Winslow, and Buddeus), the cavity had merely been filled up by the deposition of a spongy bone.

Of the former opinion, which holds that the sinus is always present, I need only quote *instar omnium*, the authority of

a As quoted above.
β Opera.
γ Comm. de Oss., p. 468.
δ De Sin. Oss. Cap.; Acta. Phys. Med. Leop. Cæs., vol. i. obs. 288.
ε Expos. Anat., Tr. des Oss. Secs., sec. 30.
ζ Obs. Anat. Sel., obs. 1.
η Ost., p. 105.
θ De Olfactu, p. 17.
ι Obs. Rar., Cent. Post., pars prior, obs. 4.
κ Anat. Descr., sect. 153, ed. 1824.
λ Elem. of Anat., i. p. 134.

Blumenbach,[a] whose illustrious reputation is in a peculiar manner associated with the anatomy of the human cranium, and who even celebrated his professional inauguration by a dissertation, in some respects the most elaborate we possess, on the Frontal Sinuses themselves. This anatomist cannot be persuaded, even on the observation of Highmore, Albinus, Haller, and the first Monro, that normal cases ever occur of so improbable a defect; " for," he says, " independently of the diseases afterwards to be considered, I can with difficulty admit, that healthy individuals are ever wholly destitute of the frontal sinus ; on the contrary, I am convinced that these distinguished men have not applied the greatest diligence and research."   In this opinion, as observed by the present Dr Monro,[β] Blumenbach is supported by the concurrence of Bertin, Portal, Sömmering, Caldani, &c.   Nor does the fiction obtain any countenance from the authors whom Blumenbach opposes. I have consulted them, and find that they are all of that class of anatomists who regard the absence of the sinus, though a possible, as a rare and memorable phenomenon. Highmore[γ] founds his assertion on the single case of a female. Albinus,[δ] on his own observation, and on that of other anatomists, declares that " the sinuses are *very rarely* absent." The first Monro,[ε] speaking of their infinite variety in size and figure, notices as a remarkable occurrence that he had " *even seen cases* in which they were absolutely wanting." And Haller[ζ] is only able to establish the exception on the case of a solitary cranium.

, My own experience is soon stated. Having examined above three hundred crania for the purpose of determining this point, I have been unable to find a single skull wholly destitute of a sinus. In crania, which were said to be examples of their absence, I found that the sinus still existed. In some, indeed, I found it only on one side, and in many not ascending to the point of the glabellar region, through which crania are usually cut round. The only instances of its total deficiency, are, I believe, those abnormal cases in which, as observed by anatomists, the

a *De Sin. Front.*, p. 5.   Oss.
β *Elem.*, vol. i. p. 133.   ε *Osteol. par Sue*, p. 54.
γ *Disq. Anat.*, lib. iii. c. 4.   ζ *Elem. Phys.*, v. p. 138.
δ *Annot. Acad.*, lib. i. c. 11, et Tab.

·original cavity has been subsequently occupied by a pumicose deposit. Of this deposit the only examples I met with occurred in males.

Authorities for fiction II.—This fiction also is in terms maintained by Gall.[a] Neither he nor any other phrenologist has adduced any proof of this paradox, nor is there, I believe, to be found a single authority for its support; while its refutation is involved in the refutation already given to fiction I. Nannoni,[β] indeed, says—"the opinion of Fallopius that the frontal sinuses are often wanting in women, is refuted by observation;" but Fallopius says nothing of the sort. It is also a curious circumstance, that the great majority of cases in which worms, &c., have been found in the sinus, have occurred in females. This is noticed by Salzmann and Honold.[γ]

My own observations, extending, as I have remarked, to above three hundred crania, confirm the doctrine of all anatomists, that in either sex the absence of this cavity is a rare and abnormal phenomenon, if not an erroneous assertion. I may notice, by the way, the opinion of some anatomists,[δ] that the sinuses are smaller in women than in men, seems to be the result of too hasty an induction; and I am inclined to think, from all I have observed, that proportionally to the less size of the female cranium, they will be found equally extensive with the male.

Authorities for fiction III.—This fiction was maintained by Spurzheim while in this country, from one of whose publications[ε] it is extracted. It is, perhaps, one of the highest flights of phrenological fancy. Nor has it failed of exciting emulation in the sect. "While a man," says Sir George Mackenzie,[ζ] "is in the prime of life, and healthy, and manifests the faculties of the frontal organs, such a cavity *very seldom* exists." (!) * * * * * "We have examined a GREAT MANY skulls, and *we have not yet seen* ONE having the sinus that could be proved to have belonged to a person in the vigour of life and mind." (!!) Did Sir George ever see any skull which belonged to any "person in the vigour

a As above.
β *Trattato di Anatomia*, 1788, p. 55.
γ *De Verme Naribus Excuso*, (Haller, *Disp. Med. Pract.*, i. n. 25.)
δ *Instar omnium*, v. Sömmering, *De*

Fabr. Corp. Humani, i. sec. 62.
ε *Answer to Objections against the Doctrines of Gall*, &c., p. 79.
ζ *Illustrations*, p. 228.

of life and mind" without a sinus ?   Did he ever see any adult
skull of any person whatever in which such a cavity was not
to be found ?

Proof of the fact, in opposition to this fiction.—This fiction
deserves no special answer.   It is already more than sufficiently
refuted under the first.

It is true, indeed, the doctrine that the frontal sinuses wax
large in old age is stated in many anatomical works.   I find it
as far back as those of Vidus Vidius and Fallopius, but I find no
ground for such a statement in nature.   This I assert on a comparative examination of some thirty aged skulls.   In fact, about
the smallest frontal sinus that I ever saw, was in the head of a
woman who was accidentally killed in her hundred and first
year.   (See also the appended Table.)   I take this indeed for
one of the instances in which anatomical authors have blindly
copied each other ; so that what originates in a blunder or a rash
induction ends in having, to appearance, almost catholic authority in its favour.   A curious instance of this sequacity occurs to
me.   The common fowl has an encephalos, in proportion to
its body, about as one to five hundred ; that is, it has a brain
less, by relation to its body, than almost any other bird or beast.
Pozzi (Puteus), in a small table which he published, gave the
proportion of the encephalos of the cock to its body, by a blunder at about half its amount ; that is, as one to two hundred and
fifty.   Haller, copying Pozzi's observation, dropped the cipher,
and records in his table, the brain of the common fowl as bearing a proportion to the body of one to twenty-five.   This double
error was shortly copied by Cuvier, Tiedemann, and, as I have
myself noticed, by some twenty other physiologists ; so that, at
the present moment, to dispute the fact of the common fowl
having a brain more than double the size of the human, in proportion to its body, would be to maintain a paradox counter to
the whole stream of scientific authority.   The doctrine of the
larger the sinus the older the skull, stands, I believe, on no better footing.   Indeed, the general opinion that the brain contracts
in the decline of life, is, to say the least of it, very doubtful, as I
may take another opportunity of showing.

As to the effect of chronic insanity in amplifying the sinuses,
I am a sceptic ; for I have seen no such effect in the crania of

madmen which I have inspected. At all events, admitting the
phrenological fancy, it could have no influence on the question,
for the statistics of insanity show, that there could not be above
one cranium in four hundred where madness could have exerted
any effect.

IV.—Extent of the Sinus.

Fact.—While the sinus is always regularly present, it, however, varies appreciably in its extent. For whilst, on the average,
it affects six or seven organs, it is, however, impossible to determine whether it be confined to one or extended to some seventeen
of these.

This fact is counter to three phrenological fictions :

Fiction I.—The frontal sinus is a small cavity.

Fiction II.—The frontal sinus, when present, affects only the
organ of Locality.

Fiction III.—When the sinus does exist, it only extends an
obstacle over two organs, (Size and Lower Individuality), or,
at most, partially affects a third, (Locality).

Authorities for fiction I.—Mr Combe[a] maintains this fiction,
that the frontal sinus " is a small cavity."

Authorities for fiction II.—Gall[β] contemplates and speaks of
the sinus as only affecting Locality ; and the same may be said of
Spurzheim, in his earlier English works.[γ]

Authorities for fiction III.—This fiction is that into which
Spurzheim modified his previous paradoxes, when, in 1825, he
published his "Phrenology."[δ] Mr Combe allows that the sinus,
in ordinary cases, extends over Locality, as well as over Size and
Lower Individuality.

All these fictions are, however, sufficiently disproved at once
by the following

Proof of the fact.—The phrenologists term the sinus, (when
they allow it being), "*a small cavity.*" Compare this with the
description given by impartial anatomists of these caverns.
Vidus Vidius[ε] characterises them by "spatium *non parvum ;*"

a *System*, p. 32.   *Obj.*, p. 79.
β As quoted above.   δ P. 115.
γ *Phys. Syst.*, p. 236, and *Exam. of*   ε *Anat.*, lib. ii. c. 2

Bauhinus[a] styles them "cavitates *insignes ;*" Spigelius[β], "cavernae satis *amplæ ;*" Laurentius,[γ] "sinus *amplissimi ;*" Bartholinus,[δ] "cavitates *amplissimæ ;*" Petit,[ε] "*grands* cavités irregulières ;" Sabatier,[ζ] "cavités *larges* et *profondes ;*" Sömmering,[η] "cava *ampla ;*" Monro, *primus,*[θ] "*great* cavities ;" and his grandson,[ι] "*large* cavities."

The phrenologists further assert, that in ordinary cases the frontal sinus covers only two petty organs and a half ; that is, extends only a few lines beyond the root of the nose. But what teach the anatomists ? "The frontal sinuses," says Portal,[κ] "are much more extensive than is generally believed." "*In general,*" says Professor Walther,[λ] "the sinuses ascend in height nearly to the *middle of the frontal bone.*" Patissier[μ] observes, that "their extent varies to infinity, is sometimes stretched upwards to the frontal protuberances, and to the sides, as far as the external orbitar apophyses, as is seen in many crania in the cabinet of the Paris Faculty of Medicine." Bichat[ν] delivers the same doctrine nearly in the same words ; which, contradicted by none, is maintained by Albinus,[ξ] Haller,[ο] Buddeus,[π] Monro *primus,*[ρ] and *tertius,*[σ] Blumenbach,[τ] Sömmering,[υ] Fife,[φ] Cloquet,[χ] Velpeau,[ψ]—and, in a word, by every osteologist ; for all represent these cavities as endless in their varieties, and extending not unfrequently to the outer angles of the eyebrow, and even to the parietal bones. To finish by a quotation from one of the last and best observers :—"In relation," says Voigtel,[ω] "to their abnormal greatness or smallness, the differences, in this respect, whether in one subject as compared with another, or in one sinus in relation to the opposite of the same skull, are of so frequent occurrence that they vary almost in every cranium. They are

a Anat., lib. iii. c. 5.
β De Fabr., lib. ii. c. 5.
γ Hist. Anat., lib. ii. c. 9.
δ Anat., lib. iv. c. 6.
ε Pulfyn An., ch. i. p. 52.
ζ Anat.
η De Fab., l. sect. 35.
θ Osteol. par Sue, p. 54.
ι Elements.
κ Anat. Med., i. pp. 102, 238.
λ Abh. v. trokn. Kn., p. 133.
μ Dict. des Sc. Med., t. li. p. 372.

ν Anat. Descr., c. i. p. 102.
ξ Annot. Acad., lib. i. c. ii. (!)
ο Elem., v. p. 138.
π Obs. Anat., sec. 8.
ρ Osteol. par Sue, p. 54.
σ Elements.
τ Anat.
υ Anat. Descr., t. i. sec. 153, edit. 3.
φ Traité d'Anat. Chir.
χ De Sin. Fr., p. 3.
ψ De Fab., c. ii. sec. 94.
ω Path. Anat., i. p. 289.

found so small, that their depth, measured from before backwards, is hardly more than a line; in others, on the contrary, a space of from four, five, to six lines, (*i.e.* half an inch), is found between the anterior and posterior wall. Still more remarkable are the variations of these cavities, in relation to their height, as they frequently rise from the trifling height of four lines to an inch at the glabella." M. Velpeau, speaking of this great and indeterminable extent of the sinus, adds : " this disposition must prevent us from being able to judge of the volume of the anterior parts of the brain by the exterior of the cranium ;—an observation sufficiently obvious in relation to Phrenology, and previously made by the present Dr Monro.[a]

On the sinus and its extent, two anatomists only, as far as I am aware, have given an articulate account of their inductions —Schulze, and the present Dr Monro.

The former,[β] who wrote a distinct treatise *On the Cavities or Sinuses of the Cranial Bones*, examined only ten skulls, and does not detail the dimension of each several sinus. After describing these cavities, which he says, " plerisque hominibus formantur," he adds, that " when of a middling size they hardly extend towards the temples beyond the centre of the eye, where the orbital vault is highest; and if you measure their height from the insertion of the nasal bones, you will find it equal to an inch. Such is the condition of this cavity when moderate. That there are sinuses far greater, was taught me by another inspection of a cranium. In this case, the vacuity on the right did not pass the middle of the orbit, but that on the left stretched so far that it only ended over the external angle of the eyebrow, forming a cavity of at least two inches in breadth. Its depth was such as easily to admit the least joint of the middle finger. Its height, measured from the root of the nose on the left side, exceeded two inches, on the right it was a little less; the left sinus was, however, shallower than the right. On the left side I have said the cavity terminated over the external angle of the orbit. From this place a bony wall ran towards the middle of the *crista Galli*, and thus separated the sinus into a posterior and an anterior cavity. The posterior extended so far towards the temples, that it reached the place

a Elem., p. 133.   *β Loc. cit.*

where the frontal and sincipetal bones and the processes of the sphenoidal meet. It covered the whole arch of the orbit, so that all was here seen hollow," &c.

After describing sundry appearances which the sinuses exhibited in another skull, he observes : "It was my fortune to see and to obtain possession of *one* cranium in which of neither of the frontal nor the sphenoidal cavities was there any vestige whatsoever. In this specimen the bones in which these vacuities are situated were thicker than usual, and more cavernous;" an observation, as we have seen, made by other anatomists. However subversive of the phrenological statement, it will soon be seen that Schulze has understated the usual extent of the impediment.

Dr Monro,[a] after mentioning that there "were forty-five crania of adults in the Anatomical Museum, cut with a view to exhibit the different sizes and forms of the frontal sinuses," says :—" I measured the breadth or distance across the forehead ; the height or distance upwards from the transverse suture, where it divides the frontal bones and bones of the nose; and also the depth of the frontal sinuses ; in nine different skulls in which these sinuses were large." Omitting the table, it is sufficient to say, that in these crania the average is as follows :— *Breadth*, within a trifle of *three inches; height, one inch and five-tenths; depth,* above *one inch.* Here the depth seems not merely the distance between the external and internal tables, but the horizontal distance from the glabella to the posterior wall of the sinus. These nine crania thus yield an average, little larger than an indifferent induction; and though the sinuses are stated to have been large, the skulls appear to have been selected by Dr Monro, not so much in consequence of that circumstance, as because they were so cut as to afford the means of measuring the cavity in its three dimensions.

By the kindness of Dr Monro and Mr Mackenzie, I was permitted to examine all the crania in the public anatomical museum and in the private collection of the Professor ; many were, for the first time, laid open for my inspection. I was thus enabled to institute an impartial induction. A random measurement of above thirty perfect crania (laying aside three skulls of old

a *Elements*, i. p. 134.

persons, in which the cavity of the sinus was almost entirely
occupied by a pumicose deposit) gave the following average
result: breadth, two inches four-tenths; height, one inch
and nearly five-tenths; depth (taken like Dr Monro), rather
more than eight-tenths of an inch. What in this induction
was probably accidental, the sinuses of the female crania exhibited an average, in all the three dimensions, almost absolutely
equal to that of the male. The relative size was consequently
greater.

Before the sinuses of the fifty crania of Dr Spurzheim's collection, (of which I am immediately to speak), were, with the
sanction of Professor Jameson, laid open upon one side, I had
measured their three dimensions by the probe. This certainly
could not ascertain their full extent, as, among other impediments, the probe is arrested by the septa, which so frequently
subdivide each sinus into lesser chambers; but the labour was
not to be undergone a second time, especially as the proportional extent of these cavities is by relation to the phrenological
organs articulately exhibited in the table. As it was, the
average obtained by the probe is as follows:—In the thirty-six
male crania (one could not be measured by the probe), the
breadth was two inches and nearly four-tenths; the height, one
inch and nearly three-tenths; the depth, rather more than one
inch. In the twelve female crania (here, also, one could not be
measured by the probe), the breadth was one inch, and rather
more than nine-tenths; the height, nearly one inch; the depth,
within a trifle of nine-tenths.

I should notice that in all these measurements, the thickness
of the external plate is included in the depth.

So true is the observation of Portal, that the *"frontal sinuses
are much more extensive than is generally believed."*

The collection of fifty crania, of which the average size of the
frontal sinuses has been given above, and of which a detailed
table of the impediment interposed by these cavities to phrenological observation now follows, was sent by M. Royer, of the
Jardin des Plantes, (probably by mistake), to the Royal Museum
of Natural History in Edinburgh; the skulls, taken from the
catacombs of Paris, having, under Dr Spurzheim's inspection,
been selected to illustrate the development of the various phreno-

logical organs, which development is diligently marked on the several crania.

Thus, though I have it in my power to afford a greatly more extensive table, the table of these fifty crania is, for the present purpose, sufficient. For—

1°, They constitute a complete and definite collection ;

2°, A collection authoritative in all points against the phrenologists ;

3°, One to which it can be objected by none, that it affords only a selected or partial induction in a question touching the frontal sinus ;

4°, It is a collection patent to the examination of the whole world ;

5°, In all the skulls a sinus has on one side been laid open to its full extent; the capacity of both is thus easily ascertained; and, at the same time, with the size of the cavity, the thickness and salience of the external frontal table remains apparent.

Table exhibiting the variable extent and unappreciable impediment, in a phrenological relation, of the Frontal Sinuses ; in a collection of fifty crania, selected, and their development marked, under the direction of Dr Spurzheim :—

Number of Skull, as here arranged, according to sex and age.	Number of Skull, according to spurzheim's former order.	Sex, as marked by apparatus.	Age, as inferred from these facts and other criteria.	Extent of the fitness, as entirely or nearly meeting (†), or as more or less affecting (*), the pretended phrenological organs, according to the late and latest enumeration. (1)																		
				20	21	22	2/19	24	1/19	23	25	L.29	26	W.19	27	30	33	28	22	7
				xxiii	xxiv	xxv	xxii	xxvii	xxx	xxxi	xxix	L.xxxii	xxxi	W.xxxiii	xxviii	xxxiv	xxxv	xxxii	xx	ix
1	viii	Male	Young	†	†	†	†	*												
2	xii			*	†	†	†	†												
3	xiii			*	†	†	†	†												
4	xvi			†	†	†	†	*	†		†	†	†	?	?			*		*
5	xxvi			†	†	†	†	*		†	†	?	†							
6	xxxiv			†	†	†	†	*	*		†		†	*						
7	xxxvi			†	†	†	†	*		†			†	*						
8	xxxvii			†	†	†	†	*		*	†	*	†							
9	xli			†	†	†	†	*		†		†		†			*	*	*	
10	xxxv			†	†	†	†	*		†	†		†		†	†	†	*	*	*?
11	xxxix	Young or Middle-aged		†	†	†	†	*		*			*							
12	ii	Middle-aged		†	†	†	†	†		*		*			*?					
13	iv			†	†	†	†	†		*?	†									
14	v			†	†	†	†	†		*?	†									
15	vi			†	†	†	†	†		*	†									
16	vii			†	†	†	†	†				*								
17	ix			†	†	†	†	†		*	†		*							
18	x			†	†	†	†	†		*	†		*							
19	xiv			†	†	†	†	†			†		*							
20	xvii			†	†	†	†	†			†		*							
21	xxi			†	†	†	†	†			*		*							
22	xxiii			†	†	†	*	†		*			*							
23	xxv			†	†	†	†	†		*			*	*						
24	xxvii			†	†	†	†	†		†	†		†							
25	xxviii			†	†	†	†	†			†		*							
26	xxix			†	†	†	†	†			†	†								
27	xxx			†	*	†	†	†			*		*					*?		
28	xlii			†	†	*	†	†			†	*	*		*					
29	xliii			†	†	†	†	†			†	*	*							
30	xliv			†	†	†	†	†			†		*							
31	xlv			†	†	†	†	†		*	*		†							
32	xlvii			†	†	†	*	†		*	*					*	*	*		
33	xlviii			†	†	†	†	†		*	*		†		†					
34	xxii	Middle-aged or Old		†	†	†	†	*												
35	xlix			†	†	†	*													
36	xxxii			†	†	*	*													
37	i	Male. Old		†	†	†	†													
38	xv	Female. Young		†	†	†	†			†			*							
39	xxxii			†	†	†	*			†			*							
40	xxxviii	Young or Middle-aged		†	†	†	*													
41	xi	Male? Middle-aged		†	†	†	†	*			†									
42	xviii			†	†	†	†	†			†									
43	xix			†	†	†	†	†												
44	xxiv			†	†	†	†	†		†										
45	xxxi			†	†	†	†	†		*	*		*							
46	xl			†	†	†	†	†		*	†		*				*			
47	xlvi			†	†	†	†	†			†		*							
48	i	Female. Middle-aged or Old		†	†	*														
49	xx			†	†	†				*										
50	iii	Female. Old		†	†															

(1) The organs denoted by these numbers :—ix. 7, Constructiveness ; xx. 22, Mirthfulness or Wit ; xxii. 19 (2), Individuality, Lower Individuality ; xxiii. 20, Configuration, Figure ; xxiv. 21, Size ; xxv. 22, Weight, Resistance ; xxvi. 23, Colour ; xxvii. 24, Locality ; xxviii. 26, Calculation, Number ; xxix. 25, Order ; xxx. 19 (1), Eventuality, Upper Individuality ; xxxi. 20, Time ; xxxii. 28, Melody, Tune ; xxxiii. 29, Language—this organ Gall divides in two, to wit, into the organ of Language and the organ of Words ; xxxiv. 30, Comparison ; xxxv. 31, Causality. The order of the numbers in this table was taken from that of a more extensive and general table ; so that whilst here xx. 32 has not been affected at all, there it was affected more frequently than ix. 7.

In these circumstances, it is to be observed—

In the first place, that, as already noticed, while the developments of all the crania have been carefully marked, the presence of the frontal sinuses has been signalised only in one skull (the male No. 19, xiv.), in which they are, however, greatly below even the average.

In the second place, that the extent of the sinus varies indeterminably from an affection of one to an affection of sixteen organs.

In the third place, in this induction of thirty-seven male and thirteen female crania, the average proportional extent of the sinuses is somewhat less in the female than in the male skulls; the sinus in the former covering 4.4, and affecting 1.2 organs; in the latter covering 5, and affecting 2.1 organs. This induction is, however, too limited, more especially in the female crania, to afford a determination of the point, even were it not at variance with other and more extensive observations.

In the fourth place, the male crania exhibit at once the largest and the smallest sinuses. The largest male sinus covers 12, and affects 4; while the largest female sinus covers 7, and affects 3 organs: whereas, whilst the smallest male sinus affects only 1, the smallest female sinus covers 2 organs.

In the fifth place, so far from supporting the phrenological assertion that the sinuses are only found, or only found in size, in the crania of the old, this their collection tends to prove the very reverse; for here we find about the smallest sinuses in the oldest heads.

END OF THE FIRST VOLUME.

PRINTED BY WILLIAM BLACKWOOD AND SONS.

CATALOGUE OF BOOKS

PUBLISHED BY

WILLIAM BLACKWOOD & SONS

EDINBURGH AND LONDON

———

EDINBURGH: 45 GEORGE STREET
LONDON: 37 PATERNOSTER ROW
MDCCCLXXVII

CATALOGUE.

ALISON. History of Europe. By Sir ARCHIBALD ALISON, Bart., D.C.L.

1. From the Commencement of the French Revolution to the Battle of Waterloo.
   LIBRARY EDITION, 14 vols., with Portraits. Demy 8vo, £10, 10s.
   ANOTHER EDITION, in 20 vols. crown 8vo, £6.
   PEOPLE'S EDITION, 13 vols. crown 8vo, £2, 11s.

2. Continuation to the Accession of Louis Napoleon.
   LIBRARY EDITION, 8 vols. 8vo, £6, 7s. 6d.
   PEOPLE'S EDITION, 8 vols. crown 8vo, 34s.

3. Epitome of Alison's History of Europe. 17th Edition, 7s. 6d.

4. Atlas to Alison's History of Europe. By A. Keith Johnston
   LIBRARY EDITION, demy 4to, £3, 3s.
   PEOPLE'S EDITION, 31s. 6d.

—— Life of John Duke of Marlborough. With some Account of his Contemporaries, and of the War of the Succession. Third Edition, 2 vols. 8vo. Portraits and Maps, 30s.

—— Essays: Historical, Political, and Miscellaneous. 3 vols. demy 8vo, 45s.

—— Lives of Lord Castlereagh and Sir Charles Stewart, Second and Third Marquesses of Londonderry. From the Original Papers of the Family. 3 vols. 8vo, £2, 2s.

—— Principles of the Criminal Law of Scotland. 8vo, 18s.

—— Practice of the Criminal Law of Scotland. 8vo, cloth boards, 18s.

—— The Principles of Population, and their Connection with Human Happiness. 2 vols. 8vo, 30s.

—— On the Management of the Poor in Scotland, and its Effects on the Health of the Great Towns. By WILLIAM PULTENEY ALISON, M.D. Crown 8vo, 5s. 6d.

ADAMS. Great Campaigns. A Succinct Account of the Principal Military Operations which have taken place in Europe from 1796 to 1870. By Major C. ADAMS, Professor of Military History at the Staff College. Edited by Captain C. COOPER KING, R.M. Artillery, Instructor of Tactics, Royal Military College. 8vo, with Maps.

AIRD. Poetical Works of Thomas Aird. Fourth Edition, fcap.
8vo, 6s.

———— The Old Bachelor in the Old Scottish Village. Fcap. 8vo,
4s.

ALEXANDER. Moral Causation ; or, Notes on Mr Mill's Notes
to the Chapter on " Freedom" in the Third Edition of his ' Examination of Sir
William Hamilton's Philosophy.' By PATRICK PROCTOR ALEXANDER, M.A.,
Author of ' Mill and Carlyle,' &c. Second Edition, revised and extended.
Crown 8vo, 6s.

ALLARDYCE. The City of Sunshine. By ALEXANDER ALLAR-
DYCE. Three vols. post 8vo, £1, 5s. 6d.

ANCIENT CLASSICS FOR ENGLISH READERS. Edited by
the Rev. W. LUCAS COLLINS, M.A. 20 vols., cloth, 2s. 6d. each. Or in 10
vols., neatly bound with calf or vellum back, £3, 10s.

CONTENTS.

HOMER : THE ILIAD. By the Editor.	GREEK ANTHOLOGY. By Lord Neaves.
HOMER : THE ODYSSEY. By the Editor.	VIRGIL. By the Editor.
HERODOTUS. By George C. Swayne, M.A.	HORACE. By Theodore Martin.
XENOPHON. By Sir Alex. Grant, Bart.	JUVENAL. By Edward Walford, M.A.
EURIPIDES. By W. B. Donne.	PLAUTUS AND TERENCE. By the Editor.
ARISTOPHANES. By the Editor.	THE COMMENTARIES OF CÆSAR. By An-
PLATO. By Clifton W. Collins, M.A.	thony Trollope.
LUCIAN. By the Editor.	TACITUS. By W. B. Donne.
ÆSCHYLUS. By R. S. Copleston, M.A.	CICERO. By the Editor.
SOPHOCLES. By Clifton W. Collins, M.A.	PLINY'S LETTERS. By the Rev. Alfred
HESIOD AND THEOGNIS. By the Rev. J.	Church, M.A., and the Rev. W. J. Brod-
Davies, M.A.	ribb, M.A.

———— Supplementary Series. Edited by the SAME. To be com-
pleted in 8 or 10 vols., 2s. 6d. each. The volumes published contain—
I. LIVY. By the Editor.   II. OVID. By the Rev. A. Church, M.A.
III. CATULLUS, TIBULLUS, AND PROPERTIUS. By the Rev. Jas. Davies, M.A.
IV. DEMOSTHENES. By the Rev. W. J. Brodribb, M.A.
V. ARISTOTLE. By Sir Alexander Grant, Bart., LL.D.

AYTOUN. Lays of the Scottish Cavaliers, and other Poems. By
W. EDMONDSTOUNE AYTOUN, D.C.L., Professor of Rhetoric and Belles-Lettres
in the University of Edinburgh. Twenty-fifth Edition. Fcap. 8vo, 7s. 6d.

———— An Illustrated Edition of the Lays of the Scottish Cavaliers.
From designs by Sir NOEL PATON. Small 4to, 21s., in gilt cloth.

———— Bothwell : a Poem. Third Edition. Fcap., 7s. 6d.

———— Firmilian ; or, The Student of Badajoz. A Spasmodic
Tragedy. Fcap., 5s.

———— Poems and Ballads of Goethe. Translated by Professor
AYTOUN and THEODORE MARTIN. Third Edition. Fcap., 6s.

———— Bon Gaultier's Book of Ballads. By the SAME. Thirteenth
Edition. With Illustrations by Doyle, Leech, and Crowquill. Post 8vo, gilt
edges, 8s. 6d.

———— The Ballads of Scotland. Edited by Professor AYTOUN.
Fourth Edition. 2 vols. fcap. 8vo, 12s.

———— Norman Sinclair. 3 vols. post 8vo, 31s. 6d.

AYTOUN. Memoir of William E. Aytoun, D.C.L. By THEODORE
MARTIN. With Portrait. Post 8vo, 12s.

BAIRD LECTURES. The Inspiration of the Holy Scriptures.
Being the Baird Lecture for 1873. By the Rev. ROBERT JAMIESON, D.D., Minister of St Paul's Parish Church, Glasgow. Crown 8vo, 7s. 6d.

———— The Mysteries of Christianity. By T. J. CRAWFORD, D.D.,
F.R.S.E., Professor of Divinity in the University of Edinburgh, &c. Being
the Baird Lecture for 1874. Crown 8vo, 7s. 6d.

———— Endowed Territorial Work : Its Supreme Importance to
the Church and Country. By WILLIAM SMITH, D.D., Minister of North Leith.
Being the Baird Lecture for 1875. Crown 8vo, 6s.

———— Theism. By ROBERT FLINT, D.D., LL.D., Professor of
Divinity in the University of Edinburgh. Being the Baird Lecture for 1876.
[In the press.

BATTLE OF DORKING. Reminiscences of a Volunteer. From
'Blackwood's Magazine.' Second Hundredth Thousand, 6d.

BY THE SAME AUTHOR.
The Dilemma. Cheap Edition. Crown 8vo, 6s.

A True Reformer. 3 vols. crown 8vo, £1, 5s. 6d.

BLACKWOOD'S MAGAZINE, from Commencement in 1817 to
December 1875. Nos. 1 to 722, forming 118 Volumes.

———— Index to Blackwood's Magazine. Vols. 1 to 50. 8vo, 15s.

———— Standard Novels. Uniform in size and legibly Printed.
Each Novel complete in one volume.

Florin Series, Illustrated Boards.

TOM CRINGLE'S LOG. Copyright Edition. By Michael Scott.	REGINALD DALTON. By J. G. Lockhart.
	PEN OWEN. By Dean Hook.
THE CRUISE OF THE MIDGE. By the Author of 'Tom Cringle's Log.'	ADAM BLAIR. By J. G. Lockhart.
	LADY LEE'S WIDOWHOOD. By Col. Hamley.
CYRIL THORNTON. By Captain Hamilton.	
ANNALS OF THE PARISH. By John Galt.	SALEM CHAPEL. By Mrs Oliphant.
THE PROVOST, AND OTHER TALES. By John Galt.	THE PERPETUAL CURATE. By Mrs Oliphant.
SIR ANDREW WYLIE. By John Galt.	MISS MARJORIBANKS. By Mrs Oliphant.
THE ENTAIL. By John Galt.	JOHN : A Love Story. By Mrs Oliphant.

Or in Cloth Boards, 2s. 6d.

Shilling Series, Illustrated Cover.

THE RECTOR, and THE DOCTOR'S FAMILY. By Mrs Oliphant.	SIR FRIZZLE PUMPKIN, NIGHTS AT MESS, &c.
THE LIFE OF MANSIE WAUCH. By D. M. Moir.	THE SUBALTERN.
	LIFE IN THE FAR WEST. By G. F. Ruxton.
PENINSULAR SCENES AND SKETCHES. By F. Hardman.	VALERIUS : A Roman Story. By J. G. Lockhart.

Or in Cloth Boards, 1s. 6d.

———— Tales from Blackwood. Forming Twelve Volumes of
Interesting and Amusing Railway Reading. Price One Shilling each in Paper
Cover. Sold separately at all Railway Bookstalls.

1. THE GLENMUTCHKIN RAILWAY, and other Tales. 2. HOW I BECAME A YEOMAN,
&c. 3. FATHER TOM AND THE POPE, &c. 4. MY COLLEGE FRIENDS, &c. 5. ADVENTURES IN TEXAS, &c. 6. THE MAN IN THE BELL, &c. 7. THE MURDERER'S LAST NIGHT,
&c. 8. DI VASARI : a Tale of Florence, &c. 9. ROSAURA : a Tale of Madrid, &c. 10.
THE HAUNTED AND THE HAUNTERS, &c. 11. JOHN RINTOUL, &c. 12. TICKLER AMONG
THE THIEVES, &c.
They may also be had bound in cloth, 18s., and in half calf, richly gilt, 30s. ;
or 12 volumes in 6, half Roxburghe, 21s., and half red morocco, 28s.

BLACKMORE. The Maid of Sker. By R. D. Blackmore, Author of 'Lorna Doone,' &c. New Edition. Crown 8vo, 7s. 6d.

BOSCOBEL TRACTS. Relating to the Escape of Charles the Second after the Battle of Worcester, and his subsequent Adventures. Edited by J. Hughes, Esq., A.M. A New Edition, with additional Notes and Illustrations, including Communications from the Rev. R. H. Barham, Author of the 'Ingoldsby Legends.' 8vo, with Engravings, 16s.

BRACKENBURY. A Narrative of the Ashanti War. Prepared from the official documents, by permission of Major-General Sir Garnet Wolseley, K.C.B., K.C.M.G. By Major H. Brackenbury, R.A., Assistant Military Secretary to Sir Garnet Wolseley. With Maps from the latest Surveys made by the Staff of the Expedition. 2 vols. 8vo, 25s.

BROUGHAM. Memoirs of the Life and Times of Henry Lord Brougham. Written by Himself. 3 vols. 8vo, £2, 8s. The Volumes are sold separately, 16s. each.

BROWN. The Forester: A Practical Treatise on the Planting, Rearing, and General Management of Forest-trees. By James Brown, Wood-Surveyor and Nurseryman. Fourth Edition. Royal 8vo, with Engravings, £1, 11s. 6d.

BROWN. A Manual of Botany, Anatomical and Physiological. For the Use of Students. By Robert Brown, M.A., Ph.D., F.L.S., F.R.G.S. Crown 8vo, with numerous Illustrations, 12s. 6d.

BROWN. Book of the Landed Estate. Containing Directions for the Management and Development of the Resources of Landed Property. By Robert C. Brown, Factor and Estate Agent. Large 8vo, with Illustrations, 21s.

BUCHAN. Handy Book of Meteorology. By Alexander Buchan, M.A., F.R.S.E., Secretary of the Scottish Meteorological Society, &c. A New Edition, being the Third. [In the press.

——— Introductory Text-Book of Meteorology. Crown 8vo, with 8 Coloured Charts and other Engravings, pp. 218. 4s. 6d.

BURBIDGE. Domestic Floriculture, Window Gardening, and Floral Decorations. Being practical directions for the Propagation, Culture, and Arrangement of Plants and Flowers as Domestic Ornaments. By F. W. Burbidge. Second Edition. Crown 8vo, with numerous Illustrations, 7s. 6d.

——— Cultivated Plants: Their Propagation and Improvement. Including Natural and Artificial Hybridisation, Raising from Seed, Cuttings, and Layers, Grafting and Budding, as applied to the Families and Genera in Cultivation. Crown 8vo, with numerous Illustrations, 12s. 6d.

BURN. Handbook of the Mechanical Arts Concerned in the Construction and Arrangement of Dwelling-Houses and other Buildings, with Practical Hints on Road-making and the Enclosing of Land. By Robert Scott Burn, Engineer. Second Edition. Crown 8vo, 6s. 6d.

BUTT. Miss Molly. By Beatrice May Butt. Third Edition. Crown 8vo, 7s. 6d.

——— Eugenie. By the Author of 'Miss Molly.' Crown 8vo, 6s. 6d.

——— Christmas Roses. Tales for Young People. By Geraldine Butt, Author of 'Lads and Lasses.' Crown 8vo, 6s.

BURTON. The History of Scotland : From Agricola's Invasion to the Extinction of the last Jacobite Insurrection. By JOHN HILL BURTON, Historiographer-Royal for Scotland. New and Enlarged Edition, 8 vols. crown 8vo, £3, 3s.

———— The Cairngorm Mountains. Crown 8vo, 3s. 6d.

———— History of the British Empire during the Reign of Queen Anne.   [*Preparing for publication.*]

CAIRD. Sermons. By JOHN CAIRD, D.D., Principal of the University of Glasgow. Thirteenth Thousand. Fcap. 8vo, 5s.

———— Religion in Common Life. A Sermon preached in Crathie Church, October 14, 1855, before Her Majesty the Queen and Prince Albert. Published by Her Majesty's Command. Price One Shilling. Cheap Edition, 3d.

CARLYLE. Autobiography of the Rev. Dr Alexander Carlyle, Minister of Inveresk. Containing Memorials of the Men and Events of his Time. Edited by JOHN HILL BURTON. 8vo. Third Edition, with Portrait, 14s.

CAUVIN. A Treasury of the English and German Languages. Compiled from the best Authors and Lexicographers in both Languages. Adapted to the Use of Schools, Students, Travellers, and Men of Business; and forming a Companion to all German-English Dictionaries. By JOSEPH CAUVIN, LL.D. & Ph.D., of the University of Göttingen, &c. Crown 8vo, 7s. 6d.

CHARTERIS. Life of the Rev. James Robertson, D.D., F.R.S.E., Professor of Divinity and Ecclesiastical History in the University of Edinburgh. By Professor CHARTERIS. With Portrait. 8vo, 10s. 6d.

CHEVELEY NOVELS, THE. A Modern Minister. To be completed in Thirteen Monthly Parts, price 1s. each, with Two Illustrations.

CHURCH SERVICE SOCIETY. A Book of Common Order : Being Forms of Worship issued by the Church Service Society. Fourth Edition, 5s.

CLIFFORD. The Agricultural Lock-Out of 1874. With Notes upon Farming and Farm Labour in the Eastern Counties. By FREDERICK CLIFFORD, of the Middle Temple. Crown 8vo, 7s. 6d.

COLQUHOUN. The Moor and the Loch. A New and Enlarged Edition. By JOHN COLQUHOUN.   [*In preparation.*]

CORKRAN. Bessie Lang : A Story of Cumberland Life. By ALICE CORKRAN. Crown 8vo, 7s. 6d.

COTTERILL. The Genesis of the Church. By the Right. Rev. HENRY COTTERILL, D.D., Bishop of Edinburgh. Demy 8vo, 16s.

COURTHOPE. The Paradise of Birds : An Old Extravaganza in a Modern Dress. By WILLIAM JOHN COURTHOPE, Author of 'Ludibria Lunæ.' Second Edition, 3s. 6d.

CRANSTOUN. The Elegies of Albius Tibullus. Translated into English Verse, with Life of the Poet, and Illustrative Notes. By JAMES CRANSTOUN, LL.D., Author of a Translation of 'Catullus.' Crown 8vo, 6s. 6d.

CRANSTOUN. The Elegies of Sextus Propertius. Translated into English Verse, with Life of the Poet, and Illustrative Notes. By the Same. Crown 8vo, 7s. 6d.

CRAWFORD. The Doctrine of Holy Scripture respecting the Atonement. By the late Thomas J. Crawford, D.D., Professor of Divinity in the University of Edinburgh. Second Edition, Revised and Enlarged. 8vo, 12s.

——— The Fatherhood of God, Considered in its General and Special Aspects, and particularly in relation to the Atonement, with a Review of Recent Speculations on the Subject. Third Edition, Revised and Enlarged. 8vo, 9s.

——— The Preaching of the Cross, and other Sermons. 8vo, 7s. 6d.

——— Mysteries of Christianity; being the Baird Lecture for 1874. Crown 8vo, 7s. 6d.

CUMMING. From Patmos to Paradise; or, Light on the Past, the Present, and the Future. By the Rev. John Cumming, D.D., F.R.S.E., Minister of the Scotch National Church, Crown Court, Covent Garden, London. Crown 8vo, 7s. 6d.

DESCARTES. On the Method of Rightly Conducting the Reason, and Seeking Truth in the Sciences; and his Meditations, and Selections from his Principles of Philosophy. Crown 8vo, 4s. 6d.

DICKSON. Japan; being a Sketch of the History, Government, and Officers of the Empire. By Walter Dickson. 8vo, 15s.

DILEMMA, THE. By the Author of the 'Battle of Dorking.' Cheap Edition, 6s.

EAGLES. Essays. By the Rev. John Eagles, A.M. Oxon. Originally published in 'Blackwood's Magazine.' Post 8vo, 10s. 6d.

——— The Sketcher. Originally published in 'Blackwood's Magazine.' Post 8vo, 10s. 6d.

ELIOT. Adam Bede. By George Eliot. A New Edition. 3s. 6d., cloth.

——— The Mill on the Floss. 3s. 6d., cloth.

——— Scenes of Clerical Life. 3s., cloth.

——— Silas Marner: The Weaver of Raveloe. 2s. 6d., cloth.

——— Felix Holt, the Radical. 3s. 6d., cloth.

——— Middlemarch. In 1 vol., 7s. 6d. Also, a Library Edition, in 4 vols. small 8vo, 21s., cloth.

——— Daniel Deronda. 4 vols., £1, 1s.

——— The Spanish Gypsy. Sixth Edition. Crown 8vo, 7s. 6d., cloth.

——— The Legend of Jubal, and other Poems. Second Edition. Fcap. 8vo, 6s., cloth.

——— Wise, Witty, and Tender Sayings, in Prose and Verse. Selected from the Works of George Eliot. Third Edition. Fcap. 8vo, 6s.

ESSAYS ON SOCIAL SUBJECTS. Originally published in the 'Saturday Review.' A New Edition. First and Second Series. 2 vols., crown 8vo, 6s. each.

EWALD. The Crown and its Advisers; or, Queen, Ministers, Lords, and Commons. By ALEXANDER CHARLES EWALD, F.S.A. Crown 8vo, 5s.

FERRIER. Philosophical Works of the late James F. Ferrier, A.B. Oxon., Professor of Moral Philosophy and Political Economy, St Andrews. New Edition. Edited by Sir ALEX. GRANT, Bart., D.C.L., and Professor LUSHINGTON. 3 vols. crown 8vo, 34s. 6d.

——— Institutes of Metaphysic. Third Edition. 10s. 6d.

——— Lectures on the Early Greek Philosophy. Second Edition. 10s. 6d.

——— Philosophical Remains, including the Lectures on Early Greek Philosophy. 2 vols., 24s.

FERRIER. Mottiscliffe; An Autumn Story. By JAMES WALTER FERRIER. 2 vols. crown 8vo, 17s.

FINLAY. History of Greece under Foreign Domination. By the late GEORGE FINLAY, LL.D., Athens. 6 vols. 8vo—viz.:

Greece under the Romans. B.C. 146 to A.D. 717. A Historical View of the Condition of the Greek Nation from its Conquest by the Romans until the Extinction of the Roman Power in the East. Second Edition, 16s.

History of the Byzantine Empire. A.D. 716 to 1204; and of the Greek Empire of Nicæa and Constantinople, A.D. 1204 to 1453. 2 vols. £1, 7s. 6d.

Greece under Othoman and Venetian Domination. A.D. 1453 to 1821. 10s. 6d.

History of the Greek Revolution of 1830. 2 vols. 8vo, £1, 4s.

FLINT. The Philosophy of History in Europe. Vol. I., containing the History of that Philosophy in France and Germany. By ROBERT FLINT, Professor of Divinity, University of Edinburgh. 8vo, 15s.

——— Theism. Being the Baird Lecture for 1876.

FORBES. The Campaign of Garibaldi in the Two Sicilies: A Personal Narrative. By CHARLES STUART FORBES, Commander, R.N. Post 8vo, with Portraits, 12s.

FOREIGN CLASSICS FOR ENGLISH READERS. Uniform with 'Ancient Classics for English Readers.' Edited by Mrs OLIPHANT.
1. DANTE. By the Editor, now published, price 2s. 6d.

In preparation:—

VOLTAIRE. By Col. Hamley.	PETRARCH. By Dr H. Reeve.
PASCAL. By Principal Tulloch.	CERVANTES. By the Editor.
GOETHE. By A. Hayward, Q.C.	MONTAIGNE. By Rev. W. L. Collins.

FRASER. Handy Book of Ornamental Conifers, and of Rhododendrons and other American Flowering Shrubs, suitable for the Climate and Soils of Britain. With descriptions of the best kinds, and containing Useful Hints for their successful Cultivation. By HUGH FRASER, Fellow of the Botanical Society of Edinburgh. Crown 8vo, 6s.

GALT. Annals of the Parish. By JOHN GALT. Fcap. 8vo, 2s.

——— The Provost. Fcap. 8vo, 2s.

——— Sir Andrew Wylie. Fcap. 8vo, 2s.

——— The Entail ; or, The Laird of Grippy. Fcap. 8vo, 2s.

GARDENER, THE : A Magazine of Horticulture and Floriculture.
Edited by DAVID THOMSON, Author of 'The Handy Book of the Flower-Garden,' &c. ; Assisted by a Staff of the best practical Writers. Published Monthly. 6d.

GENERAL ASSEMBLY OF THE CHURCH OF SCOTLAND.
——— Family Prayers. Authorised by the General Assembly of
the Church of Scotland. A New Edition, crown 8vo, in large type, 4s. 6d.
Another Edition, crown 8vo, 2s.

——— Prayers for Social and Family Worship. For the Use of
Soldiers, Sailors, Colonists, and Sojourners in India, and other Persons, at
home and abroad, who are deprived of the ordinary services of a Christian
Ministry. Second Edition, crown 8vo, 4s. Cheap Edition, 1s. 6d.

——— The Scottish Hymnal. Hymns for Public Worship. Published for Use in Churches by Authority of the General Assembly. Various
sizes—viz. : 1. Large type, cloth, red edges, 1s. 6d. ; French morocco, 2s. 6d. ;
calf, 6s. 2. Bourgeois type, cloth, red edges, 1s. ; French morocco, 2s. 3.
Minion type, limp cloth, 6d. ; French morocco, 1s. 6d. 4. School Edition, in
paper cover, 2d. No. 1, bound with the Psalms and Paraphrases, cloth, 3s. ;
French morocco, 4s. 6d. ; calf, 7s. 6d. No. 2, bound with the Psalms and Paraphrases, cloth, 2s. ; French morocco, 3s.

——— The Scottish Hymnal, with Music. Selected by the Committee on Hymns and on Psalmody. The harmonies arranged by W. H. Monk.
Cloth, 1s. 6d. ; French morocco, 3s. 6d. The same in the Tonic Sol-fa Notation,
1s. 6d. and 3s. 6d.

GLEIG. The Subaltern. By G. R. GLEIG, M.A., late Chaplain-
General of her Majesty's Forces. Originally published in 'Blackwood's Magazine.' Library Edition. Revised and Corrected, with a New Preface. Crown
8vo, 7s. 6d.

——— The Great Problem : Can it be Solved ? 8vo, 10s. 6d.

GOETHE'S FAUST. Translated into English Verse by THEODORE
MARTIN. Second Edition, post 8vo, 6s. Cheap Edition, fcap, 3s. 6d.

——— Poems and Ballads of Goethe. Translated by Professor
AYTOUN and THEODORE MARTIN. Second Edition, fcap. 8vo, 6s.

GRAHAM. Annals and Correspondence of the Viscount and First
and Second Earls of Stair. By JOHN MURRAY GRAHAM. 2 vols. demy 8vo,
with Portraits and other Illustrations. £1, 8s.

——— Memoir of Lord Lynedoch. Second Edition, crown 8vo.

GRANT. A Walk across Africa ; or, Domestic Scenes from my
Nile Journal. By JAMES AUGUSTUS GRANT, Captain H.M. Bengal Army, Fellow and Gold Medallist of the Royal Geographical Society. 8vo, with Map, 15s.

GRANT. Incidents in the China War of 1860. Compiled from the Private Journals of the late General Sir Hope Grant, G.C.B. By Henry Knollys, Captain Royal Artillery; Author of 'From Sedan to Saarbrück,' &c. Crown 8vo, with Maps, 12s.

—— Incidents in the Sepoy War of 1857-58. Compiled from the Private Journals of the late General Sir Hope Grant, G.C.B.; together with some Explanatory Chapters by Captain Henry Knollys, R.A. Crown 8vo, with Map and Plans, 12s.

GRANT. Memorials of the Castle of Edinburgh. By James Grant. A New Edition. Crown 8vo, with 12 Engravings, 7s.

HAMERTON. Wenderholme : A Story of Lancashire and York-shire Life. By Philip Gilbert Hamerton, Author of 'A Painter's Camp.' A New Edition. Crown 8vo, 6s.

HAMILTON. Lectures on Metaphysics. By Sir William Hamil-ton, Bart., Professor of Logic and Metaphysics in the University of Edinburgh. Edited by the Rev. H. L. Mansel, B.D., LL.D., Dean of St Paul's; and John Veitch, M.A., Professor of Logic and Rhetoric, Glasgow. Sixth Edition. 2 vols. 8vo, 24s.

—— Lectures on Logic. Edited by the Same. Third Edition. 2 vols. 24s.

—— Discussions on Philosophy and Literature, Education and University Reform. Third Edition, 8vo, 21s.

—— Memoir of Sir William Hamilton, Bart., Professor of Logic and Metaphysics in the University of Edinburgh. By Professor Veitch of the University of Glasgow. 8vo, with Portrait, 18s.

HAMILTON. Annals of the Peninsular Campaigns. By Captain Thomas Hamilton. Edited by F. Hardman. 8vo, 16s. Atlas of Maps to Illustrate the Campaigns, 12s.

HAMLEY. The Operations of War Explained and Illustrated. By Edward Bruce Hamley, Colonel in the Royal Artillery, Companion of the Bath, Commandant of the Staff College, &c. Third Edition, 4to, with numer-ous Illustrations, 30s.

—— The Story of the Campaign of Sebastopol. Written in the Camp. With Illustrations drawn in Camp by the Author. 8vo, 21s.

—— On Outposts. Second Edition. 8vo, 2s.

—— Wellington's Career ; A Military and Political Summary. Crown 8vo, 2s.

—— Lady Lee's Widowhood. Crown 8vo, 2s. 6d.

—— Our Poor Relations. A Philozoic Essay. With Illustra-tions, chiefly by Ernest Griset. Crown 8vo, cloth gilt, 3s. 6d.

HANDY HORSE-BOOK ; or, Practical Instructions in Riding, Driving, and the General Care and Management of Horses. By 'Magenta.' A New Edition, with 6 Engravings, 4s. 6d.

By the Same.

Our Domesticated Dogs: their Treatment in reference to Food, Diseases, Habits, Punishment, Accomplishments. Crown 8vo, 2s. 6d.

HARBORD. A Glossary of Navigation. Containing the Definitions and Propositions of the Science, Explanation of Terms, and Description of Instruments. By the Rev. J. B. Harbord, M.A., Assistant Director of Education, Admiralty. Crown 8vo. Illustrated with Diagrams, 6s.

——— Definitions and Diagrams in Astronomy and Navigation. 1s.

——— Short Sermons for Hospitals and Sick Seamen. Fcap. 8vo, cloth, 4s. 6d.

HARDMAN. Scenes and Adventures in Central America. Edited by Frederick Hardman. Crown 8vo, 2s. 6d.

HASTINGS. Poems. By the Lady Flora Hastings. Edited by her Sister, the late Marchioness of Bute. Second Edition, with a Portrait. Fcap., 7s. 6d.

HAY. The Works of the Right Rev. Dr George Hay, Bishop of Edinburgh. Edited under the Supervision of the Right Rev. Bishop Strain. With Memoir and Portrait of the Author. Complete Edition, 7 vols. crown 8vo, bound in extra cloth, £1, 11s. 6d. Or, sold separately—viz.:

——— The Sincere Christian Instructed in the Faith of Christ from the Written Word. 2 vols., 8s.

——— The Devout Christian Instructed in the Law of Christ from the Written Word. 2 vols., 8s.

——— The Pious Christian Instructed in the Nature and Practice of the Principal Exercises of Piety. 1 vol., 4s.

——— The Scripture Doctrine of Miracles Displayed. 2 vols., 10s. 6d.

HEMANS. The Poetical Works of Mrs Hemans. Copyright Editions.
One Volume, royal 8vo, 5s.
The Same, with Illustrations engraved on Steel, bound in cloth, gilt edges, 7s. 6d.
Six Volumes, fcap., 12s. 6d.
Seven Volumes, fcap., with Memoir by her Sister. 35s.
Select Poems of Mrs Hemans. Fcap., cloth, gilt edges, 3s.

——— Memoir of Mrs Hemans. By her Sister. With a Portrait, fcap. 8vo, 5s.

HOLE. A Book about Roses, how to Grow and Show Them. By the Rev. Canon Hole. Sixth Edition, Enlarged. Crown 8vo, 7s. 6d.

HOMER. The Odyssey. Translated into English Verse in the Spenserian Stanza. By Philip Stanhope Worsley. Third Edition, 2 vols. fcap., 12s.

——— The Iliad. Translated by P. S. Worsley and Professor Conington. 2 vols. crown 8vo, 21s.

HOSACK. Mary Queen of Scots and Her Accusers. Containing a Variety of Documents never before published. By John Hosack, Barrister-at-Law. A New and Enlarged Edition, with a Photograph from the bust on the Tomb in Westminster Abbey. 2 vols. 8vo, £1, 11s. 6d. The Second Volume may be had separately, price 16s. 6d.

INDEX GEOGRAPHICUS: Being a List, alphabetically arranged, of the Principal Places on the Globe, with the Countries and Subdivisions of the Countries in which they are situated, and their Latitudes and Longitudes. Applicable to all Modern Atlases and Maps. Imperial 8vo, pp. 676, 21s.

JOHNSON. The Scots Musical Museum. Consisting of upwards of Six Hundred Songs, with proper Basses for the Pianoforte. Originally published by James Johnson; and now accompanied with Copious Notes and Illustrations of the Lyric Poetry and Music of Scotland, by the late William Stenhouse; with additional Notes and Illustrations, by David Laing and C. K. Sharpe. 4 vols. 8vo, Roxburghe binding, £2, 12s. 6d.

JOHNSTON. Notes on North America: Agricultural, Economical, and Social. By Professor J. F. W. Johnston. 2 vols. post 8vo, 21s.

——— The Chemistry of Common Life. With 113 Illustrations on Wood, and a Copious Index. 2 vols. crown 8vo, 11s. 6d.

——— Professor Johnston's Elements of Agricultural Chemistry and Geology. The Tenth Edition, Revised and brought down to date. By Charles A. Cameron, M.D., F.R.C.S.I., &c. Fcap. 8vo, 6s. 6d.

KING. The Metamorphoses of Ovid. Translated in English Blank Verse. By Henry King, M.A., Fellow of Wadham College, Oxford, and of the Inner Temple, Barrister-at-Law. Crown 8vo, 10s. 6d.

KINGLAKE. History of the Invasion of the Crimea. By A. W. Kinglake. Cabinet Edition. This Edition comprises in Six Volumes, crown 8vo, at 6s. each, the contents of the Five Octavo Volumes of the present Edition, revised and prepared for the Cabinet Edition by the Author. The Volumes respectively contain:—
    I. The Origin of the War between the Czar and the Sultan.
    II. Russia Met and Invaded. With 4 Maps and Plans.
    III. The Battle of the Alma. With 14 Maps and Plans.
    IV. Sebastopol at Bay. With 10 Maps and Plans.
    V. The Battle of Balaclava. With 10 Maps and Plans.
    VI. The Battle of Inkerman. With 11 Maps and Plans.
The Cabinet Edition is so arranged that each volume contains a complete subject. Sold separately at 6s.

KNOLLYS. The Elements of Field-Artillery. Designed for the Use of Infantry and Cavalry Officers. By Henry Knollys, Captain Royal Artillery; Author of 'From Sedan to Saarbrück,' Editor of 'Incidents in the Sepoy War,' &c. With Engravings. Crown 8vo, 7s. 6d.

KNOX. John Knox's Liturgy: the Book of Common Order, and the Directory for Public Worship of the Church of Scotland. With Historical Introductions and Illustrative Notes by the Rev. George W. Sprott, B.A., and the Rev. Thomas Leishman, D.D. Handsomely printed, in imitation of the large editions of Andro Hart, on toned paper, bound in cloth, red edges, 8s. 6d.

LAVERGNE. The Rural Economy of England, Scotland, and Ireland. By Léonce de Lavergne. Translated from the French. With Notes by a Scottish Farmer. 8vo, 12s.

LEE. Lectures on the History of the Church of Scotland, from the Reformation to the Revolution Settlement. By the late Very Rev. John Lee, D.D., LL.D., Principal of the University of Edinburgh. With Notes and Appendices from the Author's Papers. Edited by the Rev. William Lee, D.D. 2 vols. 8vo, 21s.

LEWES. The Physiology of Common Life. By George H. Lewes, Author of 'Sea-side Studies,' &c. Illustrated with numerous Engravings. 2 vols., 12s.

LIE. The Pilot and His Wife. From the Norwegian of JONAS
LIE. Translated by J. L. POTTENHAM. Crown 8vo, 10s. 6d.

LOCKHART. Doubles and Quits. By Colonel L. W. M. LOCK-
HART. With Twelve Illustrations. 2 vols. post 8vo, 21s.

——— Fair to See : a Novel. New Edition, 1 volume, 6s.

LYON. History of the Rise and Progress of Freemasonry in Scotland. By DAVID MURRAY LYON, Secretary to the Grand Lodge of Scotland.
In small quarto. Illustrated with numerous Portraits of Eminent Members of
the Craft, and Facsimiles of Ancient Charters and other Curious Documents.
£1, 11s. 6d.

LYTTON. Speeches, Spoken and Unspoken. By EDWARD LORD
LYTTON. With a Memoir by his son, ROBERT LORD LYTTON. 2 volumes, 8vo,
24s.

M'COMBIE. Cattle and Cattle-Breeders. By WILLIAM M'COMBIE,
M.P., Tillyfour. A New and Cheaper Edition, 2s. 6d., cloth.

M'CRIE. Works of the Rev. Thomas M'Crie, D.D. Uniform Edition. Four vols. crown 8vo, 24s.

——— Life of John Knox. Containing Illustrations of the History of the Reformation in Scotland. Crown 8vo, 6s. Another Edition, 3s. 6d.

——— Life of Andrew Melville. Containing Illustrations of the
Ecclesiastical and Literary History of Scotland in the Sixteenth and Seventeenth Centuries. Crown 8vo, 6s.

——— History of the Progress and Suppression of the Reformation in Italy in the Sixteenth Century. Crown 8vo, 4s.

——— History of the Progress and Suppression of the Reformation in Spain in the Sixteenth Century. Crown 8vo, 3s. 6d.

——— Sermons, and Review of the 'Tales of My Landlord.' Crown
8vo, 6s.

——— Lectures on the Book of Esther. Fcap. 8vo, 5s.

M'INTOSH. The Book of the Garden. By CHARLES M'INTOSH,
formerly Curator of the Royal Gardens of his Majesty the King of the Belgians,
and lately of those of his Grace the Duke of Buccleuch, K.G., at Dalkeith Pal-
ace. Two large vols. royal 8vo, embellished with 1350 Engravings.
Vol. I. On the Formation of Gardens and Construction of Garden Edifices. 776
pages, and 1073 Engravings, £2, 10s.
Vol. II. Practical Gardening. 868 pages, and 279 Engravings, £1, 17s. 6d.

MACKENZIE. Studies in Roman Law. With Comparative Views
of the Laws of France, England, and Scotland. By Lord MACKENZIE, one of
the Judges of the Court of Session in Scotland. Fourth Edition, Edited by
JOHN KIRKPATRICK, Esq., M.A. Cantab.; Dr Jur. Heidelb.; LL.B., Edin.;
Advocate. 8vo, 12s. 6d.

MACKAY. A Manual of Modern Geography, Mathematical, Physical, and Political. By the Rev. ALEXANDER MACKAY, LL.D., F.R.G.S. New
and Greatly Improved Edition. Crown 8vo, pp. 676. 7s. 6d.

——— Elements of Modern Geography. Fifteenth Edition, revised to the present time. Crown 8vo, pp. 300, 3s.

MACKAY. The Intermediate Geography. Intended as an Intermediate Book between the Author's 'Outlines of Geography,' and 'Elements of Geography.' Third Edition, crown 8vo, pp. 224, 2s.

—— Outlines of Modern Geography. Eighteenth Edition, revised to the Present Time. 18mo, pp. 112, 1s.

—— First Steps in Geography. 18mo, pp. 56. Sewed, 4d. Cloth, 6d.

—— Elements of Physiography and Physical Geography. With Express Reference to the Instructions recently issued by the Science and Art Department. Crown 8vo.

MAJENDIE. Giannetto. By LADY MARGARET MAJENDIE. Crown 8vo, 5s.

MARSHALL. International Vanities. By FREDERIC MARSHALL. Originally published in 'Blackwood's Magazine.' 8vo, 10s. 6d.

—— French Home Life. By "an English Looker-on, who has lived for a quarter of a century in France, amidst ties and affections which have made that country his second home."—Preface. CONTENTS: Servants.—Children.—Furniture.—Food.—Manners.—Language.—Dress.—Marriage. Second Edition. 5s.

MARSHMAN. History of India. From the Earliest Period to the Close of the India Company's Government; with an Epitome of Subsequent Events. By JOHN CLARK MARSHMAN, C.S.I. Abridged from the Author's larger work. Crown 8vo, 6s. 6d.

MARTIN. Goethe's Faust. Translated by THEODORE MARTIN. Second Edition, crown 8vo, 6s. Cheap Edition, 3s. 6d.

—— The Odes of Horace. With Life and Notes. Third Edition, post 8vo, 9s.

—— Catullus. With Life and Notes. Second Edition, post 8vo, 7s. 6d.

—— The Vita Nuova of Dante. With an Introduction and Notes. Second Edition, crown 8vo, 5s.

—— Aladdin: A Dramatic Poem. By ADAM OEHLENSCHLAEGER. Fcap. 8vo, 5s.

—— Correggio: A Tragedy. By OEHLENSCHLAEGER. With Notes. Fcap. 8vo, 3s.

—— King Rene's Daughter: A Danish Lyrical Drama. By HENRIK HERTZ. Second Edition, fcap., 2s. 6d.

MERCER. Journal of the Waterloo Campaign: Kept throughout the Campaign of 1815. By General CAVALIE MERCER, Commanding the 9th Brigade Royal Artillery. 2 vols. post 8vo, 21s.

MINTO. A Manual of English Prose Literature, Biographical and Critical: designed mainly to show Characteristics of Style. By W. MINTO, M.A. Crown 8vo, 10s. 6d.

—— Characteristics of English Poets, from Chaucer to Shirley. Crown 8vo, 9s.

MITCHELL. Biographies of Eminent Soldiers of the last Four Centuries. By Major-General JOHN MITCHELL, Author of 'Life of Wallenstein.' With a Memoir of the Author. 8vo, 9s.

MOIR. Poetical Works of D. M. Moir (Delta). With Memoir by THOMAS AIRD, and Portrait. Second Edition. 2 vols. fcap 8vo, 12s.

———— Domestic Verses. New Edition, fcap. 8vo, cloth gilt, 4s. 6d.

———— Lectures on the Poetical Literature of the Past Half-Century. Third Edition, fcap. 8vo, 5s.

———— Life of Mansie Wauch. Crown 8vo, 1s. 6d.

MONTALEMBERT. Count de Montalembert's History of the Monks of the West. From St Benedict to St Bernard. Translated by Mrs OLIPHANT. 5 vols. 8vo, £2, 12s. 6d.

———— Count de Montalembert's Monks of the West. Vols. VI. and VII completing the Work. [In the press.

———— Memoir of Count de Montalembert. A Chapter of Recent French History. By Mrs OLIPHANT, Author of the 'Life of Edward Irving,' &c. 2 vols. crown 8vo, £1, 4s.

NEAVES. A Glance at some of the Principles of Comparative Philology. As illustrated in the Latin and Anglican Forms of Speech. By the Hon. Lord NEAVES. Crown 8vo, 1s. 6d.

———— Songs and Verses, Social and Scientific. By an Old Contributor to 'Maga.' Fourth Edition, fcap. 8vo, 4s.

———— The Greek Anthology. Being Vol. XX. of 'Ancient Classics for English Readers.' Crown 8vo, 2s. 6d.

NICHOLSON. A Manual of Zoology, for the Use of Students. With a General Introduction on the Principles of Zoology. By HENRY ALLEYNE NICHOLSON, M.D., F.R.S.E., F.G.S., &c., Professor of Natural History in the University of St Andrews. Fourth Edition, revised and enlarged. Crown 8vo, pp. 732, with 300 Engravings on Wood, 12s. 6d.

———— Introductory Text-Book of Zoology, for the Use of Junior Classes. A New Edition, revised and enlarged, with 136 Engravings, 3s.

———— Text-Book of Zoology, for the Use of Schools. Second Edition, enlarged. Crown 8vo, with 188 Engravings on Wood. 6s.

———— Outlines of Natural History, for Beginners; being Descriptions of a Progressive Series of Zoological Types. With 52 Engravings. 1s. 6d.

———— A Manual of Palæontology, for the Use of Students. With a General Introduction on the Principles of Palæontology. Crown 8vo, with upwards of 400 Engravings. 15s.

———— The Ancient Life-History of the Earth. An Outline of the Principles and Leading Facts of Palæontological Science. Crown 8vo, with numerous Engravings, 10s. 6d.

NICHOLSON. Redeeming the Time, and other Sermons. By the late MAXWELL NICHOLSON, D.D., Minister of St Stephen's, Edinburgh. Crown 8vo, 7s. 6d.

———— Communion with Heaven, and other Sermons. Crown 8vo, 5s. 6d.

———— Rest in Jesus. Sixth Edition. Fcap. 8vo, 4s. 6d.

NINA BALATKA. The Story of a Maiden of Prague. 2 vols. small 8vo, 10s. 6d., cloth.

OLIPHANT. Piccadilly: A Fragment of Contemporary Biography. By LAURENCE OLIPHANT. With Eight Illustrations by Richard Doyle. 5th Edition, 4s. 6d. Cheap Edition, in paper cover, 2s. 6d.

———— Narrative of Lord Elgin's Mission to China and Japan. Illustrated with numerous Engravings in Chromo-Lithography, Maps, and Engravings on Wood, from Original Drawings and Photographs. Second Edition. 2 vols. 8vo, 21s.

———— Russian Shores of the Black Sea in the Autumn of 1852. With a Voyage down the Volga and a Tour through the Country of the Don Cossacks. 8vo, with Map and other Illustrations. Fourth Edition, 14s.

OLIPHANT. Historical Sketches of the Reign of George Second. By Mrs OLIPHANT. Third Edition, 6s.

———— The Story of Valentine and his Brother. 5s., cloth.

———— Katie Stewart. 2s. 6d.

———— Salem Chapel. 2s. 6d., cloth.

———— The Perpetual Curate. 2s. 6d., cloth.

———— Miss Marjoribanks. 2s. 6d., cloth.

———— The Rector, and the Doctor's Family. 1s. 6d., cloth.

———— John : A Love Story. 2s. 6d., cloth.

OSBORN. Narratives of Voyage and Adventure. By Admiral SHERARD OSBORN, C.B. 3 vols. crown 8vo, 12s. Or separately :—

———— Stray Leaves from an Arctic Journal ; or, Eighteen Months in the Polar Regions in Search of Sir John Franklin's Expedition in 1850-51. To which is added the Career, Last Voyage, and Fate of Captain Sir John Franklin. New Edition, crown 8vo, 3s. 6d.

———— The Discovery of a North-West Passage by H.M.S. Investigator, during the years 1850-51-52-53-54. Edited from the Logs and Journals of Captain ROBERT C. M'CLURE. Fourth Edition, crown 8vo, 5s. 6d.

———— Quedah ; A Cruise in Japanese Waters : and, The Fight on the Peiho. New Edition, crown 8vo, 5s.

OSSIAN. The Poems of Ossian in the Original Gaelic. With a Literal Translation into English, and a Dissertation on the Authenticity of the Poems By the Rev. ARCHIBALD CLERK. 2 vols. Imperial 8vo, £1, 11s. 6d.

OUTRAM. Lyrics, Legal and Miscellaneous. By GEORGE OUT-RAM, Esq., Advocate. Edited, with Introductory Notice, by HENRY GLASS-FORD BELL, Esq., Advocate, Sheriff of Lanarkshire. Third Edition. Fcap. 8vo, 4s. 6d.

PAGE. Introductory Text-Book of Geology. By DAVID PAGE, LL.D., Professor of Geology in the Durham University of Physical Science, Newcastle. With Engravings on Wood and Glossarial Index. Tenth Edition, 2s. 6d.

——— Advanced Text-Book of Geology, Descriptive and Indus-trial. With Engravings, and Glossary of Scientific Terms. Sixth Edition, re-vised and enlarged, 7s. 6d.

——— Handbook of Geological Terms, Geology, and Physical Geo-graphy. Second Edition, enlarged, 7s. 6d.

——— Geology for General Readers. A Series of Popular Sketches in Geology and Palæontology. Third Edition, enlarged, 6s.

——— Chips and Chapters. A Book for Amateurs and Young Geologists. 5s.

——— The Past and Present Life of the Globe. With numerous Illustrations. Crown 8vo, 6s.

——— The Crust of the Earth : A Handy Outline of Geology. 1s.

——— Economic Geology ; or, Geology in its relation to the Arts and Manufactures. With Engravings, and Coloured Map of the British Islands. Crown 8vo, 7s. 6d.

——— Introductory Text-Book of Physical Geography. With Sketch-Maps and Illustrations. Eighth Edition, 2s. 6d.

——— Advanced Text-Book of Physical Geography. Second Edi-tion. With Engravings. 5s.

PAGET. Paradoxes and Puzzles : Historical, Judicial, and Literary. Now for the first time published in Collected Form. By JOHN PAGET, Barris-ter-at-Law. 8vo, 12s.

PATON. Spindrift. By Sir J. NOEL PATON. Fcap., cloth, 5s.

——— Poems by a Painter. Fcap., cloth, 5s.

PATTERSON. Essays in History and Art. By R. H. PATTERSON. 8vo, 12s.

PAUL. History of the Royal Company of Archers, the Queen's Body-Guard for Scotland. By JAMES BALFOUR PAUL, Advocate of the Scottish Bar. Crown 4to, with Portraits and other Illustrations. £2, 2s.

PAUL. Analysis and Critical Interpretation of the Hebrew Text of the Book of Genesis. Preceded by a Hebrew Grammar, and Dissertations on the Genuineness of the Pentateuch, and on the Structure of the Hebrew Lan-guage. By the Rev. WILLIAM PAUL, A.M. 8vo, 18s.

PETTIGREW. The Handy-Book of Bees, and their Profitable Management. By A. PETTIGREW. Third Edition, with Engravings. Crown 8vo, 3s. 6d.

POLLOK. The Course of Time : A Poem. By Robert Pollok, A.M. Small feap. 8vo, cloth gilt, 2s. 6d. The Cottage Edition, 32mo, sewed, 8d. The Same, cloth, gilt edges, 1s. 6d. Another Edition, with Illustrations by Birket Foster and others, fcap., gilt cloth, 3s. 6d., or with edges gilt, 4s.

——— An Illustrated Edition of the Course of Time. The Illustrations by Birket Foster, Tenniel, and Clayton. Large 8vo, bound in cloth, richly gilt, 21s.

PORT ROYAL LOGIC. Translated from the French : with Introduction, Notes, and Appendix. By Thomas Spencer Baynes, LL.D., Professor in the University of St Andrews. Seventh Edition, 12mo, 4s.

POTTS and DARNELL. Aditus Faciliores: An easy Latin Construing Book, with Complete Vocabulary. By A. W. Potts, M.A., LL.D., Head-Master of the Fettes College, Edinburgh, and sometime Fellow of St John's College, Cambridge ; and the Rev. C. Darnell, M.A., Head-Master of Cargilfield Preparatory School, Edinburgh, and late Scholar of Pembroke and Downing Colleges, Cambridge. Third Edition. Fcap. 8vo, 3s. 6d.

PRAYERS. Family Prayers: Authorised by the General Assembly of the Church of Scotland. A New Edition, crown 8vo, in large type, 4s. 6d. Another Edition, crown 8vo, 2s.

——— Prayers for Social and Family Worship. For the Use of Soldiers, Sailors, Colonists, and Sojourners in India, and other persons, at home and abroad, who are deprived of the ordinary services of a Christian Ministry. Cheap Edition, 1s. 6d.

PRINGLE. The Live Stock of the Farm. By Robert O. Pringle. Second Edition, Revised, crown 8vo, 9s.

PUBLIC GENERAL STATUTES AFFECTING SCOTLAND, from 1707 to 1847, with Chronological Table and Index. 3 vols. large 8vo, £3, 3s.

PUBLIC GENERAL STATUTES AFFECTING SCOTLAND, COLLECTION OF. Published Annually with General Index.

PUBLIC SCHOOLS, THE : Winchester—Westminster—Shrewsbury—Harrow—Rugby. Notes of their History and Traditions. By the Author of 'Etoniana.' Crown 8vo, 8s. 6d.

RAMSAY. Two Lectures on the Genius of Handel, and the Distinctive Character of his Sacred Compositions. Delivered to the Members of the Edinburgh Philosophical Institution. By the Very Rev. Dean Ramsay, Author of 'Reminiscences of Scottish Life and Character.' Crown 8vo, 3s. 6d.

READE. A Woman-Hater. By Charles Reade. 3 vols. crown 8vo, £1, 5s. 6d. Originally published in 'Blackwood's Magazine.'

RINK. Tales and Traditions of the Eskimo. With a Sketch of their Habits, Religion, Language, and other Peculiarities. By Dr Henry Rink, Director of the Royal Greenland Board of Trade, and formerly Inspector of South Greenland. Translated from the Danish by the Author. Edited by Dr Robert Brown, F.L.S., F.R.G.S. With numerous Illustrations, drawn and engraved by Eskimo. Crown 8vo, 10s. 6d.

ROGERS. The Geology of Pennsylvania : A Government Survey ; with a General View of the Geology of the United States, Essays on the Coal Formation and its Fossils, and a Description of the Coal-Fields of North America and Great Britain. By Professor Henry Darwin Rogers, F.R.S., F.G.S., Professor of Natural History in the University of Glasgow. With Seven large Maps, and numerous Illustrations engraved on Copper and on Wood. 3 vols. royal 4to, £8, 8s.

RUSTOW. The War for the Rhine Frontier, 1870: Its Political and Military History. By Col. W. RUSTOW. Translated from the German, by JOHN LAYLAND NEEDHAM, Lieutenant R.M. Artillery. 3 vols. 8vo, with Maps and Plans, £1, 11s. 6d.

ST STEPHENS; or, Illustrations of Parliamentary Oratory. A Poem. Comprising—Pym—Vane—Strafford—Halifax—Shaftesbury—St John —Sir R. Walpole—Chesterfield—Carteret—Chatham—Pitt—Fox—Burke— Sheridan — Wilberforce — Wyndham — Conway — Castlereagh — William Lamb (Lord Melbourne)—Tierney—Lord Gray—O'Connell—Plunkett—Shiel—Follett Macaulay—Peel. Second Edition, crown 8vo, 5s.

SANDFORD. The Great Governing Families of England. By J. LANGTON SANDFORD and MEREDITH TOWNSEND. 2 vols. 8vo, 15s., in extra binding, with richly-gilt cover.

SCHETKY. Ninety Years of Work and Play. Sketches from the Public and Private Career of JOHN CHRISTIAN SCHETKY, late Marine Painter in Ordinary to the Queen. By his DAUGHTER. Crown 8vo, 7s. 6d.

SELLAR. The Education (Scotland) Act, 1872. With Introduction, Explantory Notes, and Index. By A. C. SELLAR. New Edition. [In the press.

SELLER AND STEPHENS. Physiology at the Farm; in Aid of Rearing and Feeding the Live Stock. By WILLIAM SELLER, M.D., F.R.S.E., Fellow of the Royal College of Physicians, Edinburgh, formerly Lecturer on Materia Medica and Dietetics; and HENRY STEPHENS, F.R.S.E., Author of the 'Book of the Farm,' &c. Post 8vo, with Engravings, 16s.

SIMPSON. Paris after Waterloo: A Revised Edition of a "Visit to Flanders and the Field of Waterloo." By JAMES SIMPSON, Advocate. With 2 coloured Plans of the Battle. Crown 8vo, 5s.

SMITH. Italian Irrigation: A Report on the Agricultural Canals of Piedmont and Lombardy, addressed to the Hon. the Directors of the East India Company; with an Appendix, containing a Sketch of the Irrigation System of Northern and Central India. By Lieut.-Col. R. BAIRD SMITH, F.G.S., Captain, Bengal Engineers. Second Edition. 2 vols. 8vo, with Atlas in folio, 30s.

SMITH. Thorndale; or, The Conflict of Opinions. By WILLIAM SMITH, Author of 'A Discourse on Ethics,' &c. Second Edition. Crown 8vo, 10s. 6d.

———— Gravenhurst; or, Thoughts on Good and Evil. Second Edition, with Memoir of the Author. Crown 8vo, 8s.

———— A Discourse on Ethics of the School of Paley. 8vo, 4s.

———— Dramas. 1. Sir William Crichton. 2. Athelwold. 3. Guildone. 24mo, boards, 3s.

SOUTHEY. Poetical Works of Caroline Bowles Southey. Fcap. 8vo, 5s.

———— The Birthday, and other Poems. Second Edition, 5s.

SPEKE. What led to the Discovery of the Nile Source. By JOHN HANNING SPEKE, Captain H.M. Indian Army. 8vo, with Maps, &c., 14s.

———— Journal of the Discovery of the Source of the Nile. By J. H. SPEKE, Captain H.M. Indian Army. With a Map of Eastern Equatorial Africa by Captain SPEKE; numerous illustrations, chiefly from Drawings by Captain GRANT; and Portraits, engraved on Steel, of Captains SPEKE and GRANT. 8vo, 21s.

STARFORTH. Villa Residences and Farm Architecture: A Series of Designs. By JOHN STARFORTH, Architect. 102 Engravings. Second Edition, medium 4to, £2, 17s. 6d.

STATISTICAL ACCOUNT OF SCOTLAND. Complete, with Index, 15 vols. 8vo, £16, 16s. Each County sold separately, with Title, Index, and Map, neatly bound in cloth, forming a very valuable Manual to the Landowner, the Tenant, the Manufacturer, the Naturalist, the Tourist, &c.

STEPHENS. The Book of the Farm; detailing the Labours of the Farmer, Farm-Steward, Ploughman, Shepherd, Hedger, Farm-Labourer, Field-Worker, and Cattleman. By HENRY STEPHENS, F.R.S.E. Illustrated with Portraits of Animals painted from the life; and with 557 Engravings on Wood, representing the principal Field Operations, Implements, and Animals treated of in the Work. A New and Revised Edition, the third, in great part Rewritten. 2 vols. large 8vo, £3, 10s.

———— The Book of Farm-Buildings; their Arrangement and Construction. By HENRY STEPHENS, F.R.S.E., Author of 'The Book of the Farm;' and ROBERT SCOTT BURN. Illustrated with 1045 Plates and Engravings. Large 8vo, uniform with 'The Book of the Farm,' &c. £1, 11s. 6d.

———— The Book of Farm Implements and Machines. By J. SLIGHT and R. SCOTT BURN, Engineers. Edited by HENRY STEPHENS. Large 8vo, uniform with 'The Book of the Farm,' £2, 2s.

———— Catechism of Practical Agriculture. With Engravings. 1s.

STEWART. Advice to Purchasers of Horses. By JOHN STEWART, V.S. Author of 'Stable Economy.' 2s. 6d.

———— Stable Economy. A Treatise on the Management of Horses in relation to Stabling, Grooming, Feeding, Watering, and Working. Seventh Edition, fcap. 8vo, 6s. 6d.

STORY. Graffiti D'Italia. By W. W. STORY, Author of 'Roba di Roma.' Second Edition, fcap. 8vo, 7s. 6d.

———— Nero; A Historical Play. Fcap. 8vo, 6s.

STORMONTH. Etymological and Pronouncing Dictionary of the English Language. Including a very Copious Selection of Scientific Terms. For Use in Schools and Colleges, and as a Book of General Reference. By the Rev. JAMES STORMONTH. The Pronunciation carefully Revised by the Rev. P. H. PHELP, M.A. Cantab. Fourth Edition, crown 8vo, pp. 755. 7s. 6d.

———— The School Etymological Dictionary and Word-Book. Combining the advantages of an ordinary pronouncing School Dictionary and an Etymological Spelling-book. Fcap. 8vo, pp. 254. 2s.

STRICKLAND. Lives of the Queens of Scotland, and English Princesses connected with the Regal Succession of Great Britain. By AGNES STRICKLAND. With Portraits and Historical Vignettes. 8 vols. post 8vo, £4, 4s.

SUTHERLAND. Handbook of Hardy Herbaceous and Alpine Flowers, for general Garden Decoration. Containing Descriptions, in Plain Language, of upwards of 1000 Species of Ornamental Hardy Perennial and Alpine Plants, adapted to all classes of Flower-Gardens, Rockwork, and Waters; along with Concise and Plain Instructions for their Propagation and Culture. By WILLIAM SUTHERLAND, Gardener to the Earl of Minto; formerly Manager of the Herbaceous Department at Kew. Crown 8vo, 7s. 6d.

SWAINSON. A Handbook of Weather Folk-Lore. Being a Collection of Proverbial Sayings in various Languages relating to the Weather, with Explanatory and Illustrative Notes. By the Rev. C. SWAINSON, M.A., Vicar of High Hurst Wood. Fcap. 8vo, Roxburghe binding, 6s. 6d.

SWAYNE. Lake Victoria : A Narrative of Explorations in Search of the Source of the Nile. Compiled from the Memoirs of Captains Speke and Grant. By GEORGE C. SWAYNE, M.A., late Fellow of Corpus Christi College, Oxford. Illustrated with Woodcuts and Map. Crown 8vo, 7s. 6d.

SYMONDSON. Two Years Abaft the Mast : or, Life as a Sea Apprentice. By F. W. H. SYMONDSON. 7s. 6d.

TAYLOR. The Story of My Life. By the late Colonel MEADOWS TAYLOR, Author of the 'Confessions of a Thug,' &c. &c. Edited by his Daughter. [*In the press.*]

——— Tara ; A Mahratta Tale. 3 vols. post 8vo, £1, 11s. 6d.

——— Ralph Darnell. A Novel. 3 vols. post 8vo, £1, 11s. 6d.

THOLUCK. Hours of Christian Devotion. Translated from the German of A. Tholuck, D.D., Professor of Theology in the University of Halle. By the Rev. ROBERT MENZIES, D.D. With a Preface written for this Translation by the Author. Second Edition, crown 8vo, 7s. 6d.

THOMSON. Handy-Book of the Flower-Garden : being Practical Directions for the Propagation, Culture, and Arrangement of Plants in Flower-Gardens all the year round. Embracing all classes of Gardens, from the largest to the smallest. With Engraved and Coloured Plans, illustrative of the various systems of Grouping in Beds and Borders. By DAVID THOMSON, Gardener to his Grace the Duke of Buccleuch, K.G., at Drumlanrig. Third Edition, crown 8vo, 7s. 6d.

——— The Handy-Book of Fruit-Culture under Glass : being a series of Elaborate Practical Treatises on the Cultivation and Forcing of Pines, Vines, Peaches, Figs, Melons, Strawberries, and Cucumbers. With Engravings of Hothouses, &c., most suitable for the Cultivation of and Forcing of these Fruits. Crown 8vo, with Engravings, 7s. 6d.

THOMSON. A Practical Treatise on the Cultivation of the Grape-Vine. By WILLIAM THOMSON, Tweed Vineyards. Eighth Edition, enlarged. 8vo, 5s.

TOM CRINGLE'S LOG. A New Edition, with Illustrations. Crown 8vo, 6s.

TULLOCH. Rational Theology and Christian Philosophy in England in the Seventeenth Century. By JOHN TULLOCH, D.D., Principal of St Mary's College in the University of St Andrews; and one of her Majesty's Chaplains in Ordinary in Scotland. Second Edition. 2 vols. 8vo, 28s.

——— Some Facts of Religion and of Life. Sermons Preached before her Majesty the Queen in Scotland, 1866-76. Second Edition, crown 8vo, 7s. 6d.

——— The Christian Doctrine of Sin ; being the Croal Lecture for 1876. Crown 8vo, 6s.

——— Religion and Theology. A Sermon Preached in the Parish Church of Crathie. Second Edition. 1s.

——— Theism. The Witness of Reason and Nature to an All-Wise and Beneficent Creator. 8vo, 10s. 6d.

TRANSACTIONS OF THE HIGHLAND AND AGRICUL-
TURAL SOCIETY OF SCOTLAND. Published annually, price 5s.

TYTLER. The Wonder-Seeker; or, The History of Charles Douglas. By M. FRASER TYTLER, Author of 'Tales of the Great and Brave,' &c. A New Edition. Fcap., 3s. 6d.

VIRGIL. The Æneid of Virgil. Translated in English Blank Verse by G. K. RICKARDS, M.A., and Lord RAVENSWORTH. 2 vols. fcap. 8vo, 10s.

WALFORD. Mr Smith : A Part of his Life. By L. B. WALFORD. Cheap Edition, 3s. 6d.

WARREN'S (SAMUEL) WORKS. People's Edition, 4 vols. crown 8vo, cloth, 18s. Or separately :—

Diary of a Late Physician. 3s. 6d. Illustrated, crown 8vo, 7s. 6d.

Ten Thousand A-Year. 5s.

Now and Then. Lily and Bee. Intellectual and Moral Development of the Present Age. 4s. 6d.

Essays : Critical, Imaginative, and Juridical. 5s.

WELLINGTON. Wellington Prize Essays on "the System of Field Manœuvres best adapted for enabling our Troops to meet a Continental Army." Edited by Col. E. B. HAMLEY. 8vo, 12s. 6d.

WESTMINSTER ASSEMBLY. Minutes of the Westminster Assembly, while engaged in preparing their Directory for Church Government, Confession of Faith, and Catechisms (November 1644 to March 1649). Printed from Transcripts of the Originals procured by the General Assembly of the Church of Scotland. Edited by the Rev. ALEX. T. MITCHELL, D.D., Professor of Ecclesiastical History in the University of St Andrews, and the Rev. JOHN STRUTHERS, LL.D., Minister of Prestonpans. With a Historical and Critical Introduction by Professor Mitchell. 8vo, price 15s.

WHITE. The Eighteen Christian Centuries. By the Rev. JAMES WHITE, Author of 'The History of France.' Seventh Edition, post 8vo, with Index, 6s.

——— History of France, from the Earliest Times. Fifth Edition, post 8vo, with Index, 6s.

WHITE. Archæological Sketches in Scotland—Kintyre and Knapdale. By Captain T. P. WHITE, R.E., of the Ordnance Survey. With numerous Illustrations. 2 vols., folio, £4, 4s. Vol. I., Kintyre, sold separately, £2, 2s.

WILLS. Charles the First : An Historical Tragedy in Four Acts. By W. G. WILLS. 8vo, 2s. 6d.

——— Drawing-room Dramas for Children. By the SAME and the Hon. Mrs GREENE. Crown 8vo, 6s.

WILSON. The "Ever-Victorious Army :" A History of the Chinese Campaign under Lieut.-Col. C. G. Gordon, and of the Suppression of the Tai-ping Rebellion. By ANDREW WILSON, F.A.S.L. 8vo, with Maps, 15s.

——— The Abode of Snow : Observations on a Journey from Chinese Tibet to the Indian Caucasus, through the Upper Valleys of the Himalaya. New Edition. Crown 8vo, with Map, 10s. 6d.

WILSON. Works of Professor Wilson. Edited by his Son-in-Law, Professor FERRIER. 42 vols. crown 8vo, £2, 8s.

OTHER WORKS OF PROFESSOR WILSON.

——— Christopher in his Sporting-Jacket. 2 vols., 8s.

——— Isle of Palms, City of the Plague, and other Poems. 4s.

——— Lights and Shadows of Scottish Life, and other Tales. 4s.

——— Essays, Critical and Imaginative. 4 vols., 16s.

——— The Noctes Ambrosianæ. Complete, 4 vols., 14s.

——— The Comedy of the Noctes Ambrosianæ. By CHRISTOPHER NORTH. Edited by JOHN SKELTON, Advocate. With a Portrait of Professor Wilson and of the Ettrick Shepherd, engraved on Steel. Crown 8vo, 7s. 6d.

——— Homer and his Translators, and the Greek Drama. Crown 8vo, 4s.

WINGATE. Poems and Songs. By DAVID WINGATE. Fcap. 8vo, 5s.

——— Annie Weir, and other Poems. Fcap. 8vo, 5s.

WORSLEY. Poems and Translations. By PHILIP STANHOPE WORSLEY, M.A. Edited by EDWARD WORSLEY. Second Edition, enlarged. Fcap. 8vo, 6s.

YULE. Fortifications for the Use of Officers in the Army, and Readers of Military History. By Col. YULE, Bengal Engineers. 8vo, with numerous Illustrations, 10s. 6d.

www.ingramcontent.com/pod-product-compliance
Lightning Source LLC
Chambersburg PA
CBHW020900210326
41598CB00018B/1726